11/2015

WITHDRAWN

Undergraduate Lecture Notes in Physics

Editors

Neil Ashby
Professor Emeritus, University of Colorado, Boulder, CO, USA

William Brantley
Professor, Furman University, Greenville, SC, USA

Michael Fowler
Professor, University of Virginia, Charlottesville, VA, USA

Michael Inglis
Professor, SUNY Suffolk County Community College, Selden, NY, USA

Heinz Klose
Professor Emeritus, Humboldt University Berlin, Germany

Helmy Sherif
Professor, University of Alberta, Edmonton, AB, Canada

Undergraduate Lecture Notes in Physics (ULNP) publishes authoritative texts covering topics throughout pure and applied physics. Each title in the series is suitable as a basis for undergraduate instruction, typically containing practice problems, worked examples, chapter summaries, and suggestions for further reading.

ULNP titles must provide at least one of the following:

- An exceptionally clear and concise treatment of a standard undergraduate subject.
- A solid undergraduate-level introduction to a graduate, advanced, or non-standard subject.
- A novel perspective or an unusual approach to teaching a subject.

ULNP especially encourages new, original, and idiosyncratic approaches to physics teaching at the undergraduate level.

The purpose of ULNP is to provide intriguing, absorbing books that will continue to be the reader's preferred reference throughout their academic career.

More information about this series at http://www.springer.com/series/8917

Kerry Kuehn

A Student's Guide Through the Great Physics Texts

Volume I: The Heavens and The Earth

 Springer

Kerry Kuehn
Wisconsin Lutheran College
Milwaukee
Wisconsin
USA

ISSN 2192-4791   ISSN 2192-4805 (electronic)
ISBN 978-1-4939-1359-6   ISBN 978-1-4939-1360-2 (eBook)
DOI 10.1007/978-1-4939-1360-2
Springer New York Heidelberg Dordrecht London

Library of Congress Control Number: 2014945636

Printed on acid-free paper

Springer is part of Springer Science+Business Media (www.springer.com)

For Cindy

Preface

What is the Nature of this Book?

This four-volume book grew from a four-semester general physics curriculum which I developed and taught for the past decade to undergraduate students at Wisconsin Lutheran College in Milwaukee. The curriculum is designed to encourage a critical and circumspect approach to natural science while at the same time providing a suitable foundation for advanced coursework in physics. This is accomplished by holding before the student some of the best thinking about nature that has been committed to writing. The scientific texts found herein are considered classics precisely because they address timeless questions in a particularly honest and convincing manner. This does not mean that everything they say is true—in fact many classic scientific texts contradict one another—but it is by the careful reading, analysis and discussion of the most reputable observations and opinions that one may begin to discern truth from error.

Who is this Book for?

Like fine wine, the classic texts in any discipline can be enjoyed by both the novice and the connoisseur. For example, Sophocles' tragic play *Antigone* can be appreciated by the young student who is drawn to the story of the heroine who braves the righteous wrath of King Creon by choosing to illegally bury the corpse of her slain brother, and also by the seasoned scholar who carefully evaluates the relationship between justice, divine law and the state. Likewise, Galileo's *Dialogues Concerning Two New Sciences* can be enjoyed by the young student who seeks a clear geometrical description of the speed of falling bodies, and also by the seasoned scholar who is amused by Galileo's wit and sarcasm, or who finds in his *Dialogues* the progressive Aristotelianism of certain late medieval scholastics.[1]

---

[1] See Wallace, W. A., The Problem of Causality in Galileo's Science, *The Review of Metaphysics*, 36(3), 607–632, 1983.

Having said this, I believe that this book is particularly suitable for the following audiences. First, it could serve as the primary textbook in an introductory discussion-based physics course at the university level. It was designed to appeal to a broad constituency of students at small liberal arts colleges which often lack the resources to offer the separate and specialized introductory physics courses found at many state-funded universities (*e.g. Physics for poets, Physics for engineers, Physics for health-care-professionals, Physics of sports, etc.*). Indeed, at my institution it is common to have history and fine arts students sitting in the course alongside biology and physics majors. Advanced high-school or home-school students will find in this book a physics curriculum that emphasizes reading comprehension, and which can serve as a bridge into college-level work. It might also be adopted as a supplementary text for an advanced placement course in physics, astronomy or the history and philosophy of science. Many practicing physicists, especially those at the beginning of their scientific careers, may not have taken the opportunity to carefully study some of the foundational texts of physics and astronomy. Perhaps this is because they have (quite understandably) focused their attention on acquiring a strong technical proficiency in a narrow subfield. Such individuals will find herein a structured review of such foundational texts. This book will also likely appeal to humanists, social scientists and motivated lay-readers who seek a thematically-organized anthology of texts which offer insight into the historical development and cultural significance of contemporary scientific theories. Finally, and most importantly, this book is designed for the benefit of the teaching professor. Early in my career as a faculty member, I was afforded considerable freedom to develop a physics curriculum at my institution which would sustain my interest for the foreseeable future—perhaps until retirement. Indeed, reading and re-reading the classic texts assembled herein has provided me countless hours of enjoyment, reflection and inspiration.

How is this Book Unique?

Here I will offer a mild critique of textbooks typically employed in introductory university physics courses. While what follows is admittedly a bit of a caricature, I believe it to be a quite plausible one. I do this in order to highlight the unique features and emphases of the present book. In many university-level physics textbooks, the chapter format follows a standard recipe. First, accepted scientific laws are presented in the form of one or more mathematical equations. This is followed by a few example problems so the student can learn how to plug numbers into the aforementioned equations and how to avoid common conceptual or computational errors. Finally, the student is presented with contemporary applications which illustrate the relevance of these equations for various industrial or diagnostic technologies.

While this method often succeeds in preparing students to pass certain standardized tests or to solve fairly straightforward technical problems, it is lacking in important respects. First, it is quite bland. Although memorizing formulas and learning how to perform numerical calculations is certainly crucial for acquiring a working knowledge of physical theories, it is often the more general questions about the assumptions and the methods of science that students find particularly stimulating and enticing. For instance, in his famous *Mathematical Principles of Natural Philosophy*, Newton enumerates four general rules for doing philosophy. Now the reader may certainly choose to reject Newton's rules, but Newton himself suggests that they are necessary for the subsequent development of his universal theory of gravitation. Is he correct? For instance, if one rejects Rules III and IV—which articulate the principle of induction—then in what sense can his theory of gravity be considered universal? Questions like "is Newton's theory of gravity correct?" and "how do you know?" can appeal to the innate sense of inquisitiveness and wonder that attracted many students to the study of natural science in the first place. Moreover, in seeking a solution to these questions, the student must typically acquire a deeper understanding of the technical aspects of the theory. In this way, broadly posed questions can serve as a motivation and a guide to obtaining a detailed understanding of physical theories.

Second, and perhaps more importantly, the method employed by most standard textbooks does not prepare the student to become a practicing scientist precisely because it tends to mask the way science is actually done. The science is presented as an accomplished fact; the prescribed questions revolve largely around technological applications of accepted laws. On the contrary, by carefully studying the foundational texts themselves the student is exposed to the polemical debates, the technical difficulties and the creative inspirations which accompanied the development of scientific theories. For example, when studying the motion of falling bodies in Galileo's *Dialogues*, the student must consider alternative explanations of the observed phenomena; must understand the strengths and weaknesses of competing theories; and must ultimately accept—or reject—Galileo's proposal on the basis of evidence and reason. Through this process the student gains a deeper understanding of Galileo's ideas, their significance, and their limitations.

Moreover, when studying the foundational texts, the student is obliged to thoughtfully address issues of language and terminology—issues which simply do not arise when learning from standard textbooks. In fact, when scientific theories are being developed the scientists themselves are usually struggling to define terms which capture the essential features of their discoveries. For example, Oersted coined a term which is translated as "electric conflict" to describe the effect that an electrical current has on a nearby magnetic compass needle. He was attempting to distinguish between the properties of stationary and moving charges, but he lacked the modern concept of the magnetic field which was later introduced by Faraday. When students encounter a familiar term such as "magnetic field," they typically accept it as settled terminology, and thereby presume that they understand the phenomenon by virtue of recognizing and memorizing the canonical term. But when they encounter an unfamiliar term such as "electric conflict," as part of the scientific argument from which it derives and wherein it is situated, they are tutored into the original argument and are thus obliged to think scientifically, along with the great scientist. In other words, when reading the foundational texts, the student is led into *doing* science and not merely into memorizing and applying nomenclature.

Generally speaking, this book draws upon two things that we have in common: (i) a shared conversation recorded in the foundational scientific texts, and (ii) an innate faculty of reason. The careful reading and analysis of the foundational texts is extremely valuable in learning how to think clearly and accurately about natural science. It encourages the student to carefully distinguish between observation and speculation, and finally, between truth and falsehood. The ability to do this is essential when considering the practical and even philosophical implications of various scientific theories. Indeed, one of the central aims of this book is to help the student grow not only as a potential scientist, but as an educated person. More specifically, it will help the student develop important intellectual virtues (*i.e.* good habits), which will serve him or her in any vocation, whether in the marketplace, in the family, or in society.

How is this Book Organized?

This book is divided into four separate volumes; the plan is to publish volumes I and II concurrently in the autumn of 2014, and volumes III and IV approximately a year later. Within each volume, the readings are centered on a particular theme and proceed chronologically. For example, Volume I is entitled *The Heavens and the Earth*. It provides an introduction to astronomy and cosmology beginning with the geocentrism of Aristotle's *On the Heavens* and Ptolemy's *Almagest*, proceeding through heliocentrism advanced in Copernicus' *Revolutions of the Heavenly Spheres* and Kepler's *Epitome of Copernican Astronomy*, and arriving finally at big bang cosmology with Lemaître's *The Primeval Atom*. Volume II, *Space, Time and Motion*, provides a careful look at the science of motion and rest. Here, students engage in a detailed analysis of significant portions of Galileo's *Dialogues Concerning Two New Sciences*, Pascal's *Treatise on the Equilibrium of Fluids and the Weight of the Mass of Air*, Newton's *Mathematical Principles of Natural Philosophy* and Einstein's *Relativity*.

The forthcoming Volume III will trace the theoretical and experimental development of the electromagnetic theory of light using texts by William Gilbert, Benjamin Franklin, Charles Coulomb, André Marie Ampère, Christiaan Huygens, James Clerk Maxwell, Heinrich Hertz, Albert Michelson, and others. Volume IV will provide an exploration of modern physics, focusing on radiation, atomism and the quantum theory of matter. Selections will be taken from works by Joseph Fourier, William Thomson, Joseph Thomson, James Clerk Maxwell, Ernest Rutherford, Max Planck, James Chadwick, Niels Bohr, Erwin Schrödinger and Werner Heisenberg.

While the four volumes of the book are arranged around distinct themes, the readings themselves are not strictly constrained in this way. For example, in his *Treatise on Light*, Huygens is primarily interested in demonstrating that light can be best understood as a wave propagating through an aethereal medium comprised of tiny, hard elastic particles. In so doing, he spends some time discussing the speed of light measurements performed earlier by Ole Rømer. These measurements, in turn, relied upon an understanding of the motion of the moons of Jupiter which had recently been reported by Galileo in his *Sidereal Messenger*. So here, in this *Treatise on Light*, we find references to a variety of inter-related topics. Huygens does not artificially restrict his discussion to a narrow topic—nor does Galileo, or Newton or the other great thinkers. Instead, the reader will find in this book recurring concepts and problems which cut across different themes and which are naturally addressed in a historical context with increasing levels of sophistication and care. Science is a conversation which stretches backwards in time to antiquity.

How Might this Book be Used?

This book is designed for college classrooms, small-group discussions and individual study. Each of the four volumes of the book contains roughly thirty chapters, providing more than enough material for a one-semester undergraduate-level physics course; this is the context in which this book was originally implemented. In such a setting, one or two fifty-minute classroom sessions should be devoted to analyzing and discussing each chapter. This assumes that the student has read the assigned text before coming to class. When teaching such a course, I typically improvise—leaving out a chapter here or there (in the interest of time) and occasionally adding a reading selection from another source that would be particularly interesting or appropriate.

Each chapter of each volume has five main components. First, at the beginning of each chapter, I include a short introduction to the reading. If this is the first encounter with a particular author, the introduction includes a biographical sketch of the author and some historical context. The introduction will often contain a summary of some important concepts from the previous chapter and will conclude with a few provocative questions to sharpen the reader's attention while reading the upcoming text.

Next comes the reading selection. There are two basic criteria which I used for selecting each text: it must be *significant* in the development of physical theory, and it must be *appropriate* for beginning undergraduate students. Balancing these criteria was very difficult. Over the past decade, I have continually refined the selections so that they might comprise the most critical contribution of each scientist, while at the same time not overwhelming the students by virtue of their length, language or complexity. The readings are not easy, so the student should not feel overwhelmed if he or she does not grasp everything on the first (or second, or third) reading. Nobody does. Rather, these texts must be "grown into," so to speak.

I have found that the most effective way to help students successfully engage foundational texts is to carefully prepare questions which help them identify and understand key concepts. So as the third component of each chapter, I have prepared a study guide in the form of a set of questions which can be used to direct either classroom discussion or individual reading. After the source texts themselves, the study guide is perhaps the most important component of each chapter, so I will spend a bit more time here explaining it.

The study guide typically consists of a few general discussion questions about key topics contained in the text. Each of these general questions is followed by several sub-questions which aid the student by focusing his or her attention on the author's definitions, methods, analysis and conclusions. For example, when students are reading a selection from Albert Michelson's book *Light Waves and their Uses*, I will often initiate classroom discussion with a general question such as "Is it possible to measure the absolute speed of the earth?" This question gets students thinking about the issues addressed in the text in a broad and intuitive way. If the students get stuck, or the discussion falters, I will then prompt them with more detailed follow-up questions such as: "What is meant by the term absolute speed?" "How, exactly, did Michelson attempt to measure the absolute speed of the earth?" "What technical difficulties did Michelson encounter while doing his experiments?" "To what conclusion(s) was Michelson led by his results?" and finally "Are Michelson's conclusions then justified?" After answering such simpler questions, the students are usually more confident and better prepared to address the general question which was initially posed.

In the classroom, I always emphasize that it is critical for participants to carefully read the assigned selections before engaging in discussion. This will help them to make relevant comments and to cite textual evidence to support or contradict assertions made during the course of the discussion. In this way, many assertions will be revealed as problematic—in which case they may then be refined or rejected altogether. Incidentally, this is precisely the method used by scientists themselves in order to discover and evaluate competing ideas or theories. During our discussion, students are encouraged to speak with complete freedom; I stipulate only one classroom rule: any comment or question must be stated publicly so that all others can hear and respond. Many students are initially apprehensive about engaging in public discourse, especially about science. If this becomes a problem, I like to emphasize that students do not need to make an elaborate point in order to engage in classroom discussion. Often, a short question will suffice. For example, the student might say "I am unclear what the author means by the term *inertia*. Can someone please clarify?" Starting like this, I have found that students soon join gamely in classroom discussion.

Fourth, I have prepared a set of exercises which test the student's understanding of the text and his or her ability to apply key concepts in unfamiliar situations. Some of these are accompanied by a brief explanation of related concepts or formulas. Most of them are numerical exercises, but some are provocative essay prompts. In addition, some of the chapters contain suggested laboratory exercises, a few of which are in fact field exercises which require several days (or even months) of observations. For example, in Chap. 3 of Volume I, there is an astronomy field exercise which involves charting the progression of a planet through the zodiac over the course of a few months. So if this book is being used in a semester-long college or university setting, the instructor may wish to skim through the exercises at the end of each chapter so he or she can identify and assign the longer ones as ongoing exercises early in the semester.

Finally, I have included at the end of each chapter a list of vocabulary words which are drawn from the text and with which the student should become acquainted.

Expanding his or her vocabulary will aid the student not only in their comprehension of subsequent texts, but also on many standardized college and university admissions exams.

What Mathematics Preparation is Required?

It is sometime said that mathematics is the "language of science." This sentiment appropriately inspires and encourages the serious study of mathematics. Of course if it were taken literally then many seminal works in physics—and much of biology— would have to be considered either unintelligible or unscientific, since they contain little or no mathematics. Moreover, if mathematics is the *only* language of science, then physics instructors should be stunned whenever students are enlightened by verbal explanations which lack mathematical form. To be sure, mathematics offers a refined and sophisticated language for describing observed phenomena, but many of our most significant observations about nature may be expressed using everyday images, terms and concepts: heavy and light, hot and cold, strong and weak, straight and curved, same and different, before and after, cause and effect, form and function, one and many. So it should come as no surprise that, when studying physics *via* the reading and analysis of foundational texts, one enjoys a considerable degree of flexibility in terms of the mathematical rigor required.

For instance, Faraday's *Experimental Researches in Electricity* are almost entirely devoid of mathematics. Rather, they consist of detailed qualitative descriptions of his observations, such as the relationship between the relative motion of magnets and conductors on the one hand, and the direction and intensity of induced electrical currents on the other hand. So when studying Faraday's work, it is quite natural for the student to aim for a conceptual, as opposed to a quantitative, understanding of electromagnetic induction. Alternatively, the student can certainly attempt to connect Faraday's qualitative descriptions with the mathematical methods which are often used today to describe electromagnetic induction (*i.e.* vector calculus and differential equations). The former method has the advantage of demonstrating the conceptual framework in which the science was actually conceived and developed; the latter method has the advantage of allowing the student to make a more seamless transition to upper-level undergraduate or graduate courses which typically employ sophisticated mathematical methods.

In this book, I approach the issue of mathematical proficiency in the following manner. Each reading selection is followed by both study questions and homework exercises. In the study questions, I do not attempt to force anachronistic concepts or methods into the student's understanding of the text. They are designed to encourage the student to approach the text in the same spirit as the author, insofar as this is possible. In the homework exercises, on the other hand, I often ask the student to employ mathematical methods which go beyond those included in the reading selection itself. For example, one homework exercise associated with a selection from Hertz's book *Electric Waves* requires the student to prove that two counter-propagating waves superimpose to form a standing wave. Although Hertz casually mentions that a standing wave is formed in this way, the problem itself requires that the student use trigonometric identities which are not described in Hertz's text. In cases such as this, a note in the text suggests the mathematical methods which are required. I have found this to work quite well, especially in light of the easy access which today's students have to excellent print and online mathematical resources.

Generally speaking, there is an increasing level of mathematical sophistication required as the student progresses through the curriculum. In Volume I students need little more than a basic understanding of geometry. Euclidean geometry is sufficient in understanding Ptolemy's epicyclic theory of planetary motion and Galileo's calculation of the altitude of lunar mountains. The student will be introduced to some basic ideas of non-Euclidean geometry toward the end of Volume I when studying modern cosmology through the works of Einstein, Hubble and Lemaître, but this is not pushed too hard. In Volume II students will make extensive use of geometrical methods and proofs, especially when analyzing Galileo's work on projectile motion and the application of Newton's laws of motion. Although Newton develops his theory of gravity in the *Principia* using geometrical proofs, the homework problems often require the student to make connections with the methods of calculus. The selections on Einstein's special theory of relativity demand only the use of algebra and geometry. In Volume III, mathematical methods will, for the most part, be limited to geometry and algebra. More sophisticated mathematical methods will be required, however, in solving some of the problems dealing with Maxwell's electromagnetic theory of light. This is because Maxwell's equations are most succinctly presented using vector calculus and differential equations. Finally, in Volume IV, the student will be aided by a working knowledge of calculus, as well as some familiarity with the use of differential equations.

It is my feeling that in a general physics course, such as the one being presented in this book, the extensive use of advanced mathematical methods (beyond geometry, algebra and elementary calculus) is not absolutely necessary. Students who plan to major in physics or engineering will presumably learn more advanced mathematical methods (*e.g.* vector calculus and differential equations) in their collateral mathematics courses, and they will learn to apply these methods in upper-division (junior and senior-level) physics courses. Students who do not plan to major in physics will typically not appreciate the extensive use of such advanced mathematical methods. And it will tend to obscure, rather than clarify, important physical concepts. In any case, I have attempted to provide guidance for the instructor, or for the self-directed student, so that he or she can incorporate an appropriate level of mathematical rigor.

Figures, Formulas, and Footnotes

One of the difficulties in assembling readings from different sources and publishers into an anthology such as this is how to deal with footnotes, references, formulas and other issues of annotation. For example, for any given text selection, there may be footnotes supplied by the author, the translator and the anthologist. So I have appended a [*K.K.*] marking to indicate when the footnote is my own; I have not included this marking when there is no danger of confusion, for example in my footnotes appearing in the introduction, study questions and homework exercises of each chapter.

For the sake of clarity and consistency, I have added (or sometimes changed the) numbering for figures appearing in the texts. For example, Fig. 16.3 is the third figure in Chap. 16 of this volume; this is not necessarily how Kepler or his translator numbered this figure when it appeared in an earlier publication of his *Epitome Astronomae Copernicanae*. For ease of reference, I have also added (or sometimes changed the) numbering of equations appearing in the texts. For example, Eqs. 31.1 and 32.2 are the equations of the Lorentz and Galilei transformations appearing in the reading in Chap. 31 of Volume II, extracted from Einstein's book *Relativity*. This is not necessarily how Einstein numbered them.

In several cases, the translator or editor has included references to page numbers in a previous publication. For example, the translators of Galileo's *Dialogues* have indicated, within their 1914 English translation, the locations of page breaks in the Italian text published in 1638. A similar situation occurs with Faith Wallis's 1999 translation of Bede's *The Reckoning of Time*. For consistency, I have rendered such page numbering in bold type surrounded by slashes. So **/50/** refers to page 50 in some earlier "canonical" publication.

Acknowledgements

I suppose that it is common for a teacher to eventually mull over the idea of compiling his or her thoughts on teaching into a coherent and transmittable form. Committing this curriculum to writing was particularly difficult because I am keenly aware how my own thinking about teaching physics has changed significantly since my first days in front of the classroom—and how it is quite likely to continue to evolve. So this book should be understood as a snapshot, so to speak, of how I am teaching my courses at the time of writing. I would like to add, however, that I believe the evolution of my teaching has reflected a maturing in thought, rather than a mere drifting in opinion. After all, the classic texts themselves are formative: how can a person, whether student or teacher, not become better informed when learning from the best thinkers?

This being said, I would like to offer my apologies to those students who suffered through the birth pains, as it were, of the curriculum presented in this book. The countless corrections and suggestions that they offered are greatly appreciated; any and all remaining errors in the text are my own fault. Many of the reading selections included herein were carefully scanned, edited and typeset by undergraduate students who served as research and editorial assistants on this project: Jaymee Martin-Schnell, Dylan Applin, Samuel Wiepking, Timothy Kriewall, Stephanie Kriewall, Cody Morse, and Ethan Jahns deserve special thanks. My home institution, Wisconsin Lutheran College, provided me with considerable time and freedom to develop this book, including a year-long sabbatical leave, for which I am very grateful. During this sabbatical, I received support and encouragement from my trusty colleagues in the Department of Mathematical and Physical Sciences. Also, the Higher Education Initiatives Program of the Wisconsin Space Grant Consortium provided generous funding for this project, as did the Faculty Development Committee of Wisconsin Lutheran College. Greg Schulz has been an invaluable intellectual resource throughout this project. Aaron Jensen conscientiously translated selections of the *Almagest*, included in Volume I of this book, from Heiberg's edition of Ptolemy's Greek manuscript. And Glen Thompson was instrumental in getting this translation project initiated. Starla Siegmann and Jenny Baker, librarians at the Marvin M. Schwan Library of Wisconsin Lutheran College, were always up to the challenge of speedily procuring obscure resources from remote libraries. I would also like to thank the following individuals who facilitated the complex task of acquiring permissions to reprint the texts included in this book: Jenny Howard at Liverpool University Press, Elizabeth Sandler, Emilie David and Norma Rosado-Blake at the American Association for the Advancement of Science, Chris Erdmann at the Harvard College Observatory's Wolbach Library, Carmen Pagán at Encyclopædia Britannica, Michael Fisher and Scarlett Huffman at Harvard University Press, and Jenny Howard at Liverpool University Press. Cornelia Mutel and Kathryn Hodson very kindly provided digital images for inclusion with the Galileo and Pascal selections from the History of Hydraulics Rare Book Collection at the University of Iowa's IIHR-Hydroscience and Engineering. Also, I would like to thank Jeanine Burke, the acquisition editor at Springer who originally agreed to take on this project with me, and Robert Korec and Tom Spicer who patiently saw it through to publication. Shortly after submitting my book proposal to Springer, I received very encouraging and helpful comments from several anonymous reviewers, for whom I am thankful. I received similar suggestions from the editors of Springer's Undergraduate Lecture Notes in Physics series for which I am likewise grateful. Finally, I would especially like to thank my wife, Cindy, who has provided unwavering encouragement and support for my work from the very start.

Milwaukee, 2014   Kerry Kuehn

Contents

Chapter 1
Nature, Number and Substance

It is the unnatural which quickest passes away.

—Aristotle

1.1 Introduction

Aristotle (384–322 B.C.) was born in the Greek colonial town of Stagira, located on the Aegean Sea east of modern day Thessaloniki. He was a student of Plato and a teacher of Alexander the Great. Aristotle's writings on logic, physics, medicine, metaphysics, ethics, politics, plants, animals and the soul profoundly influenced much of western philosophy (Fig. 1.1).

The reading selections included below are from Book 1 of Aristotle's *De Caelo*, or *On the Heavens*—translated into English by J. L. Stocks. Herein, Aristotle considers the shape, size, composition and motion of both the heavens and the earth. The heavens, as Aristotle writes, refers to everything in that divine and unchangeable expanse which lies beyond the immediate vicinity of the earth.[1]

Apart from the profound enjoyment of contemplating the nature of the heavens, studying Aristotle has a number of practical benefits. First, it provides an occasion to read and analyze the arguments of a careful thinker. Try to enter Aristotle's world: see if his arguments are reasonable; try to find weaknesses in them. Is he right or wrong in his conclusions? Are you sure? If he is wrong, why? Is it because of a wrong assumption that he is making, or is it because he has reasoned incorrectly from his assumptions?

You will likely notice that Aristotle has a strong interest in classification: he likes to classify types of objects, types of motion, and types of shapes. For this reason, many beginners find reading Aristotle to be somewhat difficult. Try not to get overwhelmed—the difficulty is largely because his classification schemes are often unlike the ones we use today. For instance, he classifies all types of motion as either straight, circular, or a combination of the two. Although scientists don't usually classify motion like this today, there are good reasons for doing so. You may even discover that, in many ways, Aristotle is very modern in his thinking.

---

[1] Kepler, as we will find in Chap. 15 of the present volume, rejects Aristotle's division of the World (the universe) into two separate and utterly different domains.

K. Kuehn, *A Student's Guide Through the Great Physics Texts*,
Undergraduate Lecture Notes in Physics, DOI 10.1007/978-1-4939-1360-2_1,
© Springer Science+Business Media, LLC 2015

Fig. 1.1 A Roman copy of
Lysippos' bronze bust of
Aristotle

Second, reading Aristotle's *On the Heavens* will provide you with a framework for understanding subsequent theories of the world. Indeed, Ptolemy, Copernicus and Kepler developed their planetary theories largely in response to Aristotle's ideas. Thus, the better understanding you have of Aristotle, the more clearly you will understand their ideas.

1.2 Reading: Aristotle, *On the Heavens*

Aristotle, On the Heavens, in *Aristotle: I, Great Books of the Western World*, vol. 8, edited by Robert Maynard Hutchins, Encyclopedia Britannica, 1952.

1.2.1 Chapter 1

The science which has to do with nature clearly concerns itself for the most part with bodies and magnitudes and their properties and movements, but also with the principles of this sort of substance, as many as they may be. For of things constituted by nature some are bodies and magnitudes, some possess body and magnitude, and some are principles of things which possess these. Now a continuum is that which is divisible into parts always capable of subdivision, and a body is that which is every way divisible. A magnitude if divisible one way is a line, if two ways a surface, and if three a body. Beyond these there is no other magnitude, because the three dimensions are all that there are, and that which is divisible in three directions is divisible in all. For, as the Pythagoreans say, the world and all that is in it is determined by the number three, since beginning and middle and end give the number of an 'all', and the number they give is the triad. And so, having taken these three from nature as (so to speak) laws of it, we make further use of the number three in the worship of the Gods. Further, we use the terms in practice in this way. Of two things, or men, we say 'both', but not 'all': three is the first number to which the term 'all' has been appropriated. And in this, as we have said, we do but follow the lead which nature gives. Therefore, since 'every' and 'all' and 'complete' do not differ from one another in respect of form, but only, if at all, in their matter and in that to which they are applied, body alone among magnitudes can be complete. For it alone is determined by the three dimensions, that is, is an 'all'. But if it is divisible in three dimensions it is every way divisible, while the other magnitudes are divisible in one dimension or in two alone: for the divisibility and continuity of magnitudes depend upon the number of the dimensions, one sort being continuous in one direction, another in two, another in all. All magnitudes, then, which are divisible are also continuous. Whether we can also say that whatever is continuous is divisible does not yet, on our present grounds, appear. One thing, however, is clear. We cannot pass beyond body to a further kind, as we passed from length to surface, and from surface to body. For if we could, it would cease to be true that body is complete magnitude. We could pass beyond it only in virtue of a defect in it; and that which is complete cannot be defective, since it has being in every respect. Now bodies which are classed as parts of the whole are each complete according to our formula, since each possesses every dimension. But each is determined relatively to that part which is next to it by contact, for which reason each of them is in a sense many bodies. But the whole of which they are parts must necessarily be complete, and thus, in accordance with the meaning of the word, have being, not in some respect only, but in every respect.

1.2.2   Chapter 2

The question as to the nature of the whole, whether it is infinite in size or limited in its total mass, is a matter for subsequent inquiry.[2] We will now speak of those parts of the whole which are specifically distinct. Let us take this as our starting-point. All natural bodies and magnitudes we hold to be, as such, capable of locomotion; for nature, we say, is their principle of movement.[3] But all movement that is in place, all locomotion, as we term it, is either straight or circular or a combination of these two, which are the only simple movements. And the reason of this is that these two, the straight and the circular line, are the only simple magnitudes. Now revolution about the centre is circular motion, while the upward and downward movements are in a straight line, 'upward' meaning motion away from the centre, and 'downward' motion towards it. All simple motion, then, must be motion either away from or

---

[2] See *On the Heavens*, I.7.

[3] Cf. *Physics*, 192^b 20.

towards or about the centre. This seems to be in exact accord with what we said above: as body found its completion in three dimensions, so its movement completes itself in three forms.

Bodies are either simple or compounded of such; and by simple bodies I mean those which possess a principle of movement in their own nature, such as fire and earth with their kinds, and whatever is akin to them. Necessarily, then, movements also will be either simple or in some sort compound—simple in the case of the simple bodies, compound in that of the composite—and in the latter case the motion will be that of the simple body which prevails in the composition. Supposing, then, that there is such a thing as simple movement, and that circular movement is an instance of it, and that both movement of a simple body is simple and simple movement is of a simple body (for if it is movement of a compound it will be in virtue of a prevailing simple element), then there must necessarily be some simple body which revolves naturally and in virtue of its own nature with a circular movement. By constraint, of course, it may be brought to move with the motion of something else different from itself, but it cannot so move naturally, since there is one sort of movement natural to each of the simple bodies. Again, if the unnatural movement is the contrary of the natural and a thing can have no more than one contrary, it will follow that circular movement, being a simple motion, must be unnatural, if it is not natural, to the body moved. If then (1) the body, whose movement is circular, is fire or some other element, its natural motion must be the contrary of the circular motion. But a single thing has a single contrary; and upward and downward motion are the contraries of one another. If, on the other hand, (2) the body moving with this circular motion which is unnatural to it is something different from the elements, there will be some other motion which is natural to it. But this cannot be. For if the natural motion is upward, it will be fire or air, and if downward, water or earth. Further, this circular motion is necessarily primary. For the perfect is naturally prior to the imperfect, and the circle is a perfect thing. This cannot be said of any straight line:—not of an infinite line; for, if it were perfect, it would have a limit and an end: nor of any finite line; for in every case there is something beyond it, since any finite line can be extended. And so, since the prior movement belongs to the body which is naturally prior, and circular movement is prior to straight, and movement in a straight line belongs to simple bodies—fire moving straight upward and earthy bodies straight downward towards the centre—since this is so, it follows that circular movement also must be the movement of some simple body. For the movement of composite bodies is, as we said, determined by that simple body which preponderates in the composition. These premises clearly give the conclusion that there is in nature some bodily substance other than the formations we know, prior to them all and more divine than they. But it may also be proved as follows. We may take it that all movement is either natural or unnatural, and that the movement which is unnatural to one body is natural to another—as, for instance, is the case with the upward and downward movements, which are natural and unnatural to fire and earth respectively. It necessarily follows that circular movement, being unnatural to these bodies, is the natural movement of some other. Further, if, on the one hand, circular movement is *natural* to something, it must surely be some simple and primary body which is ordained to move with a natural circular motion, as fire is ordained to fly up and earth down. If, on the other hand, the movement of the rotating bodies about the centre is *unnatural*, it would be remarkable and indeed quite inconceivable that this movement alone should be continuous and eternal, being nevertheless contrary to nature. At any rate the evidence of all other cases goes to show that it is the unnatural which quickest passes away. And so, if, as some say, the body so moved is fire, this movement is just as unnatural to it as downward movement; for any one can see that fire moves in a straight line away from the centre. On all these grounds, therefore, we may infer with confidence that there is something beyond the bodies that are about us on this earth, different and separate from them; and that the superior glory of its nature is proportionate to its distance from this world of ours.

1.2.3   Chapter 3

In consequence of what has been said, in part by way of assumption and in part by way of proof, it is clear that not every body either possesses lightness or heaviness. As a preliminary we must explain in what sense we are using the words 'heavy' and 'light', sufficiently, at least, for our present purpose: we can examine the terms more closely later, when we come to consider their essential nature. Let us then apply the term 'heavy' to that which naturally moves towards the centre, and 'light' to that which moves naturally away from the centre. The heaviest thing will be that which sinks to the bottom of all things that move downward, and the lightest that which rises to the surface of everything that moves upward. Now, necessarily, everything which moves either up or down possesses lightness or heaviness or both—but not both relatively to the same thing: for things are heavy and light relatively to one another; air, for instance, is light relatively to water, and water light relatively to earth. The body, then, which moves in a circle cannot possibly possess either heaviness or lightness. For neither naturally nor unnaturally can it move either towards or away from the centre. Movement in a straight line certainly does not belong to it *naturally*, since one sort of movement is, as we saw, appropriate to each simple body, and so we should be compelled to identify it with one of the bodies which move in this way. Suppose, then, that the movement is *unnatural*. In that case, if it is the downward movement which is unnatural, the upward movement will be natural; and if it is the upward which is unnatural, the downward will be natural. For we decided that of contrary movements, if the one is unnatural to anything, the other will be natural to it. But since the natural movement of the whole and of its part—of earth, for instance, as a whole and of a small clod—have one and the same direction, it results, in the first place, that this body can possess no lightness or heaviness at all (for that would mean that it could move by its own nature either from or towards the centre, which, as we know, is impossible); and, secondly, that it cannot possibly move in the way of locomotion by being forced violently aside in an upward or downward direction. For neither naturally nor unnaturally can it move with any other motion but its own, either itself or any part of it, since the reasoning which applies to the whole applies also to the part.

It is equally reasonable to assume that this body will be ungenerated and indestructible and exempt from increase and alteration, since everything that comes to be comes into being from its contrary and in some substrate, and passes away likewise in a substrate by the action of the contrary into the contrary, as we explained in our opening discussions.[4] Now the motions of contraries are contrary. If then this body can have no contrary, because there can be no contrary motion to the circular, nature seems justly to have exempted from contraries the body which was to be ungenerated and indestructible. For it is in contraries that generation and decay subsist. Again, that which is subject to increase increases upon contact with a kindred body, which is resolved into its matter. But there is nothing out of which this body can have been generated. And if it is exempt from increase and diminution, the same reasoning leads us to suppose that it is also unalterable. For alteration is movement in respect of quality; and qualitative states and dispositions, such as health and disease, do not come into being without changes of properties. But all natural bodies which change their properties we see to be subject without exception to increase and diminution. This is the case, for instance, with the bodies of animals and their parts and with vegetable bodies, and similarly also with those of the elements. And so, if the body which moves with a circular motion cannot admit of increase or diminution, it is reasonable to suppose that it is also unalterable.

The reasons why the primary body is eternal and not subject to increase or diminution, but unaging and unalterable and unmodified, will be clear from what has been said to any one who believes in our assumptions. Our theory seems to confirm experience and to be confirmed by it. For all men have some conception of the nature of the gods, and all who believe in the existence of gods at all, whether barbarian or Greek, agree in allotting the highest place to the deity, surely because they suppose that immortal is linked with immortal and regard any other supposition as inconceivable. If then there is, as there certainly is, anything divine, what we have just said about the primary bodily substance was well said. The mere evidence of the senses is enough to convince us of this, at least with human certainty. For in the whole range of time past, so far as our inherited records reach, no change appears to have taken place either in the whole scheme of the outermost heaven or in any of its proper parts. The common name, too, which has been handed down from our distant ancestors even to our own day, seems to show that they conceived of it in the fashion which we have been expressing. The same ideas, one must believe, recur in men's minds not once or twice but again and again. And so, implying that the primary body is something else beyond earth, fire, air, and water, they gave the highest place a name of its own, *aither*, derived from the fact that it 'runs always' for an eternity of time. Anaxagoras, however, scandalously misuses this name, taking *aither* as equivalent to fire.

It is also clear from what has been said why the number of what we call simple bodies cannot be greater than it is. The motion of a simple body must itself be simple, and we assert that there are only these two simple motions, the circular and

---

[4] *Physics*, I. 7–9.

the straight, the latter being subdivided into motion away from and motion towards the centre.

1.3 Study Questions

QUES. 1.1 What is the significance of the *triad*, the number three, in Aristotle's world-view?

a) How many ways can time be divided? How many ways can objects be divided?
b) Are all divisible objects continuous? Are all continuous objects divisible?
c) How does Aristotle describe the grouping of things and of men?
d) In what way is the triad significant for the worship of the Gods?

QUES. 1.2 How many forms of simple motion are there?

a) What is meant by locomotion? What is capable of locomotion?
b) What two general classes of locomotion are there?
c) What does Aristotle mean by the terms "upward" and "downward"?
d) Why do you think Aristotle classifies simple motion as he does? Is there a better method of classification?

QUES. 1.3 Are the heavenly and earthly objects made of the same substance?

a) How does Aristotle classify types of bodies? Provide an example of each.
b) What is the relationship between the type of body and its motion? As an example, consider *fire*. What is its *natural* motion? What would be an *unnatural* motion for fire?
c) Can bodies ever exhibit unnatural motion? If so, under what circumstance(s)?
d) How does Aristotle infer the existence of a hitherto unknown type of substance?

QUES. 1.4 Are heavenly bodies heavy or light?

a) Define the terms "heavy" and "light." Are lightness and heaviness absolute, or relative properties?
b) What is the contrary, or opposite, of circular motion? What is the contrary of natural motion? And what do these considerations imply about heavenly objects?

QUES. 1.5 Are the heavenly bodies subject to alteration and decay?

a) Does a simple body undergoing circular motion have a contrary? Why or why not? What does this imply about the generation and decay of such bodies?
b) More generally, what is the role of contraries in the generation and decay of substances?
c) How does Aristotle use the term "motion"? Specifically, is all motion necessarily "locomotion"? In what sense are alteration and decay also types of "motion"?
d) Are Aristotle's views on the age of the heavens consistent with the observations of the ancients?
e) What, then, is the substance of the heavenly bodies? Is it one of the four earthly substances?

1.4 Exercises

Ex. 1.1 (SCIENCE AND SPECULATION ESSAY). How does Aristotle infer the existence of *aether*? Does he have any direct experimental evidence of aether? Is his reasoning sound? What about today: do modern scientists ever postulate the existence of things for which they have no direct experimental evidence? Can you cite an example? More generally, *should* scientists and philosophers engage in such speculation? And if so, are there any limitations on the types of entities which should be postulated?

1.5 Vocabulary

1. Magnitude	15. Compelled
2. Continuum	16. Ungenerated
3. Divisible	17. Substrate
4. Appropriate	18. Contrary
5. Continuity	19. Kindred
6. Virtue	20. Disposition
7. Accordance	21. Diminution
8. Subsequent	22. Supposition
9. Locomotion	23. Inconceivable
10. Compound	24. Aither
11. Infinite	25. Assert
12. Composite	26. Allot
13. Ordain	27. Recur
14. Proportionate	28. Infer

Chapter 2
The Shape and Motion of the Heavens

*Nothing which concerns the eternal can be a matter of chance
or spontaneity.*

— Aristotle

2.1 Introduction

At the outset of Book I of *On the Heavens*, Aristotle argued that the celestial bodies
cannot be comprised of one of the four commonly known elements—earth, water,
air and fire. This is because these four elements, when left to themselves, obviously
exhibit a linear motion either upward or downward. Witness, for example, the natural
motion of a dropped rock or the ascent of tongues of fire. By contrast, the heavenly
bodies obviously exhibit a naturally circular motion. The sun, moon, planets and
stars rise and set daily. This natural motion has persisted without alteration or decay
throughout recorded history—indeed from eternity, he argues—a fact which proves
that the celestial bodies are comprised of a fifth element, which Aristotle calls *æther*.
This is where the previous reading selection left off—at the end of Chap. 3 of Book
I. In the reading below, we will jump into Chap. 4 of Book II. But before doing so,
what did we skip over?

As it turns out, quite a lot. At the outset of Chap. 4 of Book I, Aristotle asks
whether the world (*i.e.* the universe) is finite or infinite in size. After exploring the
difficulties that the rotational motion of an infinite body would entail, he concludes
that "it is clear that the body of the universe is not infinite." According to Aristotle, the
world is bounded by the sphere of fixed stars which rotates once per day around the
earth. At the beginning of Chap. 8, he then asks whether there might be not just one,
but perhaps many universes. While this may seem absurd, Aristotle considers such
questions to be "all-important" in his search for truth. Indeed, the idea that there are
multiple universes is considered by some modern cosmologists to be a plausible, or
even necessary, explanation of the *anthropic principle*—that the exquisite suitability
of the universe in which we find ourselves can be best understood if ours is but one
of many universes which exist, or have existed. As for Aristotle, he concludes that
more than one universe is impossible, and moreover that the one in which we live
is eternal. His adversaries, however, argued that the heavens are *not* eternal, that
they had a beginning and may some day suffer destruction, and that their present
motion is transient and thus unnatural. So at the outset of Book II, Aristotle turns

K. Kuehn, *A Student's Guide Through the Great Physics Texts*,
Undergraduate Lecture Notes in Physics, DOI 10.1007/978-1-4939-1360-2_2,
© Springer Science+Business Media, LLC 2015

his attention to their arguments. In addressing them, Aristotle is led to consider the source, or origin, of movement itself. Perhaps surprisingly, in the first three chapters of Book II, Aristotle claims that the heavenly bodies are in some sense *animate*—that they possess a "principle of movement" similar to that which allows animals to move from place to place. In plants and animals, this principle of movement is linked not only to their possession of a soul, but also to a certain internal asymmetry which they possess. For example in plants, growth is upward, toward where their shoots aim; and in animals, locomotion is forward, toward where their senses are directed. Lacking any distinction of their parts, these would be incapable of initiating such movements. But what about the perfect spheres which comprise the heavens and on which the planets and stars ride? Lacking any distinction of parts, they would seem to be incapable of initiating movement. Their movement must be caused, then, by a *prime mover*—an immortal and divine celestial sphere which surrounds them, contains them, and generates (or perhaps inspires) their movement.

It is in Chap. 4 of Book II that we now pick up our reading once again. Herein, Aristotle examines how the motion of the heavens are related to, or perhaps even dictated by, their shape. He begins by ranking the planar and solid figures according to the number of edges and angles they have. How does this geometrical classification inform his opinion on the shape of the heavens? Are his arguments reasonable? Moreover, how is his theory of heavenly motion informed by his understanding of the growth and decay of animals? Perhaps most interesting are Aristotle's views, expressed in Chap. 5, on *chance* and *spontaneity*. In particular, can chance be the *cause of the motion* of the heavens? Which (if any) of his arguments pertaining to the shape and motion of the heavens would suffer if he were to admit chance or spontaneity as valid modes of explanation?

2.2   Reading: Aristotle, *On the Heavens*

Aristotle, On the Heavens, in *Aristotle: I, Great Books of the Western World*, vol. 8, edited by Robert Maynard Hutchins, Encyclopedia Britannica, 1952. Book II.

2.2.1   Chapter 4

The shape of the heaven is of necessity spherical; for that is the shape most appropriate to its substance and also by nature primary.

First, let us consider generally which shape is primary among planes and solids alike. Every plane figure must be either rectilinear or curvilinear. Now the rectilinear is bounded by more than one line, the curvilinear by one only. But since in any kind the one is naturally prior to the many and the simple to the complex, the circle will be the first of plane figures. Again, if by complete, as previously defined,[1] we mean a thing outside which no part of itself can be found, and if addition is always possible to the straight line but never to the circular, clearly the line which embraces the circle is complete. If then the complete is prior to the incomplete, it follows on this ground also that the circle is primary among figures. And the sphere holds the same position among solids. For it alone is embraced by a single surface, while rectilinear solids have several. The sphere is among solids what the circle is among plane figures. Further, those who divide bodies into planes and generate them out of planes seem to bear witness to the truth of this. Alone among solids they leave the sphere undivided, as not possessing more than one surface: for the division into surfaces is not just dividing a whole by cutting it into its parts, but division of another fashion into parts different in form. It is clear, then, that the sphere is first of solid figures.

If, again, one orders figures according to their numbers, it is most natural to arrange them in this way. The circle corresponds to the number one, the triangle, being the sum of two right angles, to the number two. But if one is assigned to the triangle, the circle will not be a figure at all.

Now the first figure belongs to the first body, and the first body is that at the farthest circumference. It follows that the body which revolves with a circular movement must be spherical. The same then will be true of the body continuous with it: for that which is continuous with the spherical is spherical. The same again holds of the bodies between those and the centre. Bodies which are bounded by the spherical and in contact with it must be, as wholes, spherical; and the bodies below the sphere of the planets are contiguous with the sphere above them. The sphere then will be spherical throughout; for every body within it is contiguous and continuous with spheres.

Again, since the whole revolves, palpably and by assumption, in a circle, and since it has been shown that outside the farthest circumference there is neither void nor place, from these grounds also it will follow necessarily that the heaven is spherical. For if it is to be rectilinear in shape, it will follow that there is place and body and void without it. For a rectilinear figure as it revolves never continues in the same room, but where formerly was body, is now none, and where now is none, body will be in a moment because of the projection at the corners. Similarly, if the world had some other figure with unequal radii, if, for instance, it were lentiform, or oviform, in every case we should have to admit space and void outside the moving body, because the whole body would not always occupy the same room.

Again, if the motion of the heaven is the measure of all movements whatever in virtue of being alone continuous and regular and eternal, and if, in each kind, the measure is the minimum, and the minimum movement is the swiftest, then, clearly, the movement of the heaven must be the swiftest of all movements. Now of lines which return upon themselves the line which bounds the circle is the shortest; and that movement is the swiftest which follows the shortest line. Therefore, if the heaven

---

[1] Aristotle's *Physics*, III. 207ᵃ 8.

moves in a circle and moves more swiftly than anything else, it must necessarily be spherical.

Corroborative evidence may be drawn from the bodies whose position is about the centre. If earth is enclosed by water, water by air, air by fire, and these similarly by the upper bodies—which while not continuous are yet contiguous with them—and if the surface of water is spherical, and that which is continuous with or embraces the spherical must itself be spherical, then on these grounds also it is clear that the heavens are spherical. But the surface of water is seen to be spherical if we take as our starting-point the fact that water naturally tends to collect in a hollow place— 'hollow' meaning 'nearer the centre'. Draw from the centre the lines AB, AC, and let their extremities be joined by the straight line BC. The line AD, drawn to the base of the triangle, will be shorter than either of the radii. Therefore the place in which it terminates will be a hollow place. The water then will collect there until equality is established, that is until the line AE is equal to the two radii. Thus water forces its way to the ends of the radii, and there only will it rest: but the line which connects the extremities of the radii is circular: therefore the surface of the water BEC is spherical.

It is plain from the foregoing that the universe is spherical. It is plain, further, that it is turned (so to speak) with a finish which no manufactured thing nor anything else within the range of our observation can even approach. For the matter of which these are composed does not admit of anything like the same regularity and finish as the substance of the enveloping body; since with each step away from earth the matter manifestly becomes finer in the same proportion as water is finer than earth.

2.2.2   Chapter 5

Now there are two ways of moving along a circle, from A to B or from A to C, and we have already explained[2] that these movements are not contrary to one another. But nothing which concerns the eternal can be a matter of chance or spontaneity, and the heaven and its circular motion are eternal. We must therefore ask why this motion takes one direction and not the other. Either this is itself an ultimate fact or there is an ultimate fact behind it. It may seem evidence of excessive folly or excessive zeal to try to provide an explanation of some things, or of everything, admitting no exception. The criticism, however, is not always just: one should first consider what reason there is for speaking, and also what kind of certainty is looked for, whether human merely or of a more cogent kind. When any one shall succeed in finding proofs of greater precision, gratitude will be due to him for the discovery, but at present we must be content with a probable solution. If nature always follows the best course possible, and, just as upward movement is the superior form of rectilinear movement, since the upper region is more divine than the lower, so forward movement is superior to

---

[2] Aristotle's *On the Heavens*, Book I. 4.

backward, then front and back exhibits, like right and left, as we said before and as the difficulty just stated itself suggests, the distinction of prior and posterior, which provides a reason and so solves our difficulty. Supposing that nature is ordered in the best way possible, this may stand as the reason of the fact mentioned. For it is best to move with a movement simple and unceasing, and, further, in the superior of two possible directions.

2.2.3 Chapter 6

We have next to show that the movement of the heaven is regular and not irregular. This applies only to the first heaven and the first movement; for the lower spheres exhibit a composition of several movements into one. If the movement is uneven, clearly there will be acceleration, maximum speed, and retardation, since these appear in all irregular motions. The maximum may occur either at the starting-point or at the goal or between the two; and we expect natural motion to reach its maximum at the goal, unnatural motion at the starting-point, and missiles midway between the two. But circular movement, having no beginning or limit or middle in the direct sense of the words, has neither whence nor whither nor middle: for in time it is eternal, and in length it returns upon itself without a break. If then its movement has no maximum, it can have no irregularity, since irregularity is produced by retardation and acceleration. Further, since everything that is moved is moved by something, the cause of the irregularity of movement must lie either in the mover or in the moved or both. For if the mover moved not always with the same force, or if the moved were altered and did not remain the same, or if both were to change, the result might well be an irregular movement in the moved. But none of these possibilities can be conceived as actual in the case of the heavens. As to that which is moved, we have shown that it is primary and simple and ungenerated and indestructible and generally unchanging; and the mover has an even better right to these attributes. It is the primary that moves the primary, the simple the simple, the indestructible and ungenerated that which is indestructible and ungenerated. Since then that which is moved, being a body, is nevertheless unchanging, how should the mover, which is incorporeal, be changed?

It follows then, further, that the motion cannot be irregular. For if irregularity occurs, there must be change either in the movement as a whole, from fast to slow and slow to fast, or in its parts. That there is no irregularity in the parts is obvious, since, if there were, some divergence of the stars would have taken place before now in the infinity of time, as one moved slower and another faster: but no alteration of their intervals is ever observed. Nor again is a change in the movement as a whole admissible. Retardation is always due to incapacity, and incapacity is unnatural. The incapacities of animals, age, decay, and the like, are all unnatural, due, it seems, to the fact that the whole animal complex is made up of materials which differ in respect of their proper places, and no single part occupies its own place. If therefore that which is primary contains nothing unnatural, being simple and unmixed and in its proper place and having no contrary, then it has no place for incapacity, nor, consequently, for retardation or (since acceleration involves retardation) for acceleration. Again, it is inconceivable that the mover should first show incapacity for an infinite time, and capacity afterwards for another infinity. For clearly nothing which, like incapacity, is unnatural ever continues for an infinity of time; nor does the unnatural endure as long as the natural, or any form of incapacity as long as the capacity. But if the movement is retarded it must necessarily be retarded for an infinite time. Equally impossible is perpetual acceleration or perpetual retardation. For such movement would be infinite and indefinite, but every movement, in our view, proceeds from one point to another and is definite in character. Again, suppose one assumes a minimum time in less than which the heaven could not complete its movement. For, as a given walk or a given exercise on the harp cannot take any and every time, but every performance has its definite minimum time which is unsurpassable, so, one might suppose, the movement of the heaven could not be completed in any and every time. But in that case perpetual acceleration is impossible (and, equally, perpetual retardation: for the argument holds of both and each), if we may take acceleration to proceed by identical or increasing additions of speed and for an infinite time. The remaining alternative is to say that the movement exhibits an alternation of slower and faster: but this is a mere fiction and quite inconceivable. Further, irregularity of this kind would be particularly unlikely to pass unobserved, since contrast makes observation easy.

That there is one heaven, then, only, and that it is ungenerated and eternal, and further that its movement is regular, has now been sufficiently explained.

2.3 Study Questions

QUES. 2.1 What is the shape of the heavens?

a) How does Aristotle classify plane figures? What shape does he rank as the first, or of the highest priority? And how do the notions of completeness, divisibility and number figure into his ranking?

b) What does all this have to do with the shape of the heavens? Is Aristotle justified in relating figures and bodies in this way?

c) What does the circular motion of the heavens imply about its shape? In particular, why does he say the heavens cannot be egg-shaped or lens-shaped? Does he consider the heavens to be a body, or a void? And what conceptual difficulty is he encountering here?

d) What does the speed of the heavens imply about its shape? Is his argument convincing? What corroborating evidence does he cite pertaining to the shape of the heavens?

QUES. 2.2 What is the motion of the heavens?

a) Is Aristotle correct that "nothing which concerns the eternal can be a matter of chance or spontaneity?" How does this fit into his argument?

b) Are the motions of the heavens *regular* or *irregular*? What does this mean?

c) What evidence does he provide that the heavens are *unchangeable*? And why, according to Aristotle, do animals decay, but the heavens do not?

2.4   Exercises

Ex. 2.1 (CAUSALITY AND CHANCE). Is *chance* a valid mode of explanation for observed phenomena? Consider: do we learn anything from positing chance as the *cause* of some thing? What does your answer to this question imply?

2.5   Vocabulary

1. Rectilinear
2. Curvilinear
3. Contiguous
4. Palpably
5. Void
6. Corroborative
7. Folly
8. Cogent
9. Posterior
10. Acceleration
11. Incorporeal
12. Divergence
13. Admissible
14. Perpetual
15. Fiction

Chapter 3
Harmony and Complexity

Nature is no wanton or random creator.

—Aristotle

3.1 Introduction

In Chaps. 4–6 of Book II, Aristotle argued that the motions of the heavenly spheres which surround the Earth are regular and eternal. That is, they rotate forever at a constant rate without acceleration or retardation. This must be the case because (i) their motion at present is seen to be regular, and (ii) any irregularity in their motion must have a discernible cause. After all, why would a cause of irregular motion suddenly arise, after being previously dormant? And similarly, why would a cause of irregular motion suddenly pass away, after having been previously active. Such unaccounted for causes can only be referred to chance or spontaneity, which Aristotle rejects as a valid mode of explanation for the motion of the heavens. So now, having discussed the substance (æther), size (large but finite), number (one), shape (spherical) and motion (regular) of the universe, Aristotle turns his attention to the composition, shape and motion of the bodies which populate the heavenly spheres. He begins by considering the fixed stars, then goes on to consider the complex motion of the wandering stars—the planets.

3.2 Reading: Aristotle, *On the Heavens*

Aristotle, On the Heavens, in *Aristotle: I, Great Books of the Western World*, vol. 8, edited by Robert Maynard Hutchins, Encyclopedia Britannica, 1952.

3.2.1 *Chapter 7*

We have next to speak of the stars, as they are called, of their composition, shape, and movements. It would be most natural and consequent upon what has been said that each of the stars should be composed of that substance in which their path lies, since, as we said, there is an element whose natural movement is circular. In so saying we

K. Kuehn, *A Student's Guide Through the Great Physics Texts*,
Undergraduate Lecture Notes in Physics, DOI 10.1007/978-1-4939-1360-2_3,
© Springer Science+Business Media, LLC 2015

are only following the same line of thought as those who say that the stars are fiery because they believe the upper body to be fire, the presumption being that a thing is composed of the same stuff as that in which it is situated. The warmth and light which proceed from them are caused by the friction set up in the air by their motion. Movement tends to create fire in wood, stone, and iron; and with even more reason should it have that effect on air, a substance which is closer to fire than these. An example is that of missiles, which as they move are themselves fired so strongly that leaden balls are melted; and if they are fired the surrounding air must be similarly affected. Now while the missiles are heated by reason of their motion in air, which is turned into fire by the agitation produced by their movement, the upper bodies are carried on a moving sphere, so that, though they are not themselves fired, yet the air underneath the sphere of the revolving body is necessarily heated by its motion, and particularly in that part where the sun is attached to it. Hence warmth increases as the sun gets nearer or higher or overhead. Of the fact, then, that the stars are neither fiery nor move in fire, enough has been said.

3.2.2 Chapter 8

Since changes evidently occur not only in the position of the stars but also in that of the whole heaven, there are three possibilities. Either (1) both are at rest, or (2) both are in motion, or (3) the one is at rest and the other in motion.

1. That both should be at rest is impossible; for, if the earth is at rest, the hypothesis does not account for the observations; and we take it as granted that the earth is at rest. It remains either that both are moved, or that the one is moved and the other at rest.
2. On the view, first, that both are in motion, we have the absurdity that the stars and the circles move with the same speed, *i.e.* that the pace of every star is that of the circle in which it moves. For star and circle are seen to come back to the same place at the same moment; from which it follows that the star has traversed the circle and the circle has completed its own movement, *i.e.* traversed its own circumference, at one and the same moment. But it is difficult to conceive that the pace of each star should be exactly proportioned to the size of its circle. That the pace of each circle should be proportionate to its size is not absurd but inevitable: but that the same should be true of the movement of the stars contained in the circles is quite incredible. For if, on the one hand, we suppose that the star which moves on the greater circle is necessarily swifter, clearly we also admit that if stars shifted their position so as to exchange circles, the slower would become swifter and the swifter slower. But this would show that their movement was not their own, but due to the circles. If, on the other hand, the arrangement was a chance combination, the coincidence in every case of a greater circle with a swifter movement of the star contained in it is too much to believe. In one or two cases it might not inconceivably fall out so, but to imagine it in every case alike

is a mere fiction. Besides, chance has no place in that which is natural, and what happens everywhere and in every case is no matter of chance.

3. The same absurdity is equally plain if it is supposed that the circles stand still and that it is the stars themselves which move. For it will follow that the outer stars are the swifter, and that the pace of the stars corresponds to the size of their circles.

Since, then, we cannot reasonably suppose either that both are in motion or that the star alone moves, the remaining alternative is that the circles should move, while the stars are at rest and move with the circles to which they are attached. Only on this supposition are we involved in no absurd consequence. For, in the first place, the quicker movement of the larger circle is natural when all the circles are attached to the same centre. Whenever bodies are moving with their proper motion, the larger moves quicker. It is the same here with the revolving bodies: for the arc intercepted by two radii will be larger in the larger circle, and hence it is not surprising that the revolution of the larger circle should take the same time as that of the smaller. And secondly, the fact that the heavens do not break in pieces follows not only from this but also from the proof already given of the continuity of the whole.

Again, since the stars are spherical, as our opponents assert and we may consistently admit, inasmuch as we construct them out of the spherical body, and since the spherical body has two movements proper to itself, namely rolling and spinning, it follows that if the stars have a movement of their own, it will be one of these. But neither is observed.

1. Suppose them to *spin*. They would then stay where they were, and not change their place, as, by observation and general consent, they do. Further, one would expect them all to exhibit the same movement: but the only star which appears to possess this movement is the sun, at sunrise or sunset, and this appearance is due not to the sun itself but to the distance from which we observe it. The visual ray being excessively prolonged becomes weak and wavering. The same reason probably accounts for the apparent twinkling of the fixed stars and the absence of twinkling in the planets. The planets are near, so that the visual ray reaches them in its full vigour, but when it comes to the fixed stars it is quivering because of the distance and its excessive extension; and its tremor produces an appearance of movement in the star: for it makes no difference whether movement is set up in the ray or in the object of vision.

2. On the other hand, it is also clear that the stars do not *roll*. For rolling involves rotation: but the 'face', as it is called, of the moon is always seen. Therefore, since any movement of their own which the stars possessed would presumably be one proper to themselves, and no such movement is observed in them, clearly they have no movement of their own.

There is, further, the absurdity that nature has bestowed upon them no organ appropriate to such movement. For nature leaves nothing to chance, and would not, while caring for animals, overlook things so precious. Indeed, nature seems deliberately to have stripped them of everything which makes self-originated progression possible, and to have removed them as far as possible from things which have organs of movement. This is just why it seems proper that the whole heaven and every star should be spherical. For while of all shapes the sphere is the most convenient for movement in one place, making possible, as it does, the swiftest and most self-contained motion, for forward movement it is the most unsuitable, least of all resembling shapes which are self-moved, in that it has no dependent or projecting part, as a rectilinear figure has, and is in fact as far as possible removed in shape from ambulatory bodies. Since, therefore, the heavens have to move in one place, and the stars are not required to move themselves forward, it is natural that both should be spherical—a shape which best suits the movement of the one and the immobility of the other.

3.2.3   Chapter 9

From all this it is clear that the theory that the movement of the stars produces a harmony, *i.e.* that the sounds they make are concordant, in spite of the grace and originality with which it has been stated, is nevertheless untrue. Some thinkers suppose that the motion of bodies of that size must produce a noise, since on our earth the motion of bodies far inferior in size and in speed of movement has that effect. Also, when the sun and the moon, they say, and all the stars, so great in number and in size, are moving with so rapid a motion, how should they not produce a sound immensely great? Starting from this argument and from the observation that their speeds, as measured by their distances, are in the same ratios as musical concordances, they assert that the sound given forth by the circular movement of the stars is a harmony. Since, however, it appears unaccountable that we should not hear this music, they explain this by saying that the sound is in our ears from the very moment of birth and is thus indistinguishable from its contrary silence, since sound and silence are discriminated by mutual contrast. What happens to men, then, is just what happens to coppersmiths, who are so accustomed to the noise of the smithy that it makes no difference to them. But, as we said before, melodious and poetical as the theory is, it cannot be a true account of the facts. There is not only the absurdity of our hearing nothing, the ground of which they try to remove, but also the fact that no effect other than sensitive is produced upon us. Excessive noises, we know, shatter the solid bodies even of inanimate things: the noise of thunder, for instance, splits rocks and the strongest of bodies. But if the moving bodies are so great, and the sound which penetrates to us is proportionate to their size, that sound must needs reach us in an intensity many times that of thunder, and the force of its action must be immense. Indeed the reason why we do not hear, and show in our bodies none of the effects of violent force, is easily given: it is that there is no noise. But not only is the explanation evident; it is also a corroboration of the truth of the views we have advanced. For the very difficulty which made the Pythagoreans say that the motion of the stars produces a concord corroborates our view. Bodies which are themselves in motion, produce noise and friction: but those which are attached or fixed to a moving body, as the parts to a ship, can no more create noise, than a ship on a river moving with the stream. Yet by the same argument one might say it was absurd that on a large vessel the motion of mast and poop should not make a great noise, and the like might be said of the movement of the vessel itself. But sound is caused when a moving body is enclosed in an unmoved body, and cannot be caused by one enclosed in, and continuous with, a moving body which creates no friction. We may say, then, in this matter that if the heavenly bodies moved in a generally diffused mass of air or fire, as every one supposes, their motion would necessarily cause a noise of tremendous strength and such a noise would necessarily reach and shatter us. Since, therefore, this effect is evidently not produced, it follows that none of them can move with the motion either of animate nature or of constraint. It is as though nature had foreseen the result, that if their movement were other than it is, nothing on this earth could maintain its character.

That the stars are spherical and are not self-moved, has now been explained.

3.2.4   Chapter 10

With their order—I mean the position of each, as involving the priority of some and the posteriority of others, and their respective distances from the extremity—with this astronomy may be left to deal, since the astronomical discussion is adequate. This discussion shows that the movements of the several stars depend, as regards the varieties of speed which they exhibit, on the distance of each from the extremity. It is established that the outermost revolution of the heavens is a simple movement and the swiftest of all, and that the movement of all other bodies is composite and relatively slow, for the reason that each is moving on its own circle with the reverse motion to that of the heavens. This at once leads us to expect that the body which is nearest to that first simple revolution should take the longest time to complete its circle, and that which is farthest from it the shortest, the others taking a longer time the nearer they are and a shorter time the farther away they are. For it is the nearest body which is most strongly influenced, and the most remote, by reason of its distance, which is least affected, the influence on the intermediate bodies varying, as the mathematicians show, with their distance.

3.2.5   Chapter 11

With regard to the shape of each star, the most reasonable view is that they are spherical. It has been shown that it is not in their nature to move themselves, and, since nature is no wanton or random creator, clearly she will have given things which possess no movement a shape particularly unadapted to movement. Such a shape is the sphere, since it possesses no instrument of movement. Clearly then their mass will have the form of a sphere. Again, what holds of one holds of all, and the evidence of our eyes shows us that the moon is spherical. For how else should the moon as it waxes and wanes show for the most part a crescent-shaped or gibbous figure, and only at one moment a half-moon? And astronomical arguments give further confirmation; for no other hypothesis accounts for the crescent shape of the sun's eclipses. One, then, of the heavenly bodies being spherical, clearly the rest will be spherical also.

3.2.6   Chapter 12

There are two difficulties, which may very reasonably here be raised, of which we must now attempt to state the probable solution: for we regard the zeal of one whose thirst after philosophy leads him to accept even slight indications where it is very difficult to see one's way, as a proof rather of modesty than of overconfidence.

Of many such problems one of the strangest is the problem why we find the greatest number of movements in the intermediate bodies, and not, rather, in each successive body a variety of movement proportionate to its distance from the primary motion. For we should expect, since the primary body shows one motion only, that the body which is nearest to it should move with the fewest movements, say two, and the one next after that with three, or some similar arrangement. But the opposite is the case. The movements of the sun and moon are fewer than those of some of the planets. Yet these planets are farther from the centre and thus nearer to the primary body than they, as observation has itself revealed. For we have seen the moon, half-full, pass beneath the planet Mars, which vanished on its shadow side and came forth by the bright and shining part. Similar accounts of other stars are given by the Egyptians and Babylonians, whose observations have been kept for very many years past, and from whom much of our evidence about particular stars is derived. A second difficulty which may with equal justice be raised is this. Why is it that the primary motion includes such a multitude of stars that their whole array seems to defy counting, while of the other stars each one is separated off, and in no case do we find two or more attached to the same motion?

On these questions, I say, it is well that we should seek to increase our understanding, though we have but little to go upon, and are placed at so great a distance from the facts in question. Nevertheless there are certain principles on which if we base our consideration we shall not find this difficulty by any means insoluble. We may object that we have been thinking of the stars as mere bodies, and as units with a serial order indeed but entirely inanimate; but should rather conceive them as enjoying life and action. On this view the facts cease to appear surprising. For it is natural that the best-conditioned of all things should have its good without action, that that which is nearest to it should achieve it by little and simple action, and that which is farther removed by a complexity of actions, just as with men's bodies one is in good condition without exercise at all, another after a short walk, while another requires running and wrestling and hard training, and there are yet others who however hard they worked themselves could never secure this good, but only some substitute for it. To succeed often or in many things is difficult. For instance, to throw 10,000 Coan throws with the dice would be impossible, but to throw 1 or 2 is comparatively easy.

In action, again, when A has to be done to get B, B to get C, and C to get D, one step or two present little difficulty, but as the series extends the difficulty grows. We must, then, think of the action of the lower stars as similar to that of animals and plants. For on our earth it is man that has the greatest variety of actions—for there are many goods that man can secure; hence his actions are various and directed to ends beyond them—while the perfectly conditioned has no need of action, since it is itself the end, and action always requires two terms, end and means. The lower animals have less variety of action than man; and plants perhaps have little action and of one kind only. For either they have but one attainable good (as indeed man has), or, if several, each contributes directly to their ultimate good. One thing then has and enjoys the ultimate good, other things attain to it, one immediately by few steps, another by many, while yet another does not even attempt to secure it but is satisfied to reach a point not far removed from that consummation. Thus, taking health as the end, there will be one thing that always possesses health, others that attain it, one by reducing flesh, another by running and thus reducing flesh, another by taking steps to enable himself to run, thus further increasing the number of movements, while another cannot attain health itself, but only running or reduction of flesh, so that one or other of these is for such a being the end. For while it is clearly best for any being to attain the real end, yet, if that cannot be, the nearer it is to the best the better will be its state. It is for this reason that the earth moves not at all and the bodies near to it with few movements. For they do not attain the final end, but only come as near to it as their share in the divine principle permits. But the first heaven finds it immediately with a single movement, and the bodies intermediate between the first and last heavens attain it indeed, but at the cost of a multiplicity of movement.

As to the difficulty that into the one primary motion is crowded a vast multitude of stars, while of the other stars each has been separately given special movements of its own, there is in the first place this reason for regarding the arrangement as a natural one. In thinking of the life and moving principle of the several heavens one must regard the first as far superior to the others. Such a superiority would be reasonable. For this single first motion has to move many of the divine bodies, while the numerous other motions move only one each, since each single planet moves with a variety of motions. Thus, then, nature makes matters equal and establishes a certain order, giving to the single motion many bodies and to the single body many motions. And there is a second reason why the other motions have each only one body, in that each of them except the last, *i.e.* that which contains the one star, is really moving many bodies. For this last sphere moves with many others, to which it is fixed, each sphere being actually a body; so that its movement will be a joint product. Each sphere, in fact, has its particular natural motion, to which the general movement is, as it were, added. But the force of any limited body is only adequate to moving a limited body.

The characteristics of the stars which move with a circular motion, in respect of substance and shape, movement and order, have now been sufficiently explained.

3.3   Study Questions

QUES.3.1 What, according to Aristotle, is the composition, shape and movement of the heavenly bodies?

a) Of what are the stars made? Why does Aristotle believe this? How is his argument (though not his conclusion) similar to the one made by his opponents?
b) What are the three conceivable motions of the stars which he considers?
c) Why does Aristotle reject the possibility that both the stars and the sphere of the heavens are at rest? Why does he reject the possibility that both are in motion? What principle does he invoke to support this conclusion?
d) Why does he reject the possibilities that the stars spin? That they roll? How, finally, do they move about?

QUES. 3.2 Does the rapid motion of the heavenly bodies make sound?

a) What is the view of the Pythagoreans? In what sense are the motions of the stars harmonious? And why might we not hear such sounds?
b) Why does Aristotle reject the view of the Pythagoreans? In particular, what is the cause of sound from moving objects? And what would be the effect of such sound upon the earth, and upon man? Do you find this "anthropic" argument convincing?

QUES. 3.3 What determines the speed of the various heavenly bodies? Which bodies move most rapidly? Most slowly?

QUES. 3.4 What is the shape of the heavenly bodies? What role do the principles of the *economy* of nature and of the *uniformity* of nature play in Aristotle's argument? Does he provide any empirical evidence for his view?

QUES. 3.5 What is the source of all movement in the heavens?

a) Which heavenly body, or bodies, exhibit the simplest type of motion? What exhibits the most complex type of movement?
b) How does the complexity of motion differ with distance from the center of the world? For instance, what exhibits the more complex motion, the moon or the planet Mars?
c) What difficulty does the variety of heavenly movements raise for Aristotle? What assumption gives rise to this apparent difficulty?
d) How does Aristotle address this difficulty? In particular, how does thinking of the planets as *animate*, like animals or man, provide a key to unlocking the mystery of their movements?
e) How does Aristotle's solution make use of his notion of the ease with which various bodies may finally achieve their purpose, or end? What does this have to do with the degree of perfection, or divinity, of each body?
f) Is Aristotle correct? How do you know?

3.4 Exercises

Ex. 3.1 (ANIMISM ESSAY). Is Aristotle correct in claiming that certain bodies in nature are, in fact, *animate*? What is meant by the term *animate*? Would you consider anything in nature to be animate? If so, then where would you draw a line between animate and inanimate objects? If not, then can you explain the behavior of people and of animals?

Ex. 3.2 (PLANETARY OBSERVATION). This exercise consists of charting the motion of one of the planets over the course of several months. With careful observation, you should be able to observe planetary progression, retrogression and stationary points. You will need to set aside 15 min once each week to make observations. Your particular choice of planet will depend upon which is visible in the night sky, either after sunset or before sunrise. Such information is available in various guides to the night sky for particular years and seasons.[1] When observing, select three bright stars near your chosen planet. These stars can be identified with the help of a sky atlas.[2] Measure the angular distance of your planet from each of your chosen stars. You can make rough angular measurements using the angular width of your fingers, or more accurate measurement using a cross-staff.[3] Make a plot of the location of your planet with respect to these stars on a photocopy of your sky atlas. Over the course of 2 or 3 months, you should be able to observe the motion of your planet through the zodiac.

3.5 Vocabulary

1. Composition
2. Presumption
3. Circumference
4. Conceive
5. Inevitable
6. Fiction
7. Vigour
8. Rectilinear
9. Grace
10. Concordance
11. Inanimate
12. Corroboration
13. Pythagorean
14. Gibbous
15. Serial
16. Coan
17. Consummation

---

[1] I have used Chapman, D. (Ed.), *Observer's Handbook*, Royal Astronomical Society of Canada, 2013.

[2] For example, Sinnott, R. W., *Sky and Telescope's Pocket Sky Atlas*, New Track Media, LLC, Cambridge, Massachusetts, 2006.

[3] For directions on constructing and using a simple, low-cost, cross-staff, see the Exercises of Chap. 7 of this volume.

Chapter 4
Earth at the Center of the World

Any body endowed with weight, of whatever size, moves toward the center.

—Aristotle

4.1 Introduction

Aristotle argues, in Book II of *On the Heavens*, that the stars are not self-moved, like animals. Rather, they are fixed to a vast celestial sphere which carries them around the Earth once each day without changing their positions relative to one another (see Fig. 4.1). Among these constellations, 12 are of particular importance—the zodiac.[1] The constellations of the zodiac all lie in the immediate vicinity of a particular great circle on the celestial sphere—the ecliptic—which itself is inclined at an angle of 23° with respect to the equator of the celestial sphere.[2] Polaris is located on the axis of the celestial sphere, directly above the north pole of Earth.[3]

The motions of the sun, the moon, and the five wandering stars—Saturn, Jupiter, Mars, Venus and Mercury—are more complicated than the motion of the fixed stars. It is true that on any given day (or night) all of these heavenly bodies are seen to rise above the eastern horizon, traverse the sky from east to west, and set below the western horizon—just like the stars. But on further examination, the sun, the moon and the planets are seen to exhibit a second, west-to-east motion, which is superimposed on the aforementioned daily, or *diurnal*, motion. That is, on successive days (or nights) the sun, the moon and the planets are seen to drift slowly eastward so that they are not found among the same constellations as on the previous night. For the sun, this eastward progression through the entire zodiac takes about 365 days. For the moon, this eastward progression through the zodiac takes about 27 days. And for the planets, this eastward progression through the zodiac not only differs from planet to planet, but also is punctuated by periods of retrogression, when the planet temporarily

---

[1] The constellations which make up the zodiac are described in more detail in Bede's *The Reckoning of Time*. See Chap. 8 of the present volume.

[2] A *great circle* on a sphere is formed when a plane surface bisects the sphere. A *small circle*, by contrast, is formed when the plane does not cut through the exact center point of the sphere. Both the ecliptic and the celestial equator are thus great circles. For more information on spherical geometry, see Waldseemüller's *Introduction to Cosmography*, included in Chap. 9 of the present volume.

[3] To familiarize yourself with the arrangement of some constellations, do Ex. 5.3.

K. Kuehn, *A Student's Guide Through the Great Physics Texts*,
Undergraduate Lecture Notes in Physics, DOI 10.1007/978-1-4939-1360-2_4,
© Springer Science+Business Media, LLC 2015

Fig. 4.1 The sphere
of fixed stars with Earth at the
center

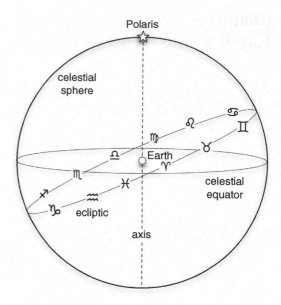

moves westward through the zodiac before continuing on its eastward progression.[4]
In order to understand these complex motions, Aristotle uses a mechanical model of
the heavens comprised of rotating homocentric spheres which are nested inside the
sphere of fixed stars.[5] How might such a system of rotating (and counter-rotating)
spheres work? Consider the planet Saturn, which is affixed to a particular sphere
labeled D in Fig. 4.2.[6] From the perspective of an Earth-bound observer at point O,
Saturn is seen to have four distinct periodic motions which must be accounted for:

(A) Sphere A, rotates daily around the axis $P_{\star}$. This governs Saturn's daily motion
    around the earth. The stars are affixed to this outermost celestial sphere.
(B) Sphere B is connected to sphere A; it rotates once every 30 years about an axis
    P_{ecl}. This governs Saturn's longitudinal motion (through the zodiac).
(C) Sphere C is connected to sphere B; it rotates about axis χ. This governs Saturn's
    anomalous longitudinal progression and retrogression (through the zodiac).

---

[4] To observe the motion of the planets and the moon through the zodiac for yourself, do Exercises 3.2
and 8.4.

[5] Aristotle based his onion-like model of the heavenly spheres on the previous astronomical work
of Eudoxus and Callippus. For a detailed account of the development of his model, see Simplicius'
commentary in Mueller, I. (Ed.), *Simplicius On Aristotle's "On the Heavens 2.10-14"*, Cornell
University Press, Ithaca, NY, 2005, pp. 33–50. Kepler summarizes Aristotle's mechanistic model
of the celestial spheres in his section entitled *Concerning the causes of movement of the planets*,
which can be found in Part II of Book IV of his *Epitome of Copernican Astronomy*. This section is
included in Chap. 16 of the present volume.

[6] Figure 4.2 is from Neugebauer, O., *A History of Ancient Mathematical Astronomy*, Springer-
Verlag, New York, Heidelberg, Berlin, 1975, p. 1360.

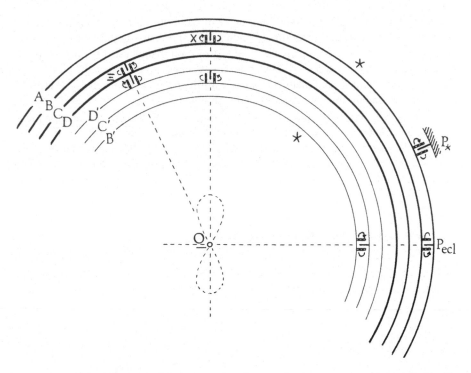

Fig. 4.2 A mechanical model of the heavens based on rotating homocentric spheres nested inside the sphere of fixed stars, P_*. (From Neugebauer's *A History of Ancient Mathematical Astronomy*, p. 1360)

(D) Sphere D is connected to sphere C; it rotates about axis $\varXi$. This governs Saturn's motion in latitude (above and below the plane of the ecliptic).

The inner three spheres, D', C' and B', are counter-rotating so as to exactly counteract all but the first of the aforementioned motions. Thus, the innermost sphere, B' has the exact same motion as the outermost sphere, A. Although not depicted, a similar set of concentric spheres can now be added inside sphere B' so as to govern the motion of the next planet, Jupiter. This procedure can be continued all the way down to Earth's moon.

While this mechanical model qualitatively describes many of the basic features of the motions of the heavens, it is lacking in several respects. For example, it does not account for the non-uniform motion of the sun as it progresses through the zodiac. Nor can it account for the fact that the planets occasionally draw nearer to, and farther from, the earth.[7] Moreover, Aristotle himself notes in the previous reading that the complexity of the planetary motions do not seem to follow a rational pattern. One

---

[7] In his *Almagest*, Ptolemy attempts to account for the the apogees and perigees of the planets by introducing non-homocentric *epicycles* upon which the planets are carried. See Chaps. 5 and 6 of the present volume.

might expect—based on the divine simplicity of the motion of the outermost sphere of fixed stars and the chaotic motion of bodies here on Earth—that the complexity of the planetary motions would increase gradually as their distances from the outermost sphere of fixed stars is increased. This, however, is not the case. Since the planets are all spherical in shape—and therefore lack the ability to move themselves—Aristotle is at pains to explain the source of their movement.

Let us now turn to our final reading selection by Aristotle. Herein, he considers the position, shape and motion of the earth in relation to the surrounding heavens. In so doing, he carefully recounts, and attempts to refute, the opinions of other thinkers such as the Pythagoreans, the Platonists and the atomists. Do you find Aristotle's arguments convincing?

4.2   Reading: Aristotle, *On the Heavens*

Aristotle, On the Heavens, in *Aristotle: I, Great Books of the Western World* (vol. 8) edited by Robert Maynard Hutchins, Encyclopedia Britannica, 1952. Book II.

4.2.1   Chapter 13

It remains to speak of the earth, of its position, of the question whether it is at rest or in motion, and of its shape.

1. As to its *position* there is some difference of opinion. Most people—all, in fact, who regard the whole heaven as finite—say it lies at the centre. But the Italian philosophers known as Pythagoreans take the contrary view. At the centre, they say, is fire, and the earth is one of the stars, creating night and day by its circular motion about the centre. They further construct another earth in opposition to ours to which they give the name counter-earth. In all this they are not seeking for theories and causes to account for observed facts, but rather forcing their observations and trying to accommodate them to certain theories and opinions of their own. But there are many others who would agree that it is wrong to give the earth the central position, looking for confirmation rather to theory than to the facts of observation. Their view is that the most precious place befits the most precious thing: but fire, they say, is more precious than earth, and the limit than the intermediate, and the circumference and the centre are limits. Reasoning on this basis they take the view that it is not earth that lies at the centre of the sphere, but rather fire. The Pythagoreans have a further reason. They hold that the most important part of the world, which is the centre, should be most strictly guarded, and name it, or rather the fire which occupies that place, the 'Guardhouse of Zeus', as if the word 'centre' were quite unequivocal, and the centre of the mathematical

figure were always the same with that of the thing or the natural centre.[8] But it is better to conceive of the case of the whole heaven as analogous to that of animals, in which the centre of the animal and that of the body are different. For this reason they have no need to be so disturbed about the world, or to call in a guard for its centre: rather let them look for the centre in the other sense and tell us what it is like and where nature has set it. That centre will be something primary and precious; but to the mere position we should give the last place rather than the first. For the middle is what is defined, and what defines it is the limit, and that which contains or limits is more precious than that which is limited, seeing that the latter is the matter and the former the essence of the system.

2. As to the position of the earth, then, this is the view which some advance, and the views advanced concerning its *rest or motion* are similar. For here too there is no general agreement. All who deny that the earth lies at the centre think that it revolves about the centre, and not the earth only but, as we said before, the counter-earth as well. Some of them even consider it possible that there are several bodies so moving, which are invisible to us owing to the interposition of the earth. This, they say, accounts for the fact that eclipses of the moon are more frequent than eclipses of the sun: for in addition to the earth each of these moving bodies can obstruct it. Indeed, as in any case the surface of the earth is not actually a centre but distant from it a full hemisphere, there is no more difficulty, they think, in accounting for the observed facts on their view that we do not dwell at the centre, than on the common view that the earth is in the middle. Even as it is, there is nothing in the observations to suggest that we are removed from the centre by half the diameter of the earth. Others, again, say that the earth, which lies at the centre, is 'rolled', and thus in motion, about the axis of the whole heaven, So it stands written in the *Timaeus*.

3. There are similar disputes about the *shape* of the earth. Some think it is spherical, others that it is flat and drum-shaped. For evidence they bring the fact that, as the sun rises and sets, the part concealed by the earth shows a straight and not a curved edge, whereas if the earth were spherical the line of section would have to be circular. In this they leave out of account the great distance of the sun from the earth and the great size of the circumference, which, seen from a distance on these apparently small circles appears straight. Such an appearance ought not to make them doubt the circular shape of the earth. But they have another argument. They say that because it is at rest, the earth must necessarily have this shape. For there are many different ways in which the movement or rest of the earth has been conceived.

The difficulty must have occurred to every one. It would indeed be a complacent mind that felt no surprise that, while a little bit of earth, let loose in mid-air, moves and will not stay still, and the more there is of it the faster it moves, the whole earth,

---

[8] For an exposition of Pythagorean cosmology, see "The Development of the Pythagorean Doctrine," by Theodor Gomperz, reprinted in Munitz, M. K. (Ed.), *Theories of the Universe*, The Free Press of Glencoe, 1957.—[*K.K.*]

free in midair, should show no movement at all. Yet here is this great weight of earth, and it is at rest. And again, from beneath one of these moving fragments of earth, before it falls, take away the earth, and it will continue its downward movement with nothing to stop it. The difficulty then, has naturally passed into a commonplace of philosophy; and one may well wonder that the solutions offered are not seen to involve greater absurdities than the problem itself.

By these considerations some have been led to assert that the earth below us is infinite, saying, with Xenophanes of Colophon, that it has 'pushed its roots to infinity',[9]—in order to save the trouble of seeking for the cause. Hence the sharp rebuke of Empedocles, in the words 'if the deeps of the earth are endless and endless the ample ether—such is the vain tale told by many a tongue, poured from the mouths of those who have seen but little of the whole'.[10] Others say the earth rests upon water. This, indeed, is the oldest theory that has been preserved, and is attributed to Thales of Miletus. It was supposed to stay still because it floated like wood and other similar substances, which are so constituted as to rest upon water but not upon air. As if the same account had not to be given of the water which carries the earth as of the earth itself! It is not the nature of water, any more than of earth, to stay in mid-air: it must have something to rest upon. Again, as air is lighter than water, so is water than earth: how then can they think that the naturally lighter substance lies below the heavier? Again, if the earth as a whole is capable of floating upon water, that must obviously be the case with any part of it. But observation shows that this is not the case. Any piece of earth goes to the bottom, the quicker the larger it is. These thinkers seem to push their inquiries some way into the problem, but not so far as they might. It is what we are all inclined to do, to direct our inquiry not by the matter itself, but by the views of our opponents: and even when interrogating oneself one pushes the inquiry only to the point at which one can no longer offer any opposition. Hence a good inquirer will be one who is ready in bringing forward the objections proper to the genus, and that he will be when he has gained an understanding of all the differences.

Anaximenes and Anaxagoras and Democritus give the flatness of the earth as the cause of its staying still. Thus, they say, it does not cut, but covers like a lid, the air beneath it. This seems to be the way of flat-shaped bodies: for even the wind can scarcely move them because of their power of resistance. The same immobility, they say, is produced by the flatness of the surface which the earth presents to the air which underlies it; while the air, not having room enough to change its place because it is underneath the earth, stays there in a mass, like the water in the case of the water-clock. And they adduce an amount of evidence to prove that air, when cut off and at rest, can bear a considerable weight.

Now, first, if the shape of the earth is not flat, its flatness cannot be the cause of its immobility. But in their own account it is rather the size of the earth than its flatness

---

[9] Diels, *Vorsokratiker*[3], II A 47 (53, 38 ff.), B 28 (63, 8). Ritter and Preller, 103 b. Cf. Burnet, E.G.P.[3] § 60.

[10] Diels, *Vors.*[3] 21 B 39 (241, 16). Ritter and Preller, 103 b. Burnet, E.G.P.[3] p. 212.

that causes it to remain at rest. For the reason why the air is so closely confined that it cannot find a passage, and therefore stays where it is, is its great amount: and this amount is great because the body which isolates it, the earth, is very large. This result, then, will follow, even if the earth is spherical, so long as it retains its size. So far as their arguments go, the earth will still be at rest.

In general, our quarrel with those who speak of movement in this way cannot be confined to the parts; it concerns the whole universe. One must decide at the outset whether bodies have a natural movement or not, whether there is no natural but only constrained movement. Seeing, however, that we have already decided this matter to the best of our ability, we are entitled to treat our results as representing fact. Bodies, we say, which have no natural movement, have no constrained movement; and where there is no natural and no constrained movement there will be no movement at all. This is a conclusion, the necessity of which we have already decided,[11] and we have seen further that rest also will be inconceivable, since rest, like movement, is either natural or constrained. But if there is any natural movement, constraint will not be the sole principle of motion or of rest. If, then, it is by constraint that the earth now keeps its place, the so-called 'whirling' movement by which its parts came together at the centre was also constrained. (The form of causation supposed they all borrow from observations of liquids and of air, in which the larger and heavier bodies always move to the centre of the whirl. This is thought by all those who try to generate the heavens to explain why the earth came together at the centre. They then seek a reason for its staying there; and some say, in the manner explained, that the reason is its size and flatness, others, with Empedocles, that the motion of the heavens, moving about it at a higher speed, prevents movement of the earth, as the water in a cup, when the cup is given a circular motion, though it is often underneath the bronze, is for this same reason prevented from moving with the downward movement which is natural to it.) But suppose both the 'whirl' and its flatness (the air beneath being withdrawn) cease to prevent the earth's motion, where will the earth move to then? Its movement to the centre was constrained, and its rest at the centre is due to constraint; but there must be some motion which is natural to it. Will this be upward motion or downward or what? It must have some motion; and if upward and downward motion are alike to it, and the air above the earth does not prevent upward movement, then no more could air below it prevent downward movement. For the same cause must necessarily have the same effect on the same thing.

Further, against Empedocles there is another point which might be made. When the elements were separated off by Hate, what caused the earth to keep its place? Surely the 'whirl' cannot have been then also the cause. It is absurd too not to perceive that, while the whirling movement may have been responsible for the original coming together of the parts of earth at the centre, the question remains, why *now* do all heavy bodies move to the earth. For the whirl surely does not come near us. Why, again, does fire move upward? Not, surely, because of the whirl. But if fire is naturally such as to move in a certain direction, clearly the same may be supposed to hold of

---

[11] I. 2–4.

earth. Again, it cannot be the whirl which determines the heavy and the light. Rather that movement caused the pre-existent heavy and light things to go to the middle and stay on the surface respectively. Thus, before ever the whirl began, heavy and light existed; and what can have been the ground of their distinction, or the manner and direction of their natural movements? In the infinite chaos there can have been neither above nor below, and it is by these that heavy and light are determined.

It is to these causes that most writers pay attention: but there are some, Anaximander, for instance, among the ancients, who say that the earth keeps its place because of its indifference. Motion upward and downward and sideways were all, they thought, equally inappropriate to that which is set at the centre and indifferently related to every extreme point; and to move in contrary directions at the same time was impossible: so it must needs remain still. This view is ingenious but not true. The argument would prove that everything, whatever it be, which is put at the centre, must stay there. Fire, then, will rest at the centre: for the proof turns on no peculiar property of earth. But this does not follow. The observed facts about earth are not only that it remains at the centre, but also that it moves to the centre. The place to which any fragment of earth moves must necessarily be the place to which the whole moves; and in the place to which a thing naturally moves, it will naturally rest. The reason then is not in the fact that the earth is indifferently related to every extreme point: for this would apply to any body, whereas movement to the centre is peculiar to earth. Again it is absurd to look for a reason why the earth remains at the centre and not for a reason why fire remains at the extremity. If the extremity is the natural place of fire, clearly earth must also have a natural place. But suppose that the centre is not its place, and that the reason of its remaining there is this necessity of indifference—on the analogy of the hair which, it is said, however great the tension, will not break under it, if it be evenly distributed, or of the men who, though exceedingly hungry and thirsty, and both equally, yet being equidistant from food and drink, is therefore bound to stay where he is—even so, it still remains to explain why fire stays at the extremities. It is strange, too, to ask about things staying still but not about their motion,—why, I mean, one thing, if nothing stops it, moves up, and another thing to the centre. Again, their statements are not true. It happens, indeed, to be the case that a thing to which movement this way and that is equally inappropriate is obliged to remain at the centre. But so far as their argument goes, instead of remaining there, it will move, only not as a mass but in fragments. For the argument applies equally to fire. Fire, if set at the centre, should stay there, like earth, since it will be indifferently related to every point on the extremity. Nevertheless it will move, as in fact it always does move when nothing stops it, away from the centre to the extremity. It will not, however, move in a mass to a single point on the circumference—the only possible result on the lines of the indifference theory—but rather each corresponding portion of fire to the corresponding part of the extremity, each fourth part, for instance, to a fourth part of the circumference. For since no body is a point, it will have parts. The expansion, when the body increased the place occupied, would be on the same principle as the contraction, in which the place was diminished. Thus, for all the indifference theory shows to the contrary, earth also would have moved in this manner away from the centre, unless the centre had been its natural place.

We have now outlined the views held as to the shape, position, and rest or movement of the earth.

4.2.2   Chapter 14

Let us first decide the question whether the earth moves or is at rest. For, as we said, there are some who make it one of the stars, and others who, setting it at the centre, suppose it to be 'rolled' and in motion about the pole as axis. That both views are untenable will be clear if we take as our starting-point the fact that the earth's motion, whether the earth be at the centre or away from it, must needs be a constrained motion. It cannot be the movement of the earth itself. If it were, any portion of it would have this movement; but in fact every part moves in a straight line to the centre. Being, then, constrained and unnatural, the movement could not be eternal. But the order of the universe is eternal. Again, everything that moves with the circular movement, except the first sphere, is observed to be passed, and to move with more than one motion. The earth, then, also, whether it move about the centre or as stationary at it, must necessarily move with two motions. But if this were so, there would have to be passings and turnings of the fixed stars. Yet no such thing is observed. The same stars always rise and set in the same parts of the earth.

Further, the natural movement of the earth, part and whole alike, is the centre of the whole—whence the fact that it is now actually situated at the centre—but it might be questioned, since both centres are the same, which centre it is that portions of earth and other heavy things move to. Is this their goal because it is the centre of the earth or because it is the centre of the whole? The goal, surely, must be the centre of the whole. For fire and other light things move to the extremity of the area which contains the centre. It happens, however, that the centre of the earth and of the whole is the same. Thus they do move to the centre of the earth, but accidentally, in virtue of the fact that the earth's centre lies at the centre of the whole. That the centre of the earth is the goal of their movement is indicated by the fact that heavy bodies moving towards the earth do not move parallel but so as to make equal angles, and thus to a single centre, that of the earth. It is clear, then, that the earth must be at the centre and immovable, not only for the reasons already given, but also because heavy bodies forcibly thrown quite straight upward return to the point from which they started, even if they are thrown to an infinite distance. From these considerations then it is clear that the earth does not move and does not lie elsewhere than at the centre.

From what we have said the explanation of the earth's immobility is also apparent. If it is the nature of earth, as observation shows, to move from any point to the centre, as of fire contrariwise to move from the centre to the extremity, it is impossible that any portion of earth should move away from the centre except by constraint. For a single thing has a single movement, and a simple thing a simple: contrary movements cannot belong to the same thing, and movement away from the centre is the contrary of movement to it. If then no portion of earth can move away from the centre, obviously still less can the earth as a whole so move. For it is the nature of the whole to move to the point to which the part naturally moves. Since, then, it would require a force greater than itself to move it, it must needs stay at the centre. This view is further supported by the contributions of mathematicians to astronomy, since the observations made as the shapes change by which the order of the stars is determined, are fully accounted for on the hypothesis that the earth lies at the centre. Of the position of the earth and of the manner of its rest or movement, our discussion may here end.

Its shape must necessarily be spherical. For every portion of earth has weight until it reaches the centre, and the jostling of parts greater and smaller would bring about not a waved surface, but rather compression and convergence of part and part until the centre is reached. The process should be conceived by supposing the earth to come into being in the way that some of the natural philosophers describe. Only they attribute the downward movement to constraint, and it is better to keep to the truth and say that the reason of this motion is that a thing which possesses weight is naturally endowed with a centripetal movement. When the mixture, then, was merely potential, the things that were separated off moved similarly from every side towards the centre. Whether the parts which came together at the centre were distributed at the extremities evenly, or in some other way, makes no difference. If, on the one hand, there were a similar movement from each quarter of the extremity to the single centre, it is obvious that the resulting mass would be similar on every side. For if an equal amount is added on every side the extremity of the mass will be everywhere equidistant from its centre, *i.e.* the figure will be spherical. But neither will it in any way affect the argument if there is not a similar accession of concurrent fragments from every side. For the greater quantity, finding a lesser in front of it, must necessarily drive it on, both having an impulse whose goal is the centre, and the greater weight driving the lesser forward till this goal is reached. In this we have also the solution of a possible difficulty. The earth, it might be argued, is at the centre and spherical in shape: if, then, a weight many times that of the earth were added to one hemisphere, the centre of the earth and of the whole will no longer be coincident. So that either the earth will not stay still at the centre, or if it does, it will be at rest without having its centre at the place to which it is still its nature to move. Such is the difficulty. A short consideration will give us an easy answer, if we first give precision to our postulate that any body endowed with weight, of whatever size, moves towards the centre. Clearly it will not stop when its edge touches the centre. The greater must prevail until the body's centre occupies the centre. For that is the goal of its impulse. Now it makes no difference whether we apply this to a clod or common fragment of earth or to the earth as a whole. The fact indicated does not depend upon degrees of size but applies universally to everything that has the centripetal impulse. Therefore earth in motion, whether in a mass or in fragments, necessarily continues to move until it occupies the centre equally every way, the less being forced to equalize itself by the greater owing to the forward drive of the impulse.

If the earth was generated, then, it must have been formed in this way, and so clearly its generation was spherical; and if it is ungenerated and has remained so always, its character must be that which the initial generation, if it had occurred, would have given it. But the spherical shape, necessitated by this argument, follows also from the fact that the motions of heavy bodies always make equal angles, and are not parallel. This would be the natural form of movement towards what is naturally spherical. Either then the earth is spherical or it is at least naturally spherical. And it is right to call anything that which nature intends it to be, and which belongs to it, rather than that which it is by constraint and contrary to nature. The evidence of the senses further corroborates this. How else would eclipses of the moon show segments shaped as we see them? As it is, the shapes which the moon itself each month shows are of every kind—straight, gibbous, and concave—but in eclipses the outline is always curved: and, since it is the interposition of the earth that makes the eclipse, the form of this line will be caused by the form of the earth's surface, which is therefore spherical. Again, our observations of the stars make it evident, not only that the earth circular, but also that it is a circle of no great size. For quite a small change of position to south or north causes a manifest alteration of the horizon. There is much change, I mean, in the stars which are overhead, and the stars seen are different, as one moves northward or southward. Indeed there are some stars seen in Egypt and in the neighbourhood of Cyprus which are not seen in the northerly regions; and stars, which in the north are never beyond the range of observation, in those regions rise and set. All of which goes to show not only that the earth is circular in shape, but also that it is a sphere of no great size: for otherwise the effect of so slight a change of place would not be so quickly apparent. Hence one should not be too sure of the incredibility of the view of those who conceive that there is continuity between the parts about the pillars of Hercules and the parts about India, and that in this way the ocean is one. As further evidence in favour of this they quote the case of elephants, a species occurring in each of these extreme regions, suggesting that the common characteristic of these extremes is explained by their continuity. Also, those mathematicians who try to calculate the size of the earth's circumference arrive at the figure 400,000 stades. This indicates not only that the earth's mass is spherical in shape, but also that as compared with the stars it is not of great size.

4.3   Study Questions

QUES. 4.1 What are Aristotle's opponents' views on the position, shape and motion of the earth?

a) Where did the Pythagoreans place the earth? The sun? What was their rationale? Does Aristotle agree?
b) Why does Aristotle compare the heavens to an animal? What is his point?
c) Do the Pythagoreans believe that the earth moves? How so? Are there other bodies in motion? What evidence do they cite to support this view?
d) What evidence do the advocates of a flat-earth advance? How does Aristotle answer this?
e) What were the opinions of Xenophanes, of Thales and of the atomists—Anaximenes, Anaxagoras and Democritus—on the shape of the earth?

f) How does Aristotle answer their assertions? In particular, how do these opinions violate Aristotle's principle of natural motion?

g) What is the "indifference theory" of Anaximander as to why the earth is motionless and at the center world? Does Aristotle agree with Anaximander's reasoning? With his conclusion?

QUES. 4.2 What is Aristotle's view on the position and motion of the earth? Is he correct?

a) What is the observed natural motion of pieces of the earth? Given this evidence, is it possible that the whole earth might be in eternal, circular motion?

b) Does Earth lie at the center of the World? What evidence does Aristotle provide? Are his arguments reasonable? Upon what assumption(s) are they based?

c) Does a lifted or thrown clod of dirt experience natural or constrained motion? What does this imply about the position and motion of the earth?

QUES. 4.3 What is the shape and size of the earth, according to Aristotle? Is he correct?

a) What does the weight of fragments of earth imply about the shape of the earth?

b) What does the shape of lunar eclipses imply about the shape of the earth?

c) What does the angle of a star when viewed from different latitudes imply about the shape and size of the earth?

d) How big is the earth?

4.4   Exercises

Ex. 4.1 (MEASURING THE EARTH). Suppose that you and your friend wish to determine the size of the earth. You are both located at the same meridian line, but at different latitudes: you stand in the city of Swenet (now Aswan), which is located (approximately) on the tropic of Cancer, so that the sun is directly overhead at noon on the day of the summer solstice; your friend stands about 520 miles north, in the city of Alexandria. On the day of the summer solstice, you each measure the length of the shadow cast by a 10 ft long vertically oriented stick, or *gnomon*, at noon. The length of your friend's gnomon's shadow is $16\frac{1}{2}$ in.

a) What is the length of your gnomon's shadow?

b) From the length of these shadows, calculate the difference in latitude of Swenet and Alexandria. How does this compare to modern values based on the latitudes of these two cities?

c) From your measurements, calculate the circumference of the earth. How does this compare to modern values? To those reported by Aristotle?

d) What do you suppose is the largest source of uncertainty in this type of measurement of the size of the earth?

4.5   Vocabulary

1. Finite
2. Analogous
3. Essence
4. Interposition
5. Causation
6. Indifferent
7. Extremity
8. Whence
9. Contrariwise
10. Centripetal
11. Potential
12. Equidistant
13. Concurrent
14. Postulate
15. Endow
16. Impulse
17. Generation
18. Gibbous
19. Concave
20. Stade

Chapter 5
The World of Ptolemy

Only mathematics can provide sure and unshakeable knowledge to its devotees.

—Claudius Ptolemy

5.1 Introduction

Claudius Ptolemy (ca. 100–165 A.D.) is the most influential astronomer of the ancient world. He lived and worked in the city of Alexandria, carrying out astronomical observations and writing about diverse topics such as optics, geography, music, mathematics, philosophy and astronomy. Ptolemy's most famous work, the *Almagest*, served as a textbook and a practical reference for calculating astronomical events for the next 1500 years. The name of the book, literally "the greatest," derives from the title given it by Arab scholars and translators from the original greek text during the ninth century.

The *Almagest* itself consists of 13 books. In the first book Ptolemy begins with his theory of the earth: its size, shape and place in the universe. The remainder of Books I and II provide careful demonstrations of how to perform numerical calculations using plane and spherical trigonometry. For instance, he demonstrates how to determine the right ascension (angular distance from the vernal equinox) of any point on the ecliptic (I, 16), how to divide the surface of the earth into geographical zones using parallels (II, 6),[1] and how to calculate the length of shadows, the position and time of sunrise, and the angle between the ecliptic and the horizon (II, 5, 2, 7 and 11, respectively). These are some of the methods that will prove useful when he develops his theories of motion of the sun (III), the moon (IV), the fixed stars (VII–VIII), and the planets (IX–XIII).

The reading selection included below is comprised of the first several chapters of Book I of the *Almagest*, translated from Greek by Aaron Jensen. Ptolemy begins in Chap. 1 by explaining the difference between theoretical and practical philosophy. (Where does Ptolemy situate the study of astronomy?) Next, after explaining the outline of the *Almagest* in Chap. 2, Ptolemy offers a summary of his theory of the heavens and the earth. Thus, Chaps. 3–8 provide a framework for understanding

---

[1] The German cartographer Martin Waldseemüller makes extensive use of Ptolemy's astronomical and geographical work in his *Introduction to Cosmography*. See Chap. 9 in the present volume.

K. Kuehn, *A Student's Guide Through the Great Physics Texts*,
Undergraduate Lecture Notes in Physics, DOI 10.1007/978-1-4939-1360-2_5,
© Springer Science+Business Media, LLC 2015

the *Almagest* as a whole. He argues against those who would claim that the earth itself is in motion. This, after all, was an ancient opinion espoused by the Pythagoreans but later rejected by Aristotle in his book *On the Heavens*. The astute reader of the *Almagest* will recognize many of Aristotle's arguments and proofs in these chapters. (Can you identify them?)

The Ptolemaic worldview came under renewed and vigorous attack by thinkers during the sixteenth and seventeenth centuries. Indeed understanding the *Almagest* is essential in understanding the astronomical works of Copernicus, Kepler and Galileo, who wrote in reaction to the geocentric worldview, which Ptolemy had so clearly and meticulously laid out in the *Almagest*.

5.2   Reading: Ptolemy, *The Almagest*

Ptolemy, C., *The Almagest*, translated for this volume by Aaron Jensen based on J. L. Heiberg's greek text in Ptolemy, C., *Claudii Ptolemaei Opera Quae Exstant Omnia*, vol. I, Lipsiae in aedibus B. G. Teubneri, 1898.

5.2.1   Chapter 1: Preface

The true philosophers, Syrus, were entirely right, it seems to me, to distinguish the theoretical aspect of philosophy from the practical. For even if it turns out that that practical, before it is practical, is theoretical, it would no less be found that there is a large difference between them. This is because while many people possess some of the moral virtues even without being taught them, it is impossible to happen upon a theory of the universe without being instructed. Not only that, but it is also because, while in the first case the greatest benefit is gained from continual practice, in the second case it is from progressing in the theories. So we order our actions by the impulses of the things we picture so that, even in everyday life, we do not forget to seek after a good, well-ordered condition. But we thought it fitting to give most of our time to teaching the theories, as they are many and beautiful, and especially those called "Mathematics" in particular.

And furthermore, Aristotle also entirely suitably divides the theoretical into three primary categories: Physics, Mathematics, and Theology. For everything which exists has its existence from matter, form, and motion. And while none of these can be seen by itself in its substratum, they can each be thought about without the others.

The first cause of the first motion of the universe, if you understand it simply, can be considered an invisible and motionless god. The category investigating this is Theology, because its activity, up above somewhere around the highest regions of the Cosmos, can only be thought about and is absolutely separated from sensory objects.

The category examining the material and ever-changing qualities—concerning itself with "white," "hot," "sweet," "soft," and such things—could be called Physics, because its object is located for the most part among the corruptible things and below the lunar sphere.

The category explaining the qualities with respect to the forms and the motions from one place to another, and examining shape, number, size, and also place, time, and similar things, can be defined as Mathematics. For its object falls, so to speak, between those other two, not only because it can be considered both with and without the senses, but also because it is a characteristic of simply everything which exists, both mortal and immortal—for those things which are always changing in their inseparable form it changes with them, and for those things which are eternal and of an ethereal nature it keeps their unchanging form unchanged.

We know from this that those first two categories of theoretical philosophy should be called conjecture rather than comprehension—Theology because it is entirely invisible and undetectable and Physics because matter is unstable and uncertain. So because of this there could never be any hope that the philosophers will be in agreement about them. But only Mathematics, if you approach it with careful examination, provides firm and credible knowledge to those who pursue it. This is because its proofs occur through indisputable methods—both Arithmetic and Geometry.

We have been drawn to be very much engaged in all such theory that we can. But because of this we have been especially drawn to that theory which considers divine and heavenly things, since only this is concerned with the inspection of things which are eternal and constant. And because of this it too, in keeping with what it properly apprehends, which is not unclear or disorderly, is able to be eternal and constant. And this is a proper characteristic of knowledge.

And Mathematics also can make contributions towards the other two categories no less than they themselves. It can most of all prepare the way for Theology. For it alone is able to make strong inferences about motionless, abstract activity because of its acquaintance with the characteristics of moving and moved objects, which are perceptible to the senses, and also with the eternal and immutable—things concerning the motions and the arrangements of motions.

And it can be of no small help to Physics, for almost every property of a material object becomes apparent from the peculiar quality of its motion from one place to another—the corruptible and the incorruptible from their respective linear and circular paths, and the heavy and light, or the passive and the active, from their respective paths towards the center and away from the center.

Certainly Mathematics most of all can prepare clear-sighted men for goodness in actions and disposition because of the uniformity, orderliness, symmetry, and calm which the divine things have. It makes those who follow it closely into lovers of this divine beauty and makes it customary and, so to speak, second nature for them to have their soul in a similar condition.

Indeed, we ourselves continuously try to increase our love for the theory of the things which are eternal and unchanging. We do this by studying the aspects of these kinds of sciences which have already been comprehended by those who have truly applied themselves to them with careful investigation. We also ourselves choose to contribute as much advancement as the time which has passed between them and us could provide. And indeed those things which we think became clear to us at the present time we will try to record both as briefly as possible and so that it may be followed by those who have already made some progress. For the sake of the treatise's completeness we will set forth everything which is useful for the theory of the heavens in its proper order. But to keep the discourse from becoming long, those things which the ancients have described accurately we will only pass through quickly. But those things which our predecessors did not grasp either as completely or as usefully we will work out as best we can.

5.2.2   Chapter 2: On the Order of the Theorems

Indeed, before entering the treatise set in front of us, we must see the relationship as a whole of the whole earth to the whole heaven. Of the individual things which follow, first would be to go through the topic of the position of the ecliptic and the places in our part of the inhabited world, and also the differences which exists between each of them at every horizon, in order, because of the changed latitude. For if the theory of these things is considered first, then it makes the investigation of everything else easier.

Second would be to go through the motions of the sun and the moon and the characteristics of these motions. For without first grasping these things it would not be possible to look into the matter of the stars in a detailed manner.

And lastly for this method is the matter of the stars. Here first the sphere of the so-called "fixed stars" can be appropriately ordered, and following that the spheres of the five so-called "planets."

We will try to explain each of these by using clear and undisputed phenomena, observed both by the ancients and in our time, as the beginnings and, so to speak, the foundations for our investigation. We will also apply the following things we comprehend through proofs by means of geometrical methods.

We would take up these kinds of general topics first: the heaven is a sphere and moves as a sphere. The earth too is generally spherical in shape to perception. As for position, it is found at the middle—nearly the center—of the entire heaven. And as for size and distance, it is the size of a point compared to the sphere of the fixed stars, and it makes no motion at all from one place to another.

For the sake of reminder we will go through each of these things briefly.

5.2.3   Chapter 3: That the Heaven Moves as a Sphere

So it seems likely that the initial conceptions about these things came to the ancients through observations such as this: they saw that the sun and the moon and the other stars move from their risings to their settings always in circles parallel to each other.

They saw them begin to ascend from below from the lower region and, so to speak, out of the earth itself, and rise up little by little to the heights, and then continue to go around in proportion and descend until finally, as if falling into the earth, they disappear. Then, after remaining hidden for some time, they once again, as from another beginning, rise and set. And the times and locations of their rising and setting, were generally both fixed and the same.

It was especially the revolution of those stars which are always visible which led them to the concept of the sphere, because it is seen to be circular and revolving around one and the same center. For of necessity that point became the pole of the heavenly sphere—the ones closer to it rotated in smaller circles and the ones farther from it traced larger circles in proportion to their distance—such is the case up until the distance is so great that they are hidden from sight. And looking at these rotating stars, they saw that the ones near those which are always visible remain hidden only a short time while the ones farther from them remain hidden longer, again in proportion.

So at first it was only because of these kinds of things that they received this concept of the sphere. But in their observations from then on, they noticed that everything else fit together with it. For simply all the phenomena disprove those ideas which are incorrect:

For if someone would suppose the motion of the stars were linear and moved towards infinity, as some have thought, what way could be contrived by which all of them would be seen as moving from the same starting point each day? For how could the stars go back when they are moving towards infinity? Or if they did turn back, how would this not be seen? Or how would they not diminish little by little before disappearing? But quite the opposite—they are seen as larger until their disappearances, when little by little they are covered up and, so to speak, cut off by the surface of the earth.

But indeed, it would also appear to be most nonsensical for them to be kindled in rising from the earth and in turn extinguished in returning to it. For suppose someone conceded that so great an arrangement with its sizes and numbers, as well as its intervals and places and times, were produced randomly and by chance. And suppose he also conceded that one whole part of the earth had a kindling nature and another an extinguishing nature, or, rather, that the same part kindles for some people and extinguishes for others, and that the same stars happen to already be kindled or extinguished for some people but not yet for others. Even if someone, I say, would concede all these things even though they are so ridiculous, what would we be able to say about the stars which are always visible, neither rising nor setting? For what reason don't the ones kindled and extinguished both rise and set for all places? And why aren't the ones not undergoing this always visible from all places on earth? For certainly the same ones will not be kindled and extinguished for some people but not at all for others. It is, however, altogether clear that the same stars do rise and set for some people but not at all for others.

In summary, if someone would suppose any kind of shape for the motion of the heavens other than spherical, it would be necessary for the distances from the earth to the higher regions—wherever and however they are situated—to be varied. In that case then also both the sizes of the stars and the distances between them ought to appear varied to the same people during the course of each rotation because they are sometimes coming to a greater and sometimes a lesser distance. This is not seen to occur. For the fact that their sizes appear larger near the horizons is not caused by the distance being smaller but by the exhalation of the moisture surrounding the earth between our viewpoint and the stars, just as things thrown into water appear larger, and the deeper they sink the larger they appear.

We are also led towards the concept of sphericity by these things: the devices constructed to observe time cannot harmonize with any hypothesis other than this one alone. And just as the motion of the heavens is unhindered and the most agile of all motions, so too the circle is the most agile of two-dimensional shapes and the sphere is the most agile of three-dimensional shapes. Likewise, in the case of different shapes which have the same perimeter/surface area, the ones which have more angles are greater in size. So the circle is the greatest in size of two-dimensional objects and the sphere is the greatest of three-dimensional objects. And the heaven is the greatest of the other bodies.

Why, indeed, we also can be spurred on to this notion by some physical things. For example, of all the bodies there are, the ether is composed of the smallest and most homogenous particles. And the surfaces of homogenous things are themselves homogenous. But the only homogenous surfaces are, among two-dimensional objects, the circular and, among three-dimensional objects, the spherical. And since the ether is not two- but three-dimensional, it follows that it is spherical. And similarly, nature has composed all earthly and corruptible bodies wholly out of shapes which are round but not homogenous and in turn all aethereal and divine bodies out of shapes that are homogenous and spherical. For if these were flat or disk-shaped, then the shape could not appear circular to everyone who looks at it at the same time from different places of the earth. Because of this, it makes sense that also the ether surrounding them, since it is of the same nature, is spherical and, because of its homogeneity, moves in a circular and uniform manner.

5.2.4   Chapter 4: The Earth Too is Generally Spherical to Perception

That the earth, too, is spherical to perception (when taken generally) we understand especially in this way: it can again be seen that the sun, moon, and other stars do not rise and set at the same time for everyone on earth but always first for those who live to the east and later for those who live to the west. For we find that the phenomenon of eclipses, and especially lunar eclipses, which occur everywhere at the same time, are not recorded by everyone as happening for them at the same hours, that is, the same intervals from their noon. Instead, the hours recorded by the more eastern observers are always later than those recorded by the more western observers. And since the difference of the hours is found to be proportional to the distances between the places, someone could appropriately suppose that the surface of the earth is spherical.

For its homogenous curvature (when taken generally) always hides objects succes-sively to successive people in proportion (to the east–west distance between them). And if the shape was something else, this would not happen, as anyone can see from the following: if it were concave, the stars would appear to rise first for those more to the west. If it were flat, they would both rise and set for everyone on earth together and at the same time. If it were triangular or square or some other polygonal shape, again they would rise and set similarly and at the same time for everyone on the same plane. This is not at all seen to happen.

And neither could it be cylindrical such that the rounded surface was turned to the east and west and the flat base sides were at the poles of the world, which some would suppose rather plausible. Here is why that is clearly the case: none of the stars would always be visible to those who lived on the curved surface. Instead, either they all would both rise and set for everyone, or the same stars would always become invisible to everyone the same distance from each of the poles. Now the further we go to the north, the more of the southern stars are hidden and the more of the northern stars appear, so that it is clear that here also the curvature of the earth, by cutting off the polar regions proportionately, shows the shape is spherical on all sides. Additionally there is the fact that, if we sail towards mountains or some higher places from any direction to any direction, their sizes are seen to grow little by little as if coming out of the sea itself after previously being sunk because of the curvature of the surface of the water.

5.2.5   Chapter 5: The Earth Is in the Middle of the Heaven

When this has been understood, if someone would next consider the position of the earth, he would understand that the phenomena surrounding it could only occur in this way if we suppose that it is in the middle of the heaven, as the center of a sphere. For if this were not so, the earth would have to be either off the axis but equidistant from each pole, or on the axis but displaced towards either of the poles, or neither on the axis nor equidistant from each pole.

Now here is what combats the first of the three positions: imagine the earth were displaced upwards or downwards in relation to a particular set of observers. In that case, at *sphæra recta*[2] it would happen for them that there would never be an equinox because the heavens, the one above the earth and the one below the earth, would always be divided by the horizon into unequal parts. For those at *sphæra obliqua*, either again there would be no equinox at all or it would happen not in the middle of passage between the summer and winter solstices. For these distances would of

---

[2] The case of *sphæra recta* refers to observations made by a person standing on the equator of the earth, so that the axis of the celestial sphere is at a right angle to the zenith (the point directly overhead). The case of *sphæra obliqua*, on the other hand, refers to observations made by a person standing at any other latitude on the earth's surface, so that the axis of the celestial sphere is at an oblique angle to the zenith. See Ex. 5.1, below.—[*K.K.*]

necessity be unequal since it is no longer the equator, the greatest of the parallel circles drawn by the rotating poles, but one of the parallels either north or south of it which would be bisected by the horizon. And it is simply confessed by everyone that these distances happen to be equal everywhere because the prolongation of the longest day at the summer solstices over the equinox is equal to the curtailment of the shortest days at the winter solstices.

But if, in turn, this displacement is supposed to be westward or eastward in relation to a particular set of observers, it would happen for them that the sizes of the stars and the distances between them would not appear equal and the same at the horizons at morning and night. And also the time from rising to culmination would not be equal to the time from culmination to setting. These are clearly altogether in opposition with the phenomena.

To the second of the positions—that the earth is considered to be on the axis and displaced towards one of the poles—it can in turn be replied that, if this were so, the plane of the horizon would always be forming the part of the heavens over the earth and the part of the heavens under the earth differently—the parts formed at different latitudes would not be unequal and at every latitude the two parts would be unequal to each other. Only in the case of *sphæra recta* would the horizon bisect it. In the case of a *sphæra obliqua*: making the nearer pole always visible, the horizon would always make what was above the earth smaller and what was under the earth larger. In that case, it would happen that also the ecliptic would be divided into unequal parts by the plane of the horizon. This is not at all seen to be so. Six of the signs of the Zodiac are always seen by everyone on earth and the other six are unseen. Then at another time those ones are all seen on earth at the same time and the first ones are not seen. So it is clearly the case that also the signs of the Zodiac are bisected by the horizon from the fact that it cuts off all the same semi-circles, at times above the earth and at times below it.

And in general, if the earth were not positioned beneath the very equator of the heaven but leaning towards its north or south, towards either of its poles, it would happen at the equinoxes that the shadows of the sundial at sunrise would not be in a straight line with the shadows at sunset in a plane parallel to the horizon, not even to perception. Yet this is clearly seen to be constant everywhere.

It is clear from this that also the third of the positions cannot be possible, for each of the things objected against the first two will also work against it.

In summary, the entire order seen concerning the prolongation and curtailment of nights and days would be thoroughly confused if the earth were not located at the middle. Additionally, there is the fact that the eclipses of the moon would not occur in every part of the heaven when it is diametrically opposite the sun. For often the earth would come between them when they were diametrically opposite each other but at distances less than a semicircle.

5.2.6 Chapter 6: The Earth Has the Ratio of a Point to the Heavenly Things

Yet certainly a great proof that, to perception, the earth also has the ratio of a point to the distance to the sphere of the so-called fixed stars is that from all parts of the earth the sizes of the stars and the distances between them appear equal and similar everywhere at the same times. The observations of the same things from different latitudes are not found to vary in the least. Why, indeed, it must be accepted that the sundials set up in every part of the earth, as well as the centers of armillary spheres, act the same as if they were at the true center of the earth. They preserve the observances of angles and the revolving of shadows in such great agreement with the hypotheses concerning the phenomenon that it is as if they happened to be the point at the middle of the earth.

Another clear sign that this is so is that everywhere the planes which are produced by our vision, which we call horizons, always bisect the whole sphere of the heaven. This would not happen if the size of the earth were perceptible in comparison with the distance of the heavens. In that case only the plane produced by the point at the center of the earth could bisect the sphere while those planes produced by any point of the surface of the earth would always make the sections below the earth greater than the ones above it.

5.2.7 Chapter 7: The Earth Does not Make any Motion from One Place to Another

It will be shown by these same preceding things that the earth cannot make any motion in any of the aforementioned directions or ever be moved at all from its place at the center. For the same things would then occur which would have if it happened to be at a position other than the middle. Therefore it seems to me that it would be superfluous for someone to seek for the causes of motion towards the middle, at least once it has been clear from the phenomena themselves that the earth occupies the middle place of the Cosmos and that all weighty things are moved towards it. And this understanding would be most readily apprehended in this way: because earth, as we said, has been shown in simply all its parts to be spherical and in the middle of the universe, both the directions and motions of bodies which have weight (I am speaking of their natural properties) are always and everywhere at right angles to the fixed plain produced tangential to the impact. Because this is so, it is clear that also, if they were not to be hindered by the surface of the earth, they would absolutely arrive at the very center. For the straight line leading towards the center is always at a right angle to the plane tangent to the sphere at the place where it intersects the tangent.

Those who think it a paradox that so heavy a thing as the earth has neither been held up by something or moved seem to me to be mistaken in making judgment by looking at their own experiences and not at the natural properties of the universe. For this kind of thing would not, I think, be so extraordinary to them if they paid attention to the fact that the size of the earth, when judged with the surrounding body, has the ratio of a point to it. For then it will seem possible that that which is smallest in ratio is both held fast and pressed against equally from all sides and in equilibrium by what is homogenous and altogether largest.

For there is no "down" or "up" in the Cosmos with respect to itself, just as no one could think of such a thing in a sphere. And as for the compound bodies in it, insofar as their proper and natural motion, the light things, those composed of the smallest particles, are blown up towards the outside and the surrounding surface and seem to move upwards for everyone. For, for all of us, that which is overhead, and is called "up," tends towards the surrounding surface. And the heavy things, those composed of large particles, move towards the middle and as to the center, and seem to fall downward. For again, for all of us, that which is towards the feet, and is called "down," tends towards the center of the earth. These things would appropriately be made to collapse around the middle by the equal and similar hindrance and resistance from all sides towards each other. Therefore, also appropriately, that whole solid of the earth, which is so large in reference to the things moved against it, is kept motionless even under the motion of much smaller weights because they come from all sides and, so to speak, it receives the things that fall to it.

And at any rate, even if the earth shared one and the same motion with the other weights, clearly it would have moved down ahead of everything else because of its so greatly excessive size. Both living things and individual weights would have been left behind to float on the air. And very soon the earth would have fallen even out of the heaven itself. But such things would appear most ridiculous of all even just to think them.

Now some people have, as they think it at least, a more plausible idea. While agreeing with these things because they have nothing with which to contradict them, they think that nothing could disprove them if they were to suppose, for instance, that the heaven were motionless and the earth were turned around the same axis from west to east approximately one turn each day. Likewise if they supposed that both of them move some amount as long as it was around the same axis, as we said, and in proportion to how they overtake each other.

But what has escaped their notice is that while nothing would prevent this from being the case for the sake of the phenomenon concerning the stars, at least not in the simpler notions, this kind of thing can be seen to be altogether ridiculous from those things which concern us and from the occurrences in the air. For suppose we concede to them that unnatural things are so. Suppose we concede that the things which are lightest and composed of the smallest particles are either wholly unmoved or moved no differently than those things of an opposite nature. (Although the things in the air composed of larger particles so clearly move much faster than all those more earthy.) And suppose we concede that the things which are heaviest and composed of the largest particles have as their proper motion one which is so swift and even. (Although the earthy things, again, are well-admitted to at times not be readily moved by other things.) If we conceded this, then they would have to admit, however, that the turning of the earth was the simply most violent of all the motions around it for it to make so great a revolution in a short time. Then the things not held up by it would all appear to be moving the opposite way as the earth. Neither clouds nor anything else which is flying or thrown could be shown to be passing to the east because the earth would always move faster and overtake their motion to the east. Therefore everything else which the earth leaves behind would seem to be displaced towards the west.

For if they would say that the air is carried around with it in the same direction and with equal speed, the compound bodies which are in it would no less always seem to be left behind by the motion of both. Or, if they too were carried along as if united to the air, they would no longer ever appear to either go forward or be left behind but would always remain and never wander or shift, whether they were flying or thrown. But we so clearly see that all of these things do occur, with the result that they are not sped up or slowed down by the earth not standing still.

5.2.8   Chapter 8: There Are Two Different Primary Motions in the Heaven

Certainly it was sufficient for the individual treatment, and those which follow them, to outline in their chief points these hypotheses, which were taken first of necessity. These will be both confirmed and entirely proven by the agreement of the things subsequently demonstrated in reference to the phenomena. In addition to these, someone would be right to think that we should also first grasp the general idea that there are two different primary motions in the heaven:

The first is that by which everything is always making its rotation from east to west at a constant and equal speed along the circles drawn parallel to each other by the poles, of course, of this sphere which rotate everything evenly. The largest of these circles is called the equator because it alone is always bisected by the largest horizon and because the revolution the sun makes when on it makes an equinox everywhere to perception.

The second is that by which the spheres of the stars, in a manner opposite the aforementioned motion, make some movement around poles which are different (and not the same as) those of the first rotation.

And we suppose that these things are so for this reason: what we view throughout any one day is that all things in general which are in the heaven are seen to perception to make their risings and high-points and settings in places uniform and parallel to the equator. For this kind of thing is the natural property of the first motion. But what we then observe over successive days is that all the other stars appear to maintain both their distances from each other and for the most part their natural properties when compared to the locations belonging to the first motion. The sun, moon, and planets, however, make some shifts which, though diverse and unequal to each other, are all a general movement in the direction east of and left behind by the stars, which preserve their distances from each other and, so to speak, revolve on one sphere.

Now if also this kind of shift of the planets were to happen on circles parallel to the equator, that is, around the poles which make the first revolution, it would be sufficient to think that they all have one and the same revolution similar to the first one. For it would then appear plausible that their shift occurred by being variously left behind and not by an opposite motion. But besides the fact that the shifts are to the east, they always appear to be displaced to the north and the south. And the size of this displacement is an anomaly when explained so that the occurrences happening seem to be from some outward forces around them. But while it is an anomaly to this kind of conjecture, it falls into place when drawn up as being produced by the ecliptic, a circle slanted with reference to the equator. From this we understand one and the same such circle to be the natural property of the planets. For it is made with precision and, so to speak, drawn by the motion of the sun. And it is revolved upon by both the moon and the planets as they always turn around it and do not by chance fall in either direction from the displacement marked for each one. And this circle is seen to be a great circle since the sun is to an equal degree north of and south the equator and because the eastward shifts of all the planets happen, as we said, around one and the same circle. Because of this it was necessary to suppose that this second, different general motion is one which happens around the poles of the aforementioned ecliptic and opposite to the first motion.

Let us indeed consider a great circle drawn through the poles of both of the aforementioned circles, that is, the equator and the ecliptic slanted with reference to it.[3] It will of necessity bisect each of them, forming right angles. Then there will be four points for the ecliptic.

Two of them are made by the equator and are on the same diameter as each other and called the equinoxes. The one which has the passage from south to north is called the vernal equinox, and the other is called the autumnal equinox.

The other two are made by the circle drawn through both poles. These are clearly on the same diameter as each other. They are called the solstices. The one south of the equator is called the winter solstice and the one to the north is the summer solstice.

The first primary motion, which encompasses all the others, will be considered traced and, so to speak, marked off by a great circle drawn through both sets of poles as it both itself rotates and rotates everything else with it from east to west around the poles of the equator. These poles are fixed, so to speak, on the meridian, which differs from the aforementioned great circle only in that it is not always drawn through the poles of the ecliptic.[4] Yet it is also called the meridian because it is

---

[3] Medieval astronomers referred to the great circle passing through both the poles of the equator and the poles of the ecliptic as the *equinoctial colure*. See, for example, Martin Waldseemüller's *Introduction to Cosmography*, included in Chap. 9 of the present volume.—[K.K.]

[4] The equinoctial colure is fixed to the rotating celestial sphere, while the meridian is not. Thus, the meridian, which is a great circle passing through both the poles of the equator and an observer's zenith, overlaps the equinoctial colure once each day. See Chap. 3 of Waldseemüller's *Introduction to Cosmography*, which is included in Chap. 9 of the present volume.—[K.K.]

always considered to be at a right angle to the horizon. For this position it has, while bisecting the hemisphere above the earth and the one below the earth, also contains the middle times of the day and the night.

The second primary motion has multiple parts. It is encompassed by the first but it encompasses the spheres of all the planets. It is moved along by the aforementioned first motion, as we said, but it revolves in the opposite direction around the poles of the ecliptic. These poles are themselves always fixed on the circle which makes the first outline, that is, the one through both sets of poles. The poles appropriately rotate with the circle, and by the second motion in the opposite direction they always preserve for the ecliptic, which is the great circle this motion draws, the same position with reference to the equator.

5.2.9   Chapter 9: On the Individual Concepts

This, then, would be how the general preliminary explanation sets forth in its chief parts the things we must lay down beforehand. We are about to begin the individual proofs, the first of which we think to be that through which it is understood how large the arc of a great circle drawn through the poles is between these poles. But we see that it is necessary to first set forth the methods for finding the size of the chords in a circle in order to at last prove everything geometrically.

5.3   Study Questions

QUES. 5.1 In what sense is the *Almagest* a work of theoretical philosophy?

a) What, according to Ptolemy, is the difference between *theoretical* and *practical* pursuits? Provide an example of each.
b) What are the three branches into which Ptolemy divides theoretical philosophy? Did he conceive of this division? With what is each branch concerned? Is his description consistent with the contemporary understanding of the subjects by the same names?
c) Which branch of theoretical philosophy does Ptolemy seem to favor? Why?
d) Is it true, as Ptolemy states, that theology is "absolutely separated from sensory objects"?

QUES. 5.2 What is the shape and motion of the heavens?

a) Upon what ancient observation is Ptolemy's belief based?
b) On what grounds does he reject the notion that the stars move in straight lines?
c) On what grounds does he reject the "nonsensical" hypothesis that the stars are kindled and extinguished daily?
d) What other evidence does he offer in support of his view? Are these reasonable? Are they correct?

QUES. 5.3 What is the shape of the earth?

a) What do the different rising and setting times of the sun, moon and stars for
   different observers tell us about the shape of the earth?
b) Why does Ptolemy here speak of eclipses? Consider: if the ancients used the
   motion of the moon as a standard to measure time, then how could they measure
   whether the moon rose at a different time for people at different locations on the
   earth?
c) How would our observations of the sun be different if the earth were, say, concave?
   planar? polygonal? cylindrical?
d) What other evidence does he offer for his belief? Are his arguments convincing?

QUES. 5.4 What is the location of the earth with respect to the celestial sphere?

a) If the earth were located at the center of the celestial sphere, then how much of
   the zodiac would appear above the horizon on any given night?
b) How would the length of daylight differ for a person standing at the equator or
   standing in northern Europe? Would the hours of daylight and darkness be equal
   throughout the year?
c) How would your answers to the previous questions differ if the earth were located
   off the axis of the celestial sphere? What about if the earth were located on the
   axis of the celestial sphere, but nearer the north pole of the sphere?
d) What does Ptolemy conclude, based upon all these considerations? Is his
   argument convincing?

QUES. 5.5 What does Ptolemy believe concerning the size and the motion of the
earth?

a) What evidence does he provide for the size of the earth?
b) If the earth were not at the center of the world, but objects were attracted to the
   center of the world, then what (contra-factual) observation would follow?
c) How does Ptolemy address the apparent paradox that the earth, being heavy, is
   supported by nothing?
d) If the earth were indeed falling, then what (contra-factual) observation would
   follow? Does his reasoning rely upon Aristotelian physics?
e) On what grounds does Ptolemy reject the notion that the earth spins on its axis?
   Are his arguments reasonable? Are they correct?

QUES. 5.6 How do the heavens move?

a) In which direction do the sun, the stars and the planets move, when viewed over
   the course of a single day? Do they all move at the same speed? Do they all move
   parallel to the celestial equator?
b) In which direction do the sun, stars and planets move, when viewed at the same
   time on successive days throughout the month, or even the year? Do they all
   move at the same speed? Do they move parallel to the celestial equator? Does the
   celestial equator always (or ever) meet the horizon at a right angle?

Fig. 5.1 Geocentric diagram depicting the horizon, *hg*, for an observer standing on the equator. The earth (♁) is at the exact center of the sphere of fixed stars which rotates around the axis *ab* daily. The sun (☉) is fixed to a different sphere which rotates yearly about the axis *cd*

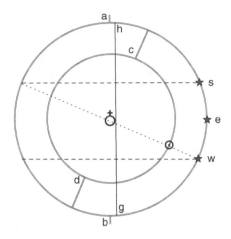

c) What is the ecliptic circle? In particular, why was it introduced? And what is its orientation with respect to the circle of the equator?

d) How are the solsticial points defined, and what is their significance? How are the equinoctial points defined and what is their significance?

e) How is an observer's meridian circle defined? What is the relationship between the meridian circle and the horizon? Are the meridian circles and horizons of all observers standing on the earth the same?

5.4 Exercises

Ex. 5.1 (PTOLEMAIC GEOCENTRISM). In Chap. 5 of the *Almagest*, Ptolemy argues that Earth must be located precisely at the center of the heavens. In so doing, he carefully considers three distinct possibilities (though not in the following order); we will refer to them as (i) the earth is located at the center of the sphere of fixed stars, (ii) the earth is located along the equator of the sphere, but off its axis, and (iii) the earth is located on the axis of the sphere but nearer one pole than the other. Let us begin by carefully analyzing case (i) together. Afterwards, you will be asked to similarly analyze the other two cases, and explain why Ptolemy finds them to be erroneous.

Case (i) is depicted schematically in Fig. 5.1. The earth is located at the center of a celestial sphere, which rotates about the axis *ab* once per day. This westward rotation (clockwise from above the axis at *a*) accounts for the daily rising and setting of the stars, which are fixed to the celestial sphere. From the perspective of a person standing on the equator of the stationary earth, the sphere of fixed stars (and hence the zodiac) is bisected by the horizon which is depicted by the vertical line *hg*. Polaris, located at *a*, lies just on (or slightly below) the northern horizon.

The inner sphere, to which the sun is fixed, rotates counterclockwise about the axis *cd* once per year. Axis *cd* is tilted at an angle of approximately 23° with respect

Fig. 5.2 Similar to Fig. 5.1,
but depicting the horizon for
an observer in the
mid-latitudes

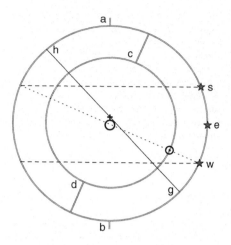

to the axis of the sphere of fixed stars.[5] The yearly rotation about *cd*, coupled with
the daily rotation about *ab*, gives rise to both the rising and setting of the sun and
also its seasonal eastward drift along the ecliptic. The limits of the sun's northern
and southern motion is indicated by the horizontal dashed lines which are formed by
the intersection of the equator of sphere *cd* with the sphere of fixed stars.

On the day of the winter solstice, the sun will appear to lie in front of the stars
at *w* from the vantage point of an observer on the earth. As the year progresses, the
sun's position will drift northward. On the day of the vernal equinox, the sun will
appear in front of the stars at *e*. On the day of the summer solstice, it will appear in
front of the stars at *s*. It will then drift back southward during the course of the year,
past the point *e* on the autumnal equinox, until it reaches the solsticial point *w* once
again. Notice that, for an observer standing on the earth's equator, the sun would
spend an equal time above and below the horizon on every day of the year. Also, the
sun would be at the zenith (directly overhead) at local noon on only two days of the
year, the vernal and autumnal equinoxes.

What about for an observer located in the temperate region, say in Milwaukee? In
such a case, the horizon *hg* would again bisect the sphere of fixed stars, but it would
do so at an oblique angle, as shown in Fig. 5.2.

Polaris would reside at an angle above the horizon equal to the observer's latitude.
During the course of the year, the motion of the sun would again be confined between
the two horizontal dashed lines. But during the days of winter, the sun would spend
far less time above the horizon than during the summer months. Only on the vernal
equinox and the autumnal equinox would the day and night be of equal duration.

Finally, for an observer located above the arctic circle, the horizon line *hg* would
bisect the heavenly sphere at a very oblique angle, as shown in Fig. 5.3. Again,

---

[5] For the sake of clarity, I have omitted the several intervening spheres which, according to Ptolemy,
govern the motion of the planets which lie between the sun and the sphere of fixed stars. For more
information, see the Introduction to Chap. 4 of the present volume.

Fig. 5.3 Similar to Fig. 5.1, but depicting the horizon for an observer in the arctic region

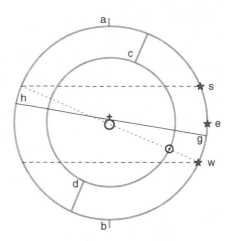

Polaris would reside at an angle above the horizon equal to the observer's latitude. During certain days in the depth of winter, the sun never rises above the horizon, and during certain days in the height of summer, the sun never sets below the horizon.

Now, finally, here are the homework exercises:

a) How would the observations differ if the earth were located off of the axis of the sphere of fixed stars, but equidistant from the poles? Draw appropriate diagrams, similar to Figs. 5.1, 5.2 and 5.3, and answer the following questions. (i) Would the horizon bisect the zodiac? (ii) Would Polaris be visible on the horizon? (iii) Would the hours of daylight and darkness be equal throughout the year? If not, would they ever be equal? When? (iv) Would the sun pass directly overhead on the equinoxes? And (v) would the equinoxes lie midway between the solstices?

b) What would be observed if the earth, instead, were located on the axis of the sphere of fixed stars, but nearer its northern pole? Draw appropriate diagrams and answer the same questions as before.

Ex. 5.2 (SHAPE OF THE EARTH). Suppose that there were no satellite photographs of the earth from space. How might you determine the shape and size of the earth? Be as clear, detailed and complete as possible in your explanation.

Ex. 5.3 (STARGAZING LABORATORY). In this field exercise, we will use both star charts and the technique of "star hopping" to find our way around the night sky.

Stargazing Preparation In order to make a stargazing experience enjoyable and productive, first make a tentative plan of what you would like to observe during your stargazing session. Next, choose a good viewing location, such as an open field or the top of a building. Be sure to dress appropriately.[6] Also, bring snacks, a folding chair,

---

[6] During cold winter observing sessions, I have used Hot Hands Air-Activated Warmers by Kobayashi, Dalton, Georgia.

and a blanket to catch dropped objects and to make viewing more relaxing. Bring a star chart,[7] a magnetic compass for orienting yourself, and a small flashlight for viewing the charts. You might also consider installing motion tracking astronomy software[8] for your handheld electronic device, but keep in mind that software is artificial: it cannot substitute for actual observations of nature—and it may not even be correct.

Mapping the Heavens   For practical purposes, the night sky may be considered a vast, dark, celestial sphere, to which the stars are affixed, and at the center of which we on Earth reside. This model is employed by both Ptolemy (who considered it to be true) and modern astronomers (who consider it to be useful). The celestial sphere rotates around the earth once per day such that, for earthbound observers, the stars rise on the eastern, and set on the western, horizon. The planets, sun and moon, as you will find, are not strictly affixed to the celestial sphere; rather, they drift about the celestial sphere with their own apparent motion which is superimposed on that of the daily, or *diurnal*, rotation of the celestial sphere. Accordingly, one can use a spherical coordinate system, similar to the longitude and latitude lines on the earth's surface, to describe locations on the celestial sphere. There are two particularly important coordinate systems: celestial and horizontal.

The *celestial coordinate system* is essentially a projection of the earth's latitude and longitude lines outward and onto the celestial sphere. The poles of the celestial sphere lie directly above the earth's poles; the celestial equator directly above the earth's equator. Only the terms, and the zero points are different: whereas on the earth, distances from the equator are measured using latitude, on the celestial sphere, distances from the equator are measured using declination (DEC). The celestial equator lies at $0°$ and the poles lie at $\pm 90°$. And whereas on the earth, distances from east to west are measured using longitude, on the celestial sphere, distances from east to west are measured using right ascension (RA). Moreover, RA is not measured from a point directly above the Prime Meridian passing through Greenwich, but rather from the first point of Aries—the point on the celestial sphere at which the celestial equator crosses the ecliptic on the vernal equinox. Furthermore, RA is not measured in degrees, but in hours and minutes. In particular, the circumference of the celestial sphere is conceptually divided not into $360°$, but into 24 h.[9] For instance, the star *Vega* is located at RA 18^h36^m.

The *horizontal coordinate system* is similar to the celestial coordinate system, except that it is measured in degrees of azimuth and altitude, rather than DEC and RA. The altitude of an object is the angular distance of that object above an observer's horizon. The zenith lies at an altitude of $90°$ above the horizon. The azimuth of an

---

[7] Such as Sinnott, R. W., *Sky and Telescope's Pocket Sky Atlas*, New Track Media, LLC, Cambridge, Massachusetts, 2006.

[8] I have been very happy with the *Luminos* Astronomy Companion for iOS by Wobbleworks LLC.

[9] This is of course the principle of operation behind old-fashioned analog clocks, which use the angular position of pointers to relay time. Newer digital clocks, which use only digits or other symbols, conceal this connection between time and astronomy, for the sake of simplicity.

object is the angular distance of that object eastward from a line joining the zenith to the observer's north point. The north point for an observer is simply the point on the horizon due north of his position. Notice that the azimuth and altitude of the same object are different for two observers at different locations on the earth's surface. For this reason, it is critical that the observer accurately record his or her own position (latitude and longitude) on the earth when employing the horizontal coordinate system.

Southern Sky Viewing This exercise is designed for winter observing in the northern hemisphere; it should be carried out in the late evening. For each recognized star, you should record the common name, the Bayer designation,[10] the celestial coordinates (right ascension and declination) and the horizontal coordinates (altitude and azimuth). You might also enjoy reading up on the myth or story behind each constellation you observe.

a) Begin by finding the constellation Orion, the hunter. Orion's belt can be recognized as three adjacent stars forming a remarkably straight line. Now find the stars *Betelgeuse* and *Rigel*, which comprise Orion's left shoulder and right foot (from your perspective).

b) Extrapolate a line from *Rigel* beyond *Betelgeuse* to find the stars *Castor* and *Pollux*, the brightest stars in the constellation Gemini.

c) Extrapolate a line rightward from Orion's belt to find *Aldebaran*, the brightest star in Taurus.

d) Continue along this same line beyond Taurus to find the *Pleiades*, a cluster of stars containing seven bright stars.

e) Extrapolate a line leftward from Orion's belt to find *Sirius*, the brightest star in Canis Major.

f) Extrapolate a line from *Bellatrix*, Orion's right shoulder, leftward through *Betelgeuse* to find *Procyon*, the brightest star in Canis Minor.

Northern Sky Viewing This exercise is also designed for winter observing in the northern hemisphere; it should be carried out in the pre-dawn hours. Again, for each recognized star, you should record the common name, the Bayer designation, the celestial coordinates and the horizontal coordinates.

a) Begin by finding the constellation Ursa Major, or the big bear, the hindquarters of which form the commonly recognizable Big Dipper. *Dubhe* and *Merak*, the two brightest stars in Ursa Major, form the edge of the saucepan opposite the long, curving handle.

---

[10] German astronomer Johannes Bayer systematically assigned names to over a thousand of the brightest stars. He assigned each a three-letter combination (indicating its parent constellation) preceded by a lower-case greek letter (indicating its relative magnitude, or brightness, among the other stars in its constellation). For example, Betelgeuse is given the Bayer designation α-Ori, since it is the brightest star in the constellation Orion. See, for example, Swerdlow, N. M., A Star Catalogue Used by Johannes Bayer, *Journal for the History of Astronomy, 17,* 189, 1986.

b) By extrapolating a line upward along the edge of the saucepan approximately five times the distance between *Dubhe* and *Merak*, one arrives at *Polaris*. *Polaris* lies at the tip of the handle of the little dipper, which forms the hindquarters of Ursa Minor, the little bear. *Polaris* also lies very near the axis about which the celestial sphere rotates daily.

c) By following a great circle from the zenith through *Polaris* and down to the horizon, one can identify the North point on the horizon. The north point on the horizon should correspond with the direction of magnetic north. Does it?

d) Returning to Ursa Major, follow the arc formed by the stars *Alioth*, *Mizar* and *Alkaid* in the handle downward to find the bright star *Arcturus*, which lies in the constellation Boötes.

e) Continue the same arc beyond *Arcturus* to find *Spica*, the brightest star in the constellation Virgo. Since Virgo does not rise above the horizon until late evening in January, it will be impossible to find *Spica* in an early evening observation session.

5.5   Vocabulary

1. Virtue	13. Concave
2. Disposition	14. Equidistant
3. Aethereal	15. Zenith
4. Corruptible	16. Equinox
5. Passive	17. *Sphæra obliqua*
6. Latitude	18. Solstice
7. Ecliptic	19. Bisect
8. Subsequent	20. Culmination
9. Pole	21. Diametrical
10. Exhalation	22. Armillary sphere
11. Sphericity	23. Tangent
12. Eclipse	24. Plausible

Chapter 6
Measuring the Tropical Year

We think it is in general fitting to demonstrate the phenomena through the simplest hypotheses of all—as long as there does not appear to be anything from the observations worth mentioning which conflicts with this supposition.

—Claudius Ptolemy

6.1 Introduction

At the outset of Book I of his *Almagest*, Ptolemy provides an overview of his world-view, describing the size, shape, position and motion of the heavens and the earth. Now in Book III, Ptolemy focuses his attention on the motion of the sun. In the reading selection that follows, he considers how the length of the year is determined. This is no simple problem. First, one must decide how the year is to be defined. How does Ptolemy do it? Is the motion of the sun uniform, or does its speed vary from year to year, or perhaps even from day to day? Next, one must find and interpret the ancient historical documents which record, to varying degrees of reliability, the time at which significant astronomical events—such as solstices and equinoxes—were observed. In fact, measuring the precise time of the summer solstice is notoriously difficult.[1]

The reader will notice that Ptolemy records the dates of the solstices and the equinoxes according to an Egyptian calendar (which is referred to the era of Nabonassar). The Egyptian year was constant in length, having 365 days. It consists of twelve 30-day months, beginning with Thoth and ending with Mesore, followed by five additional 'intercalated' days. This addition is akin to the modern practice of inserting an extra day at the end of February once every four years. Ptolemy's recorded dates may be converted to an 'astronomical' calendar (which is referred to the Christian era). Toomer has carried out these conversion;[2] I have included his astronomical dates in square brackets next to each Egyptian date reported by Ptolemy.[3]

---

[1] For a history of such measurements, even up until the seventeenth century, see Heilbron, J. L., *The Sun in the Church*, Harvard University Press, Cambridge, Massachusetts, 1999.

[2] See Toomer, G. J. (Ed.), *Ptolemy's Almagest*, Princeton University Press, Princeton, NJ, 1998.

[3] To be clear, in the astronomical calendar, the year -1 corresponds to 2 B.C. and year 1 to 1 A.D.

K. Kuehn, *A Student's Guide Through the Great Physics Texts,*
Undergraduate Lecture Notes in Physics, DOI 10.1007/978-1-4939-1360-2_6,
© Springer Science+Business Media, LLC 2015

The reader will also notice that Ptolemy reports angular measurements using a sexagesimal (base-60) system which the Greek astronomers inherited from the Babylonians. Consider, for example, one of Ptolemy's recorded angles: 0;59,8,17,13,12,31. The number before the semi-colon is just the number of degrees; the comma-delimited numbers after the semi-colon are subsequent digits in the sexagesimal system: $(0 \times 60^0) + (59 \times 60^{-1}) + (8 \times 60^{-2}) + (17 \times 60^{-3}) + (13 \times 60^{-4})$ etc. degrees. This can also be expressed as 0 degrees, 59 minutes, 8 seconds, 17 thirds, 13 fourths etc.

6.2   Reading: Ptolemy, *The Almagest*

Ptolemy, C., *The Almagest*, translated for this volume by Aaron Jensen based on J. L. Heiberg's greek text in Ptolemy, C., *Claudii Ptolemaei Opera Quae Exstant Omnia*, vol. I, Lipsiae in aedibus B. G. Teubneri, 1898. Book III.

6.2.1   Preface

In the parts of the treatise before this, we inspected the things concerning heaven and earth which need to be taken first in a general mathematical way as well as the inclination of the ecliptic. We also inspected its individual characteristics at both *sphaera recta* and the *sphaera obliqua* in each residence. After these things we think we should next discuss the sun and the moon and go through the characteristics of their motions. For none of the phenomenon concerning the stars can be completely found without explaining this first. And we find that the matter of the sun's motion goes first of these, for without it, again, it would not even be possible to obtain the things concerning the moon in detail.

6.2.2   Chapter 1: On the Size of the Time of a Year

As the first of all the things to be demonstrated about the sun is to find the time of a year, we can learn the disagreements and difficulties of the ancients in their statements about this kind of thing from their treatises, especially that of Hipparchus, a man who loved both hard work and the truth. For even he especially leads the matter into this difficulty: because of its apparent returns to the solstices and the equinoxes, the time of a year is found to be less than $1/4$ of an extra day beyond 365 days, but from returns seen concerning the fixed stars it is greater than that. From this he also comes to the conclusion that the movement of the fixed stars takes somewhat more time and that it, just like those of the wandering stars, towards the reverse of motion making the first revolution upon a circle drawn through both the poles and the equator.

But we will demonstrate that this is the case, and the way in which it happens, in the parts of the treatise about the fixed stars. For the things about them, too, would not even be possible to look at in general without first explaining the sun and the moon. But in our present observation, we think that it is necessary, when looking at the time of a solar year, to look at nothing other than the return of the sun to itself, that is, the return which happens on its ecliptic. We must also define the time of the year according to which it next arrives from some motionless point of the circle to the same point, considering the only proper beginnings for such a return are the points of the aforementioned circle marked by the points of the solstices and equinoxes.

For if we approach the matter mathematically, we will find no more proper return than one which carries the sun to the same configuration both in place and time, whether such a thing is considered in relation to the horizons, the meridian, or the durations of day and night. And we can find no other beginnings in the middle of the ecliptic, but only those which happen to be marked off by the solstice and equinox points.

And if someone examines what is proper from more of a physics perspective,[4] he will find that no return makes more sense than one which carries the sun from a condition of the atmosphere at a season to a similar condition at the same season. And he will find that there are no other beginnings than only those by which the seasons are most of all distinguished. Additionally is the fact that looking at the return in relation to the fixed stars seems out of place for several reasons, especially that their sphere also is seen to make an ordered movement to the reverse of that of the [daily motion of the] heaven. For if these things were so, nothing would prevent us from saying that the time of the solar year were as great as that in which the sun passed the star Saturn, for sake of argument, or any of the other planets, and then the times for the year would be many and various.

Because of these considerations, we think of this time of the solar year as that found from any solstice (or equinox) to the next same one, especially through observations taken at the greatest interval.

Since Hipparchus is somewhat troubled by the inequality of this return, and he became suspicious (through successive, continual observation), we will attempt to briefly show that this is not troublesome. For we have become confident that these times are not unequal, from those successive solstices and equinoxes which we ourselves have happened to observe. For we find that they have no difference worthy of mention from the addition of $\frac{1}{4}$, but at times it errs, close to the amount it can, due to both the construction and the position of the instruments [employed].

Also from the things which Hipparchus considers, we guess that his suspicion concerning the inequalities [of the times of the years] is from an error in the observations. For in *On the change of the solstice and equinox points*, he first sets forth the summer and winter solstices which he thinks have been observed accurately and in

---

[4] ... as opposed to a purely mathematical perspective. See Ptolemy's distinction between physics, mathematics and theology in his *Preface* of Book I of the *Almagest*, found in Chap. 5 of the present volume.—[*K.K.*].

succession and himself admits that the difference in them is not so much that through them inequalities of the time of the year are observed. For he considers things in this way:

> So from these things it is clear that the differences of the years have been altogether small. But in the case of the solstices I am not confident that we and Archimedes have not erred in our observation and calculation even up to one fourth of a day. The anomalies of the times of the years can be understood from the observations made in the bronze circle in Alexandria on the so-called "square stoa," which seems to indicate that the equinox is the day on which the concave surface begins to be illuminated from the other direction.

Then he sets out first the times of the autumnal equinox, observed as accurately as possible:

[1] In the 17th year of the third Kallipic Cycle, Mesore 30 [−161 Sept. 27], around sunset.

[2] Three years later, in the 20th year, on the first of the intercalated days [−158 Sept. 27], at dawn. It should have been at noon, so it differed by $\frac{1}{4}$ of a day.

[3] One year later, in the 21st year [−157 Sept. 27], at the 6th hour. This was in keeping with the observation before it.

[4] Eleven years later, in the 32nd year, at midnight between the 3rd and 4th of the intercalated days [−146 Sept. 26/27]. It should have been at dawn, so it again differed by that amount.

[5] One year later, in the 33rd year, on the 4th of the intercalated days [−145 Sept. 27], at dawn. This also was in keeping with the observation before it.

[6] Three years later, in the 36th year, on the 4th of the intercalated days [−142 Sept. 26], in the evening. This should have been at midnight, so it again differed by only that amount.

After these he sets out the vernal equinoxes which have similarly been accurately observed:

[1] In the 32nd year of the third Kallipic Cycle, Mechir 27 [−145 Mar. 24], at dawn. Also, he says, the circle in Alexandria was illuminated from both directions at about the 5th h. So already the same equinox, observed differently, is approximately 5 h different.

[2–6] He says that the next ones until the 37th year [−144 to −140] were in agreement with the addition of the $\frac{1}{4}$.

[7] 11 years later, in the 43rd year, he says, the equinox happened after the midnight between Mechir 29 and 30 [−134 Mar. 23/24]. This also was in keeping with the observation in the 32nd year, and it is again in agreement, he says, with the observations [8 to 13, −133 to −128] in each of the coming years until the 50th year [14]. This was in Phamenoth 1 [−127 Mar. 23], around sunset, approximately $1\frac{3}{4}$ days later than in the 43rd year, which also comes to the 7 intervening years.

So in these observations no difference took place worthy of mention, even though it is possible that some error happened in them even up to $\frac{1}{4}$ of a day in observations, not only of the solstices but also the equinoxes. For even if the position or the graduation of the instruments deviates by only $^1/_{360}$ of a degree from being accurate to the circle through the poles of the equator, the sun corrects such latitudinal displacement (with reference to the degrees of the equinox) by moving $^1/_4$ of a degree longitudinally on the ecliptic, such that the difference comes to approximately $^1/_4$ of a day.

There can be even greater error in the case of instruments [which were] not set up and made accurate each time for these very observations, but [instead] erected from some beginning with underlying bases so that the position would be stationary for a long time. For there may have been some movement around them over the time which has passed. One can see this, at any rate, in the case of the bronze circles we have in the Palaestra, which are thought to be positioned on the plane of the equator. When we observe it, such a great discrepancy appears in their position, especially for that of the larger and older one, that sometimes their concave surfaces undergo two changes in illumination in the same equinox.

But Hipparchus himself does not think that any of these things happen to be credible in relation to his suspicion of the inequality of the times of the years. But from considering some lunar eclipses he thinks he has found, he says, that the anomaly of the times of the years, when viewed in relation to the average, contains no difference greater than $^3/_4$ of a day.

Now this would be worthy of some attention if it were the case—and not seen to be false—from the very things which he brings forth. For through some lunar eclipses near fixed stars, he considers in each case how much the star called Spica is ahead of the autumnal point. And through these he thinks himself to have found that it is distant from its times by a maximum of $6^1/_2°$ and a minimum of $5^1/_4°$. From this he gathers that, since it is not possible to move so far in so little time, it is plausible that the sun, from which Hipparchus examines the placed of the fixed stars, does not make its return at an equal time. But he has not noticed that, since this consideration cannot wholly take place without laying down the place of the sun at the eclipse, in each case he undertakes for this [purpose of measuring the sun's position, he unwittingly makes use of] the solstices and equinoxes he has observed accurately in those [same] years. From this he makes clear that when the years are compared there is no difference at all existing from the addition of the $^1/_4$.

To give one example, from the eclipse he observed in the 32nd year of the third Kallippic Cycle which he cites, he thinks that Spica is ahead of the autumnal point by $6^1/_2°$, but in the 43rd year of the same cycle he thinks it is ahead by $5^1/_4°$. And in similar fashion he cites for the aforementioned calculations the vernal equinoxes, observed accurately in the same years, so that through these things he may receive the places of the sun during the middle of the eclipses, and from these places those of the moon, and from those of the moon those of the stars. He says that in the 32nd year it happened on Mechir 27 [−145 Mar. 24] at dawn and in the 43rd year it was after the midnight between the 29th and 30th [−134 Mar. 23/24], close to $2^3/_4$ days later than the one which happened in the 32nd year. This is how much the $^1/_4$ add to each of the 11 years in between makes. So since the sun makes its return to the underlying equinoxes in neither more nor less time than the addition of $^1/_4$, and since it is impossible for Spica to have moved $1^1/_4°$ in so few years, how is it not out of place to use the things computed using these foundations to slander the foundational things themselves? And how is it not out of place to assign the cause for the impossibly large motion of Spica, even though there are many things which can cause so large an error, to nothing else but to the underlying equinoxes—as if they were at the same time observed both accurately and inaccurately? For it would seem more possible that either the distances of the sun to the nearest of the stars in the eclipses are inferred too roughly or the considerations either of its parallaxes in relation to the examination of the apparent positions or of the motion of the sun from the equinoxes in the middle of the time of the eclipse is received either untruly or inaccurately.

But I think that even Hipparchus himself comprehended that there was no credible reason in such things to attribute some second anomaly to the sun, but because of his love for the truth he just wanted to not pass by in silence any of the things which could in some way bring some people to suspect it. So at any rate, he himself has used as his suppositions for the sun and moon that there is one and the same anomaly concerning the sun which returns with the time in relation to the solstices and equinoxes of the year. And just because we assume that the proposed cycles of the sun are equal in time we do not anywhere see the phenomena around the eclipses having any difference worth mentioning from the ones considered by the suppositions set forth. This would happen to be entirely perceptible if there were a correction concerning the inequality of the time of the year which we did not include, even if it were only one degree, or approximately two equinoctial hours.

So from both all these things and those times of the return we obtained through the successive cycles of the sun we observed, we do not find that the size of the year is unequal—provided it is viewed in relation to any one thing (and not at times in relation to the solstice and equinoctial points and at [other] times in relation to the fixed stars). We also do not find another return more proper than the one carrying the sun from any solstice (or equinox, or any other point of the ecliptic) back to the same point. And we think it is in general fitting to demonstrate the phenomena through the simplest hypotheses of all—as long as there does not appear to be anything from the observations worth mentioning which conflicts with this supposition.

So the fact that the time of the year (viewed in relation to the solstices and equinoxes) is less than the excess of $^1/_4$ of a day (to the 365 days) has been made clear to us also through the things which Hipparchus demonstrated. How much shorter it is cannot be received entirely securely, because for many years the addition of $^1/_4$ remains unchanged to perception because of how small the difference is. And because of this, in keeping with the comparison throughout a larger time, the addition of the days found, which must be distributed among the years between the interval, are viewed to be the same whether there are more or less years. Such a return could be observed more accurately as the time between the compared observations is increased. And this kind of thing happens not only in this case but also in all the cyclical returns. For the error which occurs from the weakness of their observations, even if done correctly, is to their perception small and approximately the same, both in the case of phenomenon seen over a long time and those seen over a short time. When distributed over less years it makes the error greater for both each year and that which is concluded from it for a longer time, but when distributed over more years it makes the error less.

From this it is fitting to think it sufficient if we ourselves try to contribute as much as the time between our [own] observations, and those we have from the ancients, can add to the accuracy of the cyclical hypotheses. And we will not intentionally overlook the appropriate investigation [of the records]. But we think assertions about the whole of "eternity"—or even about a time much longer than that of the observations—are foreign to a love of learning and truth.

So as for antiquity, the summer solstices observed by those around Meton and Euketemon and after them by those around Aristarchus ought to be compared with those happening in our [own] time. But since in general the solstice observations are difficult to distinguish and, in addition to this, the ones they handed down were taken rather roughly, as it seems to appear also to Hipparchus, we have excused ourselves from these. And for the sake of accuracy we instead used for the proposed comparison the observations of the equinoxes, both the ones taken by Hipparchus which he especially indicated are most secure, and the especially undoubtful ones observed by us ourselves through instruments for these kinds of things demonstrated at the beginning of the treatise. From these we find that in approximately 300 years the solstices and equinoxes happen one day earlier than the addition of $\frac{1}{4}$ to the 365 days.

For in the 32nd year of the third Kallippic Cycle, he indicated that the autumnal equinox was observed especially securely and says that it is considered to have happened in the third of the intercalated day at midnight, moving to the fourth day [−146 Sept. 26/27]. And that is the 178th year from the death of Alexander. Around 285 years later in the third year of Antoninus, which is 463 years from the death of Alexander, we again most securely observed that the autumnal equinox happened on Athyr 9 [139 Sept. 26] approximately 1 h after sunrise. So the return took, in addition to 285 complete Egyptian years, that is, 365-day years, all of $70\frac{1}{4}$ days and approximately $\frac{1}{20}$ of a day, instead of the $71\frac{1}{4}$ days received in keeping with the addition of $\frac{1}{4}$ day for the years set forth. Therefore the return has happened earlier than the addition of the $\frac{1}{4}$ of a day, short by approximately $\frac{1}{20}$ of a day.

Hipparchus again likewise says that the vernal equinox was most securely observed to have happened in the same 32nd year of the third Kallippic Cycle on Mechir 27 [−145 Mar. 24] at dawn. And that is the 178th year from the death of Alexander. We find that likewise 285 years later, in the 463rd year from the death of Alexander, the vernal equinox happened in Pachon 7 [140 Mar. 22] approximately 1 h after noon. So this cycle also took an equal number of days, $70\frac{1}{4}$ and approximately $\frac{1}{20}$ of a day, instead of the $71\frac{1}{4}$ days received for the 285 years by the $\frac{1}{4}$ of a day. So from these things the return of the vernal equinox happened earlier than the addition of the $\frac{1}{4}$ of a day, short by approximately $\frac{1}{20}$ of a day. Therefore since 300 years has the same ratio to 285 years as one day has to the day which is short by $\frac{1}{20}$ of a day, it is gathered that also in 300 years the return of the sun (happening in relation to the equinoctial points) is earlier than the addition of the $\frac{1}{4}$ by one day.

And if, for the sake of antiquity, we make comparison with the summer solstice observed and rather roughly recorded by those around Meton and Euktemon with the one calculated especially undoubtfully by us, we will find the same thing. This is recorded as having happened when Apseudes was leader of Athens, Phamenoth 21 according to the Egyptians [–431 June 27], at dawn. We securely calculated that it has happened in the same 463rd year from the death of Alexander close to 2 h after the midnight between Mesore 11 and 12 [140 June 24/25]. And there are 152 years from the summer solstice recorded in the time of Apseudes until the one observed by those around Aristarchus in the 50th year of the first Kallippic Cycle [–279], as Hipparchus also says. And there were 419 years from that 50th year, which was the 44th year from the death of Alexander, until the 463rd year, the year of our observation. Therefore in the 571 years of the whole interval, if the summer solstice observed by those around Euktemon were to have happened around the beginning of Phamenoth 21, there would have been, in addition to the whole Egyptian years, approximately 140$\frac{5}{6}$ instead of the 142$\frac{3}{4}$ taken for the 571 years in keeping with the addition of the $\frac{1}{4}$. Therefore the same return happened $\frac{1}{12}$ of a day short of 2 days earlier than the addition of the $\frac{1}{4}$. Therefore it is apparent that it likewise happened that in 600 whole years the time of the year adds approximately 2 full days to the addition of the $\frac{1}{4}$.

And through many other observations we find this same thing happening and many times we see Hippparchus agreeing with it. For in *On the size of the year* he compared the summer solstice observed by Aristarchus at the end of the 50th year of the first Kallippic Cycle [–279] with the one he himself observed, again accurately, at the end of the 43rd year of the third Kallippic Cycle [–134]. This is what he says:

> So it is clear that in 145 years the solstice has become sooner than the addition of $\frac{1}{4}$ by half the joint sum of day and night.

And again in *On intercalated months and days* he first says that according to those around Meton and Euktemon the time of the year contains 365$\frac{1}{4}$ days and $\frac{1}{76}$ of a day, but according to Kallippos is is only 365$\frac{1}{4}$ days. But then he continues, as his words goes,

> We find as many whole months contained in 19 years as they did, but we find the year to still be adding most of all $\frac{1}{300}$ of a day less than $\frac{1}{4}$, as in 300 years 5 days are missing from Meton and 1 day is missing from Kallippos.

And all but summing up his ideas through the register of his own treatises, he says this:

> I have also written a treatise about the time of the year in one book in which I demonstrate that the solar year—this is the time in which the sun moves from a solstice to the same solstice or from equinox to the same equinox—contains 365 and approximately $\frac{1}{300}$ of a day and a night less than $\frac{1}{4}$ days, and not, as the mathematicians think, the same $\frac{1}{4}$ added to the aforementioned quantity of days.

So according to the agreement of those today with those in the past I think it has become apparent that the phenomena concerning the size of the time of the year until now coincides with the aforementioned size of the return with reference to the solstice and equinoctial points. Since these things are so, if we distribute the one day to the 300 years, we attribute to each year 12 s of one day. If we subtract these from the 365;15, in keeping with the addition of the $\frac{1}{4}$, we will have the sought-for time of the year in days as 365;14,48. For this would be the best possible approximation we have received for the quantity of days.

As for the inspection of both the sun and the other ones (in relation to the cycles each of them make), the arrangement of individual astronomical tables is on hand and, so to speak, left it exposed. We think the purpose and goal for a mathematician should be to show all the phenomena in the heaven through even, circular motions. And in keeping with this purpose an astronomical chart is most fitting of all for this purpose as divides the individual even motions from the anomaly which seems to happen because of the suppositions of the circles. And it further shows their apparent cycles by mixing and combining both of these. So in order to receive this kind of thing in an easier form and on hand for the demonstrations themselves, we will hereupon set forth the individual even motions of the sun in this way:

Since one return is demonstrated in days to be 365;14,48, if we distribute it among the 360° of one circle, we will have the average daily motion of the sun at approximately 0;59,8,17,13,12,31 degrees. (It will be sufficient to carry the degrees to this many sexagesimal places.)

Furthermore, by taking $1/24$ of the daily motion, we will have the hourly motion at approximately 0;2,27,50,43,3,1 degrees. Likewise by multiplying the daily motion by 30 (the days of one month), we will have the average monthly motion at 29;34,8,36,36,15,30 degrees, and by multiplying it by the 365 days (of the Egyptian years), we will have the average yearly motion at 359;45,24,45,21,8,35 degrees.

Furthermore, by multiplying the yearly motion by 18 years, for the sake of the symmetry of the astrological tables which would appear, and subtracting the whole circles, we will have in 18 years an additional 355;37,25,36,20,34,30.

So we arranged three charts for the even motion of the sun, each of them with 45 lines and 2 parts. The first chart will contain the average motions of the 18 years. The second will contain first the yearly motions and under them the hourly motions. The third will contain first the monthly motions and underneath the daily motions. The numbers of the time are arranged in the first parts and the comparison of the degrees is in the second part in keeping with the appropriate sums for each. And these are the charts... [5]

6.3 Study Questions

QUES. 6.1 Why does Ptolemy provide a theory of the sun prior to discussing the motion and phases of the moon?

QUES. 6.2 How is the year defined, and what is its length?

a) What is the source of "disagreement and confusion" regarding the length of the year? Specifically, how has the motion of the sun been measured? And how might this give rise to many different definitions of the year?

---

[5] The tables of the mean motion of the sun have been omitted from this volume for brevity. —[K.K.].

b) What reference point does Ptolemy himself recommend? What reasons does he provide?

c) What are some important sources of error when determining the length of the year using Ptolemy's recommended method? What precaution does he recommend to ensure the highest level of accuracy?

d) On the basis of what type of observations did Hipparchus conclude that the length of the year can vary by up to $^3/_4$ of a day? To what plausible error does he attribute Hipparchus' miscalculation? Does Ptolemy believe the length of the year varies from year to year?

e) What general principle guides Ptolemy's understanding of astronomical phenomena?

f) How may the effects of errors in successive measurements of the length of the year be minimized? Does he consider all recorded measurements of equinoxes and solstices equally trustworthy? If not, does he limit his considerations to measurements which he himself performed?

g) Is the length of the year $365^1/_4$ days? If not, is it longer or shorter? What value does Ptolemy derive from the available data?

QUES. 6.3 How far does the sun move during 1 h?

a) What, according to Ptolemy, is the task and goal of the mathematician?

b) Does the sun exhibit uniform motion from the perspective of an earth-bound observer? What problem does this present? And how will Ptolemy attempt to solve it?

c) What is the mean daily, hourly and monthly motion of the sun?

6.4   Exercises

Ex. 6.1 (MEASURING THE YEAR). In 1575 in the city of Florence, a Dominican scholar named Egnatio Danti attempted to measure the duration of the tropical year by constructing a meridian line, or *meridiana*, on the floor of the church of Santa Maria Novella. By knocking a small hole in the south façade of the church, 21 m above the *meridiana* on the floor, a ray of light from the sun could be made to traverse the north-south running *meridiana*, at mid-day, in an eastward direction. By counting the number of days between successive solstices (or equinoxes) one can measure the duration of the tropical year.

a) What is the minimum distance with which the (center of the) image of the sun on the floor approaches the southern façade of the church? When does this occur? What is the maximum distance? When does this occur?

b) If the angular width of the sun is 30′ of a degree, then what is the width of the image of the sun as it traverses the meridian line on the winter solstice? How does the width of the image of the sun affect the uncertainty in the length of a tropical year using the meridian method?

Fig. 6.1 The eccentric orbit of the sun, S, about the earth, Z, from where the sun appears to progress across the sphere of fixed stars, $ABCD$

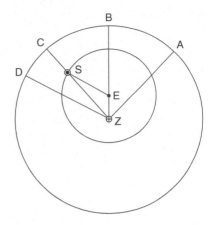

Ex. 6.2 (PTOLEMY'S THEORY OF THE SUN). In the above reading selection, Ptolemy notes that the sun does not exhibit uniform motion from the perspective of an earth-bound observer. To reconcile the sun's apparent non-uniform progression through the zodiac with Aristotle's demand that the heavenly bodies must be moving uniformly, Ptolemy employs an eccentric model, in which the sun moves around a circle which is itself offset from the center of the ecliptic (see Fig. 6.1).[6] The sun, S, moves uniformly around a circle centered at E. From the perspective of an earth-bound observer at Z, the sun appears among the zodiacal stars, C, on the ecliptic circle $ABCD$. According to this model, while the *mean* sun angle, BES (or BZD), increases at a constant rate, the *observed* sun angle BZC does not. Thus, the sun's motion appears non-uniform. At any position in its orbit, the observed angle differs from the mean angle by an *anomaly* angle, CZD. At what position(s) on the eccentric circle is the sun's anomaly angle equal to zero? At what position(s) is the sun's anomaly maximum? Can you prove this geometrically? As a challenge, can you prove that the apparent speed of the sun, when at the point(s) of maximum anomaly, is equal to the mean speed of the sun?

---

[6] The eccentric model for the sun's motion, which Ptolemy inherited from Hipparchus, is described in Chap. 3 of Book III of the *Almagest*.

6.5 Vocabulary

1. Ecliptic
2. *Sphaera obliqua*
3. *Sphaera recta*
4. Equinox
5. Solstice
6. Horizon
7. Meridian
8. Succinct
9. Explicable
10. Conjunction
11. Declination
12. Adduce
13. Impugn

Chapter 7
Geometrical Tools

> *We were led to this understanding and belief—both from the passages of the moon observed and recorded by Hipparchus, and from the things we ourselves have received through an instrument we constructed.*
>
> —Claudius Ptolemy

7.1 Introduction

In the first chapter of Book III of the *Almagest*, Ptolemy arrived at a reliable determination of the length of the year based on the time between subsequent equinoxes. In the course of this discussion, he noted that the sun's motion is not uniform throughout the year. This presents a problem for followers of Aristotle. For how can one reconcile the *apparent* non-uniform motion of the sun with *actual* uniform circular motion characteristic of Aristotle's celestial spheres. To address this problem Ptolemy considers two models, both of which involve the uniform rotation of spheres.[1] The first model makes the sun rotate uniformly on a circle about a point which is slightly offset from the center of the World (where the earth is located). This *eccentric* model is depicted in Fig. 6.1.[2] The second model is more complicated in that it involves two interpenetrating but uniformly rotating circles—a deferent circle and an epicycle (more on this shortly).[3] Ptolemy, following Hipparchus, finally decides in favor of the eccentric model to describe the motion of the sun. But he returns to a modified form of the epicyclic model in order to understand the motion of the planets. For even Hipparchus, that "great lover of truth," was unable to grasp their complex and varied movement. So how was Ptolemy able to?

---

[1] This is done in Chap. 3 of Book III of the *Almagest*, which has been omitted from the present volume.

[2] Figure 6.1 accompanies Ex. 6.2 in the previous chapter of this volume.

[3] The eccentric and the epicyclic models were both employed by Hipparchus of Nicaea (d. 127 B.C.), but were probably known to Apollonius of Perga (d. 190 B.C.). See the careful discussion of Ptolemy's skillful use of these models in Pedersen, O., *A Survey of the Almagest*, revised ed., Springer, New York, 2010.

K. Kuehn, *A Student's Guide Through the Great Physics Texts*,
Undergraduate Lecture Notes in Physics, DOI 10.1007/978-1-4939-1360-2_7,
© Springer Science+Business Media, LLC 2015

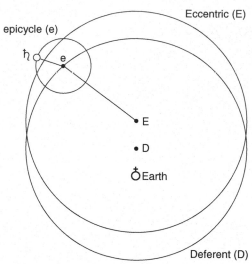

Fig. 7.1 Ptolemy's model for the motions of Venus, Mars, Jupiter and Saturn (he treats Mercury somewhat differently). The planet (♄) is attached to an epicycle (center e). Point e travels at a constant angular speed around the center of an eccentric circle (center E), but is constrained to lie on a deferent circle whose center, D, lies midway between E and the center of the World (⊕)

Ptolemy's epicyclic model of planetary motion is shown in Fig. 7.1. Here, the planet Saturn (♄) is shown fixed to an epicycle which rotates at a constant angular speed about its center, e. The center of this epicycle, in turn, rotates at a constant angular speed about the center of an eccentric circle located at E. Until now, this model seems like a simple combination of the eccentric and epicyclic models. But here is the surprising thing: the path of the center of the epicycle is constrained to lie on yet another circle, the deferent, whose center D lies midway between the aforementioned center of the eccentric circle, E, and the center of the World (where Earth is located). This means that an observer located at the center of the deferent circle, D, would see point e rotate about him at a constant distance but *not* at a constant angular speed. Ptolemy's skillful use of the epicycle, the eccentric and the deferent allowed him to very accurately describe the complex motions of the planets. Indeed, Ptolemy's models of the heavens would stand unchallenged for more than a thousand years—until the time of Copernicus.[4]

Let us linger with Ptolemy for a little while longer to look at some of the other tools that he used. In the two reading selections included in this chapter, taken from Books I and V of the *Almagest*, Ptolemy describes the design and use of three astronomical instruments which may be used to measure the angular distance between heavenly bodies. The first two instruments are very similar; they can both be used to measure

---

[4] We will come to Copernicus' *Revolutions of the Heavenly Spheres* in Chap. 11 of the present volume.

the altitude of the sun above the horizon at any time during the day. Since for a particular observer the altitude of the sun at local noon varies over the course of the year (in the northern latitudes the sun is much closer to the horizon at noon during the winter than during the summer), these types of instruments were employed to measure the length of the year itself. This, in turn, played a significant role in the development of the calendar.[5] The third instrument, which Ptolemy refers to as an "astrolabe," is today called a *spherical astrolabe* or an *armillary sphere*. It is depicted in Fig. 7.2. The spherical astrolabe consists of a network of adjustable intersecting rings which can be oriented so as to measure the celestial coordinates of heavenly bodies, such as the sun and the moon.

7.2 Reading: Ptolemy, *The Almagest*

Ptolemy, C., *The Almagest*, translated for this volume by Aaron Jensen based on J. L. Heiberg's greek text in Ptolemy, C., *Claudii Ptolemaei Opera Quae Exstant Omnia*, vol. I, Lipsiae in aedibus B. G. Teubneri, 1898. Book I, Chap. 12 & Book V, Chap. 1.

7.2.1 Book I, Chapter 12: On the Arc Between the Solstices

Having set forth the size of the chords in the circle, first would be, as we said, to show how much the ecliptic, the circle through the middle of the signs of the Zodiac, is at an incline in relation to the equator: What is the ratio of the great circle through both sets of poles to the arc cut off between the poles? Clearly this arc is equal in length to the one from either solstice point to the equator. With the use of an instrument, we understand this kind of thing directly through a certain simple device like this:

We use a lathe to make a good-sized bronze circle such that its surface is accurately squared off. We use this as a meridian, graduating it into the conventional 360° of a great circle, and dividing each of the degrees into as many parts as there is room for. Then we fit another circle, smaller and thinner, inside the first one such that their sides remain in the same plane but the smaller circle can be rotated within that same plane independent of the larger one towards the north and south. And on both sides of the smaller circle, at two diametrically opposed degree-points, we add small equal-sized wood pieces pointing toward each other and the center of the circles. In the middle of the sides of these wood pieces we place thin pointers touching the side of the larger, graduated circle. Because of the individual necessities we fit this circle securely inside a good-sized pillar and we station the base of the pillar out in the open on a foundation which has no inclination in relation to the plane of the horizon.

---

[5] See Ex. 6.1 in the previous chapter of this volume.

We are quite careful that the plane of the circles is at a right angle to that of the horizon and parallel to that of the meridian. The first of these is addressed by hanging a plumb-line from the point which will be the top and watch until it is straightened by the supports and points toward the point diametrically opposite it. The second of these is addressed by clearly obtaining a meridian line in the plane below the pillar and moving the circles from side to side until their plane is seen to be parallel to the line.

When we positioned things like this, we observed the sun's displacement toward the north and south by moving the small inner circle each noon until the entire lower wood piece was enshadowed by the entire upper one. And when this happened, the tips of the pointers indicated to us at every time how many degrees away along the meridian the center of the sun was from the zenith.

We made such an observation in an even easier way by constructing instead of the circles a stone or wood block, square and rigid, with one of its sides cut accurately flat. On this side we drew a quadrant using a point near one of the corners as the circle center, connecting the center point to the drawn arc with straight lines enclosing the right angle of the quadrant. We similarly divided the arc into 90°, and their parts.

After these things, on one of the lines, the one which was going to be at a right angle to the plane of the horizon and positioned towards the south end of the block, we inserted at right angles all the way around two small cylinders of the same size, made similarly on a lathe. The one we accurately inserted with its middle on the center point itself. The other we inserted at the lower end of the line. Then we stationed this drawn-on side of the block on the meridian line drawn on the underlying surface such that its position was also parallel to the plane of the median. And by a plumb-line between the cylinders we accurately make the line through them to be at a right angle to the plane of the horizon without any inclination, again with some thin supports straightening out whatever side is short.

We likewise observed each noon the shadow from the cylinder at the center, placing something on the drawn-on arc so that its place may appear more visible. And by marking the middle of the shadow we took the degrees it falls on the arc of quadrant to indicate the latitudinal passage of the sun upon the meridian.

So from these kinds of observations, and especially when we examine the ones around the solstices over the course of many revolutions, the marking generally returns the same exact degrees of the median circle from the zenith for that solstice, both for the summer and the winter solstices. Because of this we gathered that the arc from the most northern end to the most southern, which is the arc between the degrees of the two solstices, is always more than $47\frac{2}{3}°$ and less than $47\frac{3}{4}°$. Through this is concluded the same ratio as that of Eratosthenes, which Hipparchus made use of. For when the meridian is 83, the arc between the solstices is approximately 11.

From the preceding observation it is also easy to obtain the latitudes of the locations in which the observations are made by taking the point in the middle of these two ends. This is the equinoctial point. And clearly also the arc between this middle and the zenith is equal to the distance from the poles to the horizon.

7.2.2   Book V, Chapter 1: On the Construction of an Astrolabe Instrument

So as regards the syzygies with the sun, both in conjunction and opposition, and the eclipses which take place with them, we find that the hypothesis set forth in the first simple anomaly is sufficient, even if we take it just as it is. However, for its individual passages in the other configurations with the sun it is no longer found to be sufficient because also, as we said, a second anomaly of the moon occurs arising from the distances to the sun. The second anomaly is reduced to the first anomaly at both syzygies but is greatest at both half-moons. We were led to this understanding and belief—both from the passages of the moon observed and recorded by Hipparchus, and from the things we ourselves have received through an instrument we constructed for these kinds of things. It is comprised in this manner (see Fig. 7.2):[6] We took two good-sized circles, accurately made on a lathe and squared-off in their surfaces, equal and similar to each other all the way around. We connected them at diametrically opposite places so that respective surfaces were at right angles to each other. So the one [3] is understood to be the ecliptic and the other [4] to be the meridian made through its poles and [the poles of the] equator.[7] From the side of the squares we have [identified] points determining the poles of the ecliptic [e, e]; we took those points and inserted pegs into both of them so that they were sticking out of both the inner and the outer surface.

On their outsides we inserted another circle [5] whose concave surface fit accurately to the convex surfaces of the two circles [3, 4] joined together all the way around. It could move longitudinally around the aforementioned poles [e, e] of the ecliptic.

On their insides we likewise inserted another circle [2] whose convex surface fit accurately everywhere to the concave surfaces of the two circles [3, 4]. It likewise moved longitudinally around the same poles as the outer circle [5].

We divided this inner circle [2], as well as the one made to be the ecliptic [3], into the conventional 360° of a circumference and as many parts of these as it allowed.[8] We accurately attached another small thin circle [1], which had holes sticking out of it at diametrically opposite points [b, b], within the inner [2] of the two circles [3 and 4] so that it could be moved within the same plane as that one [2] towards each of the aforementioned poles [e, e] for the sake of latitudinal observation.

Once these things were so, on the circle [4] understood to be through both sets of poles [d, d and e, e] we marked off from both of the poles of the Zodiac [e, e] the

---

[6] Figure 7.2 is based on the schematic reconstruction found in Rome, A., L'Astrolabe et le Météoroscope d'aprés le commentaire de Pappus sur le 5ᵉ livre de l'Almageste, *Annales de la Société Scientifique de Bruxelles*, 47(2), 77–102, 1927.—[K.K.]

[7] In other words, the so-called meridian circle [4] represents the *solsticial colure*, which is a great circle intersecting both the poles of the equator and the poles of the ecliptic.—[K.K.]

[8] *i.e.* minutes, seconds, thirds, *etc.*—[K.K.]

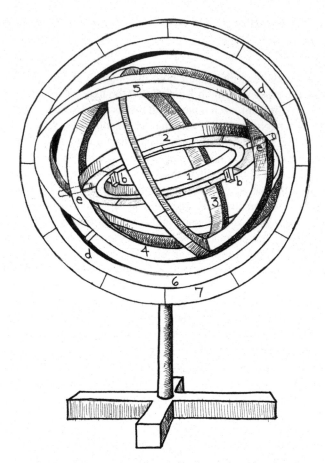

Fig. 7.2 A schematic reconstruction of Ptolemy's spherical astrolabe.—[*K.K.*]

demonstrated arc between the two poles [*e, e*] of the ecliptic [3] and of the equator.[9]
Where these ends were, we again made insertions at points diametrically opposite
each other [*d, d*], placing them into a meridian circle [6 and 7] similar to the ones
demonstrated at the beginning of the treatise for observing the arc of the meridian
between the solstices.[10] So this was constructed in the same position as that was,
that is, at a right angle to the plane of the horizon, at the elevation of the pole proper
for the underlying location, and yet parallel to the natural plane of the meridian

---

[9] So the arc *de* is made identical to the (previously) measured angle of the ecliptic (with respect to
the equator), which is about 24° according to Ptolemy.—[*K.K.*]

[10] Ptolemy is here referring to the instrument described in Book I, Chap. 12, which he used to
determine the inclination of the ecliptic by measuring the altitude of the sun above the horizon
during the solstices.—[*K.K.*]

Fig. 7.3 Orientation of the
spherical astrolabe with
respect to the celestial
sphere.—[*K.K.*]

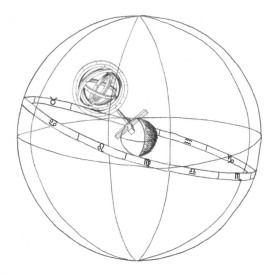

(see Fig. 7.3).[11] Because of this, the revolution of the inside circles occurred around
the poles of the equator from east to west, following the first motion of the universe.[12]

So after constructing the instrument in this way, whenever both the sun and the
moon were visible over the earth at the same time we set the outer circle [5] of the
astrolabe to the approximate degree of the sun found at that time.[13] And we rotated
the circle [4] which was through the poles [d, d] so that, by accurately turning the
intersection of the circles [3 and 5] to the sun's angle, it points at the sun, [and] both
circles enshadow themselves—both the ecliptic [3] and the one through its poles [5].[14]

Or if a star was being sighted, we rotated it so that the outer circle [5] was set
to its conventional degree point on the ecliptic. Then when one eye was placed to
one of the sides of that outer circle [5], the star was sighted through the opposite
parallel side of the circle in the plane through them, joined, so to speak, to both of
its surfaces.

---

[11] In other words, the meridian circles [6, 7] are aligned in the north-south direction. The inner
circle [6] is then rotated within the outer circle [7] until the axis [d, d] is aligned with the North
Star. Thus, the angle between the axis [d, d] and the zenith is determined by the latitude at which
the instrument is located.—[*K.K.*]

[12] The inside circles [1–5], all being attached to the meridian circle [6], can be rotated together
around an axis [d, d] which is aligned with the axis of the celestial sphere.—[*K.K.*]

[13] For example, during the month of October, the outer circle [5] is spun around until it is in front
of the constellation Libra on the ecliptic circle [3].—[*K.K.*]

[14] The circle representing the solsticial colure [4] is rotated about the poles of the equator [d, d]
until a line connecting the two intersections of the ecliptic circle [3] and the one through its poles
[5] is pointed at sun. Proper orientation is identified by noting when the sunward half of the ecliptic
circle [3] casts a shadow on its back half, and the sunward half of the other circle [5] casts a shadow
on its back half.—[*K.K.*]

We moved the other circle [2], the inner one of the astrolabe, to the moon (or to some other thing being sought) so that the moon (or that other thing sought) is seen through both holes [b, b] on the little circle [1] attached along with the sight of the sun, or some other conventional thing.

For thus we know what longitudinal degrees of the ecliptic it occupies from the graduation at which the circle corresponding to it [3] intersects the inner circle [2].[15] And we know how many degrees it is removed to the north or south upon the circle [2] through its poles [e, e] from both the graduation of the inner astrolabe [2] and the distances found between the middle of the hole over the earth on the small circle [1] rotating within it [2] and the middle line of the ecliptic [3].[16]

7.3   Study Questions

QUES. 7.1 What is the angle between the ecliptic and the equator? How can it be measured?

a) Based on the text, make a sketch of the first apparatus described by Ptolemy in Book I, Chap. 12.
b) How should the apparatus be oriented with respect to the local meridian, and with respect to the local horizon?
c) Why must the inner circle be movable? How does one know when it is oriented exactly towards the sun?
d) How can this apparatus be used to measure the altitude of the sun above the local horizon?
e) What is the ecliptic? And how do measurements of the sun allow one to measure the angle of the ecliptic with respect to the celestial equator?
f) On what day of the year is the altitude of the sun at noon most northward? Most southward? What is the size of the arc measured between the northernmost and southernmost points? And how is this related to the angle of the ecliptic?
g) Conversely, how can knowledge of the angle of the ecliptic be used to determine the latitude from which observations of the sun are made?

QUES. 7.2 Describe the construction and the various components of Ptolemy's spherical astrolabe.

a) Which ring represents the ecliptic? How is it marked, or graduated? Where is its axis and where are its poles?
b) Where is the axis of the celestial sphere? How is the angle between the axis of the ecliptic and the axis of the celestial sphere determined?

---

[15] The angular separation of the moon and the sun along the ecliptic is determined by measuring the angular separation of the inner [2] and outer [5] circles.—[K.K.]

[16] The angular separation of the moon and the sun perpendicular to the ecliptic is determined by measuring the angular separation of a line through the sights [b, b] and the ecliptic circle [3].—[K.K.].

c) How can the orientation of the axis of the celestial sphere be adjusted? How should it be oriented for an observer using the spherical astrolabe at the earth's equator? at the earth's north pole?

d) Which is the meridian ring? How should it be oriented with respect to the compass directions?

QUES. 7.3 How can the spherical astrolabe be used to measure the position of the moon with respect to the sun?

a) How must the outer astrolabe ring [5], which rotates about the poles of the ecliptic [e, e], be oriented when using the sun as a reference?

b) When using the location of the sun as a reference, what is significant about the shadows cast by the outer astrolabe ring [5] and the ecliptic ring [3]?

c) Can one use a known star, instead of the sun, as a reference point? If so, how is the outer astrolabe ring [5] oriented?

d) What is the purpose of the movable inner astrolabe ring [2]? And how is it used to measure the celestial coordinates of the moon?

7.4 Exercises

Ex. 7.1 (SUN AT ZENITH). Are there any locations on the earth where the sun is at zenith (directly overhead) on the summer solstice? If so, where, and at what time? (ANSWER: at 23.4°N latitude at local noon.) What about on the autumnal equinox?

Ex. 7.2 (ALTITUDE OF THE SUN). Using geometry and the angle of the ecliptic (look it up), calculate the altitude (in degrees above the horizon) of the sun at local noon for each of the following cases:

a) at your latitude on the vernal equinox,
b) at your latitude on the summer solstice,
c) at your latitude on the winter solstice,
d) at the equator on the summer solstice,
e) at the equator on the winter solstice.

Ex. 7.3 (SPHERICAL ASTROLABE). Carefully examine the current orientation of the rings of the spherical astrolabe depicted in Fig. 7.2. Assume that a competent astronomer has set it up in this particular orientation and is currently using it to simultaneously observe the moon and the sun.

a) Which direction is northward?
b) At what (approximate) latitude is the instrument located? (ANSWER: Approximately 45°N.)
c) What season is it?
d) What time is it? (ANSWER: morning, perhaps 10 a.m.)
e) How many degrees east or west of the sun is the moon located?
f) How many degrees north or south of the sun is the moon located? (ANSWER: approximately 30° south of the sun.)
g) What is the phase of the moon?

Fig. 7.4 A simple cross-staff
can be constructed from a
wooden dowel and a ruler

Ex. 7.4 (Epicyclic theory of Planet X). Suppose that the motion of planet X
can be accurately described by Ptolemy's epicyclic planetary model as shown in
Fig. 7.1. Also, (i) the distance from the center of the eccentric to the center of the
ecliptic is two units of distance, (ii) the diameters of the eccentric and deferent are
both eight units of distance, and the diameter of the epicycle is two units of distance,
and (iii) the epicycle undergoes six complete counterclockwise rotations during one
complete counterclockwise rotation of its center around the earth. How many cycles
of anomaly will planet X undergo during one complete cycle of longitude? At what
angular location(s) will planet X appear to be moving most quickly?

Ex. 7.5 (Cross-staff laboratory). A cross staff is a simple and inexpensive tool
that allows one to consistently measure the angular separation of two distant objects.
The two objects might be celestial objects, such as two stars in a constellation or two
craters on the moon, or they might be terrestrial objects, such as the top and bottom
of a mountain or the ears on each side of a face. Historically, the cross-staff was used
extensively in both astronomy and navigation, but it has been largely replaced by
more sophisticated devices and techniques, such as the sextant and global positioning
system.

Constructing the Cross-staff A cross-staff consists of a scale fixed perpendicu-
larly to the end of a long straight rod. By holding the rod just below the eye and
looking down the rod, as shown in Fig. 7.4, the transverse scale of the cross-staff
can be used to measure the angular separation between two objects, such as the
sun and the horizon. Two sliding markers on the transverse scale are used to facilitate such measurements. In order to construct a cross-staff you will need the
following materials: a straight wooden dowel at least 2 ft long, a flexible plastic
or wooden ruler at least 1 ft long, at least 4 ft of string, two rectangular pieces of
cardboard which measure about $1^{1}/_{2} \times 3$ inches, a razor blade or precision utility
knife, a small wood-cutting saw, a wood screw or nail, and a small drill to make
a starter hole for the screw. Begin by cutting the wooden dowel to an appropriate
length. What is an appropriate length? By choosing wisely, you can design your
cross-staff so that one centimeter on the transverse scale corresponds to an angular
separation of $1°$. Once you have your dowel cut to an appropriate length, you will
need to drill a hole in the far end and attach your plastic ruler so that it forms a T.
Next, cut two lengthwise slits, about a centimeter apart, in your rectangular pieces
of cardboard. These can be slid over the ends of the ruler and will serve as markers.

In order to make your cross-staff more precise, you may wish to bend the ruler into an arc, so that all points on the ruler are equidistant from the tip of the dowel near the observer's eye. This can be accomplished with a length of string; it may be helpful to cut a shallow groove in the near end of the dowel in which to lay the string.

Practicing with the Cross-staff Begin by using your cross-staff to measure the angular width of an object across the room, such as a chalkboard or a window. Adjust your markers to measure the angular width of the object. How confident are you in the precision of your angular measurement? In other words, by how much could you change the positions of the markers and still feel that you have done a reasonable angular measurement? This is the (angular) precision of your cross-staff. Now go outside during the daytime and measure the angular size of various distant objects—towers, trees, mountains—so as to become acquainted with using your cross-staff. The more comfortable you are with your device, the easier it will be for you to use it later, in the dark. When you do take the cross-staff outside at night, you may need a small red-tinted light source to assist in reading your angular scale.

7.5 Vocabulary

1. Chord
2. Ecliptic
3. Horizon
4. Zenith
5. Meridian
6. Plumbline
7. Anomaly
8. Syzygies
9. Quadrature
10. Colure
11. Astrolabe

Chapter 8
The Sun, the Moon and the Calendar

> *The rule of catholic teaching commands that Easter is not to be celebrated before the passing of the spring equinox.*
> —The Venerable Bede

8.1 Introduction

The Venerable Bede (*ca.* 673–735 A.D.) was a scholar and monk who was born in Northumbria. He was ordained as a priest and then spent the remainder of his days at the monastery of Wearmouth-Jarrow studying the Scriptures. His writings include several lives of the saints, an *Ecclesiastical History*, various liturgical writings, poetry, books on grammar and rhetoric, a book referred to as *On the nature of things*, and two books on the subject of chronology. The longer of his books on chronology came to be known as *De temporum ratione* or *The Reckoning of Time*. It dealt with the topic of *computus*—the medieval name given to the technique of constructing a Christian calendar from detailed mathematical analysis of lunar and solar cycles and interpretation of Biblical texts. Such analysis relied upon centuries of astronomical observation and was instrumental in the establishment of modern Western Christian calendars.

In the reading selections that follow, translated by Faith Wallis from the latin text of *The Reckoning of Time*, Bede instructs the reader regarding the motion of the sun and moon through the zodiac (Chaps. 16 and 17), the orientation of the crescent moon in the sky during various times of the year (Chap. 25), the effect of the moon on the tides (Chap. 29), and finally the relationship between the liturgical calendar and celestial events such as solstices and equinoxes (Chap. 30). On this latter point, Bede states that the dates for certain Christian festivals (such as the Incarnation and the Annunciation) were chosen for astronomical and allegorical reasons, rather than for historical reasons.

When reading Bede, you will come across some terms with which you are probably unfamiliar. For instance, Bede uses the terms *punctus* and *partes* to refer to intervals of 12 and 4 min, respectively. Bede also writes of *Kalends*, *Nones* and *Ides*. Readers of Shakespeare will perhaps recall the soothsayer's warning to Julius Caesar, "Beware the ides of March." But when is the Ides of March? In early Roman calendars, the new moon marked the day of *Kalends*. The lunar cycle being predictable, the date of the first quarter moon was anticipated to be in about 6 or 7 days. This was referred to as *Nones*. So the days after Kalends and leading up to Nones were referred to as

K. Kuehn, *A Student's Guide Through the Great Physics Texts,*
Undergraduate Lecture Notes in Physics, DOI 10.1007/978-1-4939-1360-2_8,
© Springer Science+Business Media, LLC 2015

6 Nones, 5 Nones, 4 Nones, 3 Nones, 2 Nones and finally, Nones. After Nones, the date of the full moon—referred to as *Ides*—was anticipated. The days leading up to Ides were referred to as 7 Ides, 6 Ides, and so on. After Ides, and anticipating the next new moon, the Roman calendar begins counting down to the Kalends of the next month. So the days immediately after the Ides of March were known as the 17 Kalends of April, 16 Kalends of April, and so on.

In 45 B.C., Julius Caesar implemented a reform of the early Roman calendar. This newer system came to be known as the Julian calendar. From here, the development of modern calendars is a truly interesting story.[1] Bede himself played a significant role in calendrical reforms during the eighth century. Several centuries later, the Holy Roman Empire adopted a set of calendrical reforms which were initiated by the Council of Trent in 1563 and introduced *via* papal bull by Pope Gregory XIII in 1582. Despite its late adoption by many non-papal states, today the Gregorian calendar is the most widely used civil calendar in the world.

8.2   Reading: Bede, *The Reckoning of Time*

Wallis, F. (Ed.), *Bede: The Reckoning of Time*, Liverpool University Press, 1999.

8.2.1   Chapter 16: The Signs of the Twelve Months

Each of the months has its own sign in which it receives the Sun: *April [the sign of] Aries; May, Taurus; June, Gemini; July, Cancer; August, Leo; September, Virgo; October, Libra; November, Scorpio; December, Sagittarius; January, Capricorn; February, Aquarius; March, Pisces. As one of the ancients explained* in heroic verse: /333/

> The Phrygian Ram looks back on the kalends of April.
> May marvels at the horns of Agenor the Bull.
> June sees the Laconian pair travel the heavens.
> At the solstice July brings in the constellation of fiery Cancer.
> Burning Leo scorches the month of August with flame.
> O Virgo, September swells Bacchus under your star.
> In the sowing season, October balances Libra.
> The diving Scorpion bids wintry November come in.
> The Archer brings his sign to a close in mid-December.
> The tropic of Capricorn marks off the commencement of January.
> The star of stout Aquarius stands in the midst of Numa's month.
> The twin fish go forth in the season of Mars.[2]

---

[1] For a very enjoyable account of the use of astronomy in calendrical reforms, refer to Chap. 1 of Heilbron, J. L., *The Sun in the Church*, Harvard University Press, Cambridge, Massachusetts, 1999.—[*K.K.*]

[2] Ausonius, *Ecloga* 16, ed. Sextus Prete (Leipzig: Teubner, 1978):108–109. Bede derived the eclogue and its introduction from a text entitled *De causis quibus nomina acceperunt duodecim*

He shows that what he states explicitly about December should be understood of the others in every case, namely, that each sign ends in the middle of its month, and begins in the middle of the previous month. I beseech those who are knowledgeable not to consider it an imposition to instruct those who do not know concerning the precise position [of the signs].

The gyre of the heavens, perfectly round at every point, is bound by the line of the zodiacal circle, like the discrete settings of twelve gems adjacent to each other on a sort of girdle wrapped around a very large sphere.[3] /334/ They are of such great size that they cannot rise or set or move from a position in less than 2 h. Thirty parts are ascribed to each sign, because of the 30 days in which the Sun illuminates each of them. *The* $10^{1}/_{2}$ h left over are not added in right away,[4] because they do not make up a full measure of 24 h. Nonetheless, since the full solar year is finished, not just in 360 days, but with an additional $5^{1}/_{4}$ days, has been added to the 30 parts.[5]

Now *the first sign of Aries begins* in that part of the heavens where the Sun stands *in the middle of March.*[6] It ends in that part which the Sun has traversed in the middle of April. Therefore it indicates the location of the vernal equinox, according to some,[7] in its eighth part, but according to *the Egyptians, who are skilled in calculation*

---

signa, ed. by Jones in *BOD* 665–667. This edition omits the poem proper, but it is included in Jones' earlier transcription in *Bedae pseudepigrapha* 103. This same text was the primary source for *The Nature of Things* 17. Its presence in the "Bobbio *computus*" (Milan Ambrosiana H 150 inf; PL 129.1324–1325) suggests an Irish origin. On this, and other possible sources for this chapter, see Jones, *BOT* 351–2. It should be noted that Bede has rearranged the poem to begin in April rather than January. This deliberate displacement away from the civic calendar and towards the natural phenomenon of the equinox may reflect the positive attraction of Easter for the computist, but it may also be in part inspired by a long standing Christian polemic against the persistent pagan associations of the January New Year: see Chap. 12, n. 147.

[3] Bede's image of the zodiac signs as jewels is striking. He may have been thinking of the 12 jewels in the foundations of the New Jerusalem (*Apoc* 2.19–20), which were closely connected with the twelve gates of the city, facing the four cardinal directions (*ibid.* 21.21). However, unlike later writers, Bede did not, in his *Explnatio Apocalypsis,* compare the foundations or gates of the New Jerusalem to the zodiac, and there is no indication that his immediate lapidary sources would have suggested such a link: *cf.* Peter Kitson, "Lapidary Traditions in Anglo-Saxon England: Part I, the Background; Old English Lapidary" *Anglo-Saxon England* 7 (1978): 9–60; "Part II, Bede's *Explanatio Apocalypsis* and Related Works", *Anglo-Saxon England* 12 (1983): 73–123.

[4] *De causis quibus nomina acceperunt duodecim signa,* 667.60–61.

[5] This is an interesting illustration of a basic principal of calendar construction, which is that one cannot calculate with less than a whole day. The Sun actually spends 30 days and $10^{1}/_{2}$ h in each zodiac sign—as is evident from the length of the tropical year—but if one is going to mark its entry into each sign on a calendar, one will have to choose one or another day, not a part of a day. In this chapter, Bede only states that the Sun enters each zodiac sign "in the middle" of the calendar month, but his main source, *De causis quibus nomina acceperunt duodecim signa,* gives the exact dates and we have included these in our reconstruction of Bede's solar calendar in Appendix 1. The extra 5 days are absorbed by the additional day in each of the seven 31 day months, minus two for the 2 days' shortfall in February. The quarter-day, of course, is cumulated into the leap-year day.

[6] *De causis quibus nomina acceperunt duodecim signa,* 666.49–50.

[7] *E.g.* Isidore, *Etym.* 5.34.

beyond all others,[8] [the equinox] is more correctly [said to be] in the fourth part.[9] The second [sign], Taurus, starting from that part of the celestial circle where the Sun is borne about in mid-April, finishes in [the part] which [the Sun] occupies in mid-May. The fourth [sign], Cancer, begins in that [part] which it illumines in the middle of June. Therefore, according to some, it receives the solstice in its eighth part,/335/ but in fact, if one follows a more careful line of research, it is a few parts earlier. From this point I take a downward course towards the bottom of the zodiac,[10] that is, the circle of the signs. The fifth [sign], Leo, [begins] in the part where the Sun is in the middle of July; the sixth, Virgo, from where it is in the middle of August. The seventh [sign], Libra, begins to rise from where [the Sun] orbits in the middle of September; hence according to common opinion it furnishes a place in its eighth part for the autumn equinox,[11] but to a more unclouded judgement it occurs rather before this. Then, as the zodiac tips towards the winter region, the eighth sign, Scorpio, begins from where [the Sun is] in the middle of October; the ninth, Sagittarius, from where it is in the middle of November. The tenth [sign], Capricorn, takes its start from where the Sun is in mid-December. Hence it consecrates a mansion for the winter solstice in its eighth part, as is commonly believed; but as Egypt, *the mother of the arts*[12] teaches, it is a few days before this. Then, when the course of the zodiac returns to higher regions, the 11th constellation, Aquarius, [starts] from the Sun's position in the middle of January. The 12th [sign], Pisces, beginning in that same part of February, comes to an end in the middle of March, where Aries, following after, shows the beginning of its rise. As all the signs agree in the fashion which they are measured and the conjunction of their zone[13]—albeit not in their visible form—the poet has said of them:

> thus the golden Sun holds sway over the sphere, divided into fixed section, by the twelve stars.[14]

/336/ The zodiac touches the Milky Way in Sagittarius and Gemini.[15] Much can be said about this, but in can be done to better effect by someone speaking than through the written word.

---

[8] Dionysius Exiguus, *Ep. ad Petronium* 65.22.

[9] Pliny, *HN* 2.17.81 (2.224).

[10] The "bottom" of the zodiac *(ad inferiora situ zodiaci)* comprises the signs through which the Sun will pass from summer solstice to winter solstice, that is, from the point at which its daily orbit will begin to travel southwards. It represents not only the "bottom" of the year, but also the fact that the Sun will be heading towards its lowest point in the sky, closest to the horizon.

[11] Isidore, *Etym.* 3.71.29; Pliny, *HN* 2.224.

[12] Macrocius, *Saturnalia* 1.15.1 (69.13–14). *cf.* Proterius, *Ep. ad Leonem* 2(271), who refers to the Egyptian church *quae mater huiuscemodi laboris extitit.*

[13] *regionum tamen suarum coniunction:* Jones thinks that this obscure phrase refers to the meeting-point of Pisces and Aries, and cites in this regard Hyginus 3.29. But Bede clearly states that "all the signs" (and not just Pisces and Aries) "agree" in this. What is meant is not exactly clear: one possibility is that Bede means to say that all the signs are of the same width.

[14] Vergil, *Georgics* 1.231–232; *cf. Letter to Wicthed* 6.

[15] *Cf. The Nature of Things* 18, which draws on the same source, namely Pliny, *HN* 18.281 (5.367).

Because we discussed above[16] the order and periodicity [*tempore*] of the planets which course through the zodiac, we will merely repeat here that the Sun illumines the circuit of the zodiac in 365 days and 6 h, the Moon in 27 days and 8 h. The Sun slips through each of the signs in 30 days and $10^{1}/_{2}$, the Moon in 2 days and $6^{2}/_{3}$ [*bisse*] h. Should you ask what *bisse* means, I shall now repeat briefly what I stated at the beginning of this little work about measures.[17] A *bisse* is smaller than a whole hour in the same proportion as 8 is smaller than 12, 20 than 30, 10 than 15. Take away a third part, and two parts remain every time; these two parts are called the *bisse*, and the third is called *triens*. Therefore they are mistaken who say that the Moon traverses as much of the heavens in 30 days as the Sun does in 365 days, since (as we stipulated earlier) the plain truth demonstrates that the Moon is held to complete as much of the course in $27^{1}/_{3}$ days as the Sun completes in $365^{1}/_{4}$ **/337/** days, and that the Moon covers as much territory in one of its [synodic] months as the orbit of the Sun does in 13 of its months.[18]

8.2.2   Chapter 17: The Course of the Moon Through the Signs

Every day the Moon either retreats from the Sun by 4 *puncti* with respect to [its location on] the preceding day when it is waxing, or approaches the Sun by 4 *puncti*, when waning.[19] As we observed above,[20] *every sign has 10 "puncti", that is, 2 h,*[21] *for 5 "puncti" make 1 h.*[22] So if you want to know what sign the Moon is in, take whatever [day of the] Moon **/338/** you wish—the fifth, for example. Multiply it by 4,[23] which makes 20. Divide by 10[24] (10 times 2 is 20); therefore the fifth Moon is

---

[16] Chapter 8.

[17] Chapter 4.

[18] It is Isidore who is in error: *DNR* 19.1 (247.5–8), and *Etym.* 5.36.3. But so also is Ambrose, *Hexaemeron* 4.5.24 (131.17–21); cf. Chap. 18 below. The Moon returns to the point in the zodiac where it was new in $27^{1}/_{3}$ days, but it must "catch up" with the Sun, which has meanwhile moved into the next zodiac sign, before conjunction can take place. Hence, as Bede puts it, the zodiacal distance covered by a synodic month is the equivalent of 13 solar months.

[19] As Bede explained above, the Moon circles the zodiac at a much faster rate than does the Sun. At conjunction, Moon and Sun will occupy the same point on the zodiac, but with each day it waxes, the Moon will pull ahead of the Sun at the rate of about $^{4}/_{5}$ of a zodiac sign per day, until it is full—or in other words, until it is exactly opposite the Sun on the zodiacal band. Then, as it wanes, it continues along the zodiac, but now returning towards the Sun, again at the rate of about $^{4}/_{5}$ of a sign per day.

[20] Chapter 3.

[21] "every sign ... hours": *De causis quibus nomina acceperunt duodecim signa*, 667.64–65. What Bede means is that each sign represents $^{1}/_{12}$ of the daily revolution of the heavenly sphere about the Earth, or in other words, 2 h.

[22] *De causis quibus nomina acceperunt duodecim signa* 667.64–65. Note that Bede is speaking of "lunar *puncti*" here; ordinarily an hour has 4 *puncti*, as he points out in Chap. 3.

[23] i.e. the number of *puncti* by which the Moon advances towards or retreats from the Sun each day.

[24] i.e. the number of *puncti* in a zodiac sign.

separated from the Sun by two signs. Then again, take the eighth [day of the] Moon. Multiply by 4 and that makes 32. Divide by 10. Three times 10 yields 30, with 2 remaining. Therefore the eighth moon is separated from the Sun by three signs and 2 *puncti*. You know that 2 *puncti* equal 6 *partes*, that is, the distance covered by the Sun in its journey through the zodiac in 6 days. A *punctus* has 3 *partes*, so each sign has 10 *puncti* or 30 *partes*. Once again, take the 19th Moon. Multiply by 4, which makes 76. Divide by 10: 7 times 10 yields 70 with 6 remaining. Therefore, on the journey which it began away from the Sun, the Moon has parted company with it by seven signs, plus 1 h, which is half a sign and a *punctus*, that is, 3 *partes*. And lest you suspect a faulty argument, find the point which no one can doubt is diametrically opposite [the position] of the fifteenth Moon. Multiply 15 by 4, which makes 60. Divide by 10: 60 divided by 10 is 6. Now the 15th Moon can always be seen six signs away from the Sun, that is, half of the celestial sphere away, whether you look ahead or behind.

Therefore you see the orb of the Moon to be completely full whenever it is opposite the Sun; /339/ you see it low [in the sky] when the Sun is high, and high when it is low.[25] For to be sure, when the Sun is in the summer circle [*i.e.* the tropic of Cancer], the Moon, when it is full is in the winter circle; when it is full and the Sun has dipped down into the winter circle, the longest night [of the year] sends [the Moon] forth to climb up the [summer] solstitial circle. And when [the Sun] keeps to one equinox, the Moon, when she is full, keeps to the other. And the distance by which the Sun has passed the equinox or solstice which it has most recently illumined obviously corresponds to the distance by which the Moon has passed the opposite solstice or equinox.[26]

8.2.3   Chapter 25: When and Why the Moon Appears to be Facing Upwards, Facing Downwards, or Standing Upright

Those who have attempted to investigate the upper air say that whenever a new Moon is seen with the crescent lying flat out, it portends a stormy month, and when it is upright, a fair one.[27] Natural reason [*naturalis ratio*] shows that this is far from the case. How is this so? Is it really credible that the position of the Moon, which remains

---

[25] "Therefore you see the orb ... when it is low." *Cf.* Chap. 6. What Bede means is that when the Sun is farthest to the north (*i.e.* "high" in the sky, or closest to zenith) at the summer solstice in Cancer, the Moon at full will be in Capricorn, "low" in the sky to the south, and vice versa. See Chap. 26 and Commentary.

[26] "For to be sure ... solstice or equinox." *Cf.* Anatolius 2, as reported in Rufinus' trans. of Eusebius, *HE* 7.32.18 (725.8–14); and Victor of Capua, *De pascha*. On the relationship of Bede's text to that of Victor of Capua and Anatolius, see Jones, *BOT* 353.

[27] Bede's immediate source is not known, but Jones has assembled a number of likely classical and post-classical references, *BOT* 360. Isidore, *DNR* 38, relates a number of weather superstitions about the Moon, but not this particular one. The same is true of Vegtius, *Epitoma rei militarius* 4.40–41, ed. K. Lang (Leipzig: Teubner, 1869): 158.13–160.17.

fixed in the ether, could be altered under the influence of a change in the winds or clouds which lie beneath it, and that it should lift up its horns any higher than nature [*naturae ordo*] dictates, as if it dreaded bad weather to come, particularly when such a blast of wayward wind would not occur everywhere on earth? The rotation of the Moon's position ought to be constant with respect to its varying degree of separation from the Sun. As St. Augustine says in his exegesis of Psalm 10: they say that *[the Moon] does not have its own light but is lit up by the Sun. But when it is [in conjunction] with [the Sun], it turns towards us the part which is not lit up, and so no light can be seen. But when it begins to draw away from [the Sun], it is lit up in the part which faces the earth, beginning (as it must) with a crescent, until the fifteenth Moon is in opposition to the Sun. At that point, [the Moon] rises when the Sun sets, so that anyone /358/ who is observing the setting Sun, if he turns to the east when he can no longer see [the Sun], may see the Moon rise. Then, when it begins to approach the Sun in the opposite direction, [the Moon] turns towards us that part which is not lit up, until it is reduced to a crescent. Eventually, it does not appear at all, because the part which is lit up is the upper part, the one facing towards the heavens, and the part which the Sun cannot shine upon is facing towards the Earth.*[28]

So when the Sun gradually ascends from the southern clime to the northern regions, and daylight increases, it is necessary that the Moon which appears at that time[29] should precede the Sun towards the northern signs at a more leisurely gait. Therefore the new Moon which is seen about to set after sunset is situated, without a doubt, not beside the Sun, but above the Sun, by which it is lit up from below (see Fig. 8.1). [The Moon's] horns extend virtually parallel, and it seems to move along flat on its back, like a ship.[30] But after the summer solstice, when the Sun's course turns back downwards towards the south, the Moon born in these months must likewise make haste towards the lower regions [of the sky]. Hence it happens that when [the Moon] is about to set in the southern regions, where the Sun has just set, without a doubt it will not still appear [above] the Sun, but beside it, positioned towards the south, when it first appears after sunset (see Fig. 8.2). Thus [the Moon's] northern flanks, facing the Sun, will seem to advance in an erect position, for with its horns turned away from the Sun, /359/ the Moon always presents its round part to [the Sun].[31]

---

[28] Augustine, *Enarrationes in Psalmos* 10. 3 (76.30-42), where this is one of a number of possible explanations advanced; *cf.* Augustine, *Ep.* 55.6–7, CSEL 34 (175.15–177.16), where he is rather more committed to the explanation proposed here.

[29] *i.e.* the crescent Moon on the first day of the lunation.

[30] This is a rather convoluted way of saying that in the northern hemisphere, the crescent Moon at sunset at the spring equinox will be to the north of the celestial equator. Since the equator slants to the south, this means that the Moon will seem to be directly above the setting Sun. The Moon's underside, that is, the side directly opposite the horizon, will be illuminated from below by the setting Sun. The crescent will look like a boat, with the two horns pointing up. For illustration, see Commentary.

[31] "for with its horns ... round part to him": *cf.* Augustine, *Ep.* 55.7 (176.20–177.7). Again, what Bede is saying is that at the autumn equinox the crescent Moon at sunset is to the south of the

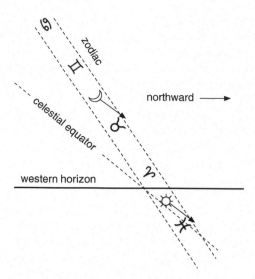

Fig. 8.1 Looking toward the western horizon just after sunset shortly before the vernal equinox. The celestial equator makes an angle of 55° with respect to the horizon in northern England. The arrows represent the trajectories of the sun and the moon as they set.—[*K.K.*]

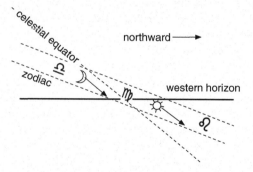

Fig. 8.2 Looking toward the western horizon just after sunset shortly before the autumnal equinox. The sun is leaving Leo and entering Virgo.—[*K.K.*]

By this logic, it follows that the longer the day is, the higher the new Moon will be; and the shorter and more inclined towards the south the day is, the lower down the new Moon will be seen. Hence the opinion of the untutored came to the conclusion that when the new Moon appeared higher up and supine, it betokened whirlwinds of storm, and when erect and lower down towards the south, fair weather. For such,

---

celestial equator. Since in the northern hemisphere the equator slants to the south, this means that the crescent Moon will seem to be to the south and left of the setting Sun. Its northern side will be illuminated, forming a crescent standing upright on one of its horns. See commentary.

indeed, is the condition of the revolving year: the motion of the atmosphere is much more gentle in the 6 months during which the day decreases than during the other six. The same logic explains why the waning Moon in its morning rising appears now upright, and now supine. It also explains why she will often emerge during the day bent downwards [*prona*],[32] since the rays of the sun, of course, are striking her upper part. Therefore the alteration of the Moon, which is natural and fixed, cannot portend the condition of the month to come. But those who are inquisitive about this sort of thing will often pronounce on atmospheric conditions by the colour of [the Moon] or sun, or from the heavens themselves, the stars or the shifting shapes of clouds, or other omens.[33] Hence they claim that if the fourth moon is unblemished and its horns unblunted, it is a sign of fair weather for the rest of the month, and things of this sort.[34]

8.2.4   Chapter 29: The Harmony of the Moon and the Sea

But more marvellous than anything else is the great fellowship that exists between the ocean and the course of the Moon.[35] For at [the Moon's] every rising and setting, [the ocean] sends forth the strength of his ardour, *which the Greeks call "rheuma,"*[36] to cover the coasts far and wide; and when it retreats, it lays them bare. The sweet streams of the rivers it abundantly mingles together and covers over with salty waves. When the Moon passes on, [the ocean] retreats and restores the rivers to their natural sweetness and level, without delay. *It is as if [the ocean] were dragged forwards* against its will *by certain exhalations of the Moon,* and when her power ceases, *it is poured back again into his proper measure.*[37]

   For as we taught above,[38] the Moon /367/ rises and sets each day 4 *puncti* later than it rose or set the previous day; likewise, both ocean tides, be they by day or by night, morning or evening never cease to come and go each day at a time which is later by almost the same interval. Now a *punctus* is one-fifth of an hour, and 5 *puncti* make and hour. Furthermore, because the Moon in 2 lunar months (that is, in 59 days) goes around the globe of the earth 57 times,[39] therefore the ocean tide during this same period of time, surges up to its maximum twice this number of times, that is, 114, and sinks back again to its bed the same number of times. For in the course

---

[32] *i.e.* with the horns pointing downwards, like an eyebrow.

[33] Bede is alluding to Isidore, *DNR* 38 (299–303) and Vegetius (see note 229 above).

[34] Isidore, *DNR* 38.3 (301. 27–28); Pliny, *HN* 2.48.128 (1.268).

[35] *Cf.* ps.-Isidore, *Liber de ordine creaturarum* 9.4 (148.28–30); Virgilius Maro Grammatiucs, *Epitoma* 4 (17.6–7).

[36] Vegetius 4.42 (161.4); by contrast, Isidore uses the word *"rheuma"* only in its medical sense of "morbid discharge": *Etym.* 4.7.11.

[37] Ambrose, *Hexaemeron* 4.7.30 (136.3–5). *Cf.* Isidore, *DNR* 40.1 (307.8–11).

[38] Chapter 17.

[39] See n. 42 below.

of 29 days, the Moon lights up the confines of the Earth 28 times, and in the 12 h which are added on to make up the fullness of the natural month, it circles half the globe of the Earth, so that, for example, /368/ the new Moon which emerged [from conjunction] last month above the Earth at noon, will this month meet up with the Sun to be kindled at midnight beneath the Earth. Through this length of time, the tides will come twice as often, and 57 times

> the high seas swell
> breaking their barriers and once again retreat unto themselves.[40]

Because the Moon in half a month (that is, in 15 days and nights) circles about the Earth 14 times, and once besides around half the Earth, it happens that it is in the east at evening when it is full, while earlier it was in the west at evening when it was new. In this period of time, the sea ebbs and flows 29 times. Just as the Moon in 15 days (as we said) is flung back by its natural retardation towards the east from its position in the west at evening,[41] [so that while it] occupies the east today at morning, it will be in the west at morning 15 days from now, so also the ocean tide which now occurs at evening, will after 15 days be in the morning. On the other hand, the morning tide, impeded by this daily drag, will then rise at evening. And because the Moon in one year (that is, in 12 of its months, which makes 354 days) circles the globe of the Earth 12 fewer times (that is 342 times), /369/ so the ocean tide within the same period washes up against the land and recoils 684 times.[42]

The sea reflects the course of the Moon not only by sharing its comings and goings, but also by a certain augmenting and diminishing of its size, so that the tide recurs not only at a later time than it did yesterday, but also to a greater or lesser extent. When the tides are increasing, they are called *malinae*, and when they are decreasing, *ledones*.[43] In alternating periods of 7 or 8 days, they divide up every month among themselves in their fourfold diversity of change. Sometimes by equal shares both fill up their course in $7\frac{1}{2}$ days; sometimes they will arrive earlier or later, or be more or less strong than usual when they are pushed onwards or forced back by the winds,

---

[40] Vergil, *Georgics* 2.479–480, ed. R.A.B. Mynors (Oxford: Clarendon Press, 1990): li.

[41] The Moon moves across the sky in two ways. Each day, it rises in the east and sets in the west: from the medieval perspective, this is because it is borne above by the daily east–west rotation of the heavenly sphere. But the Moon also progresses eastwards around the zodiac once each month. Bede imagines the Moon "falling behind" the daily rotation of the outermost heavenly sphere, so that it rises each day a little later. The Moon that rose and set with the Sun at conjunction, rises and sets 12 h later, and half the sky away, when it is full.

[42] Bede imagines that the Moon every day "runs ahead" of the Sun by four *puncti*, or $\frac{4}{5}$ of an hour (48 min), so that it can meet up with the Sun at conjunction 29$\frac{1}{2}$ days later (*cf.* Chap. 17.) Over 354 days, this 50-min advance amounts to about 12 days ($50 \times 354 = 17,700$ min $\div 60 = 295$ h $\div 24 =$ approximately 12 days). The "gain" is therefore about 1 day each month. Hence, as Bede said earlier, the Moon circles the Earth 28 times each 29 days, 57 times each 59 days, *etc*. From a modern heliocentric perspective, we would say that the Earth rotates 29 times, but the retardation of the Moon means that it seems to circle our planet one less time.

[43] These terms are usually translated as "spring tide" and "neap tide" respectively. For further discussion, see Commentary.

or by the pressure of some other phenomenon or natural force, with the result that sometimes their order is upset, and the *malina* tide claims more in this month and less in the next. Hence both directions will begin now in the evening, now in the morning. Indeed, when an evening tide occurs at the full or new Moon, it will be a *malina*, and for the next 7 days this same *malina* tide will be greater and stronger than the morning tide. Similarly if a *malina* tide starts in the morning, the morning tide for days afterwards will cover /370/ the land with more sea. The evening tide, confined to the boundaries laid down by the morning tide, refrains from extending its course further, although in some months both tides grow at utterly different rates.

The more the stronger tide covers the shore and lands and fills up the rivers and straits, the more it is wont to leave these same coastlines empty and bare as it recedes. So let him who is capable, see if what Philip says is true or no: *There are those who claim and affirm that an enormous outpouring of the ocean takes place in all the streams of every region and land at one and the same time.*[44] But we who live at various places along the coastline of the British Sea know that where the tide begins to run in one place, it will start to ebb at another at the same time. Hence it appears to some that the wave, while retreating from one place, is coming back somewhere else; then leaving behind the territory where it was, it swiftly seeks again the region where it first began. Therefore at a given time a greater *malina* deserts these shores in order to be able all the more to flood other [shores] when it arrives there.

This can easily be grasped from the Moon's course. For example, positioned at about the winter sunrise or at the [summer] solstice sunset, the Moon at any given age, whether it be above the earth or beneath it, will pull at the tide here, but will repel it when she is positioned at winter sunset or summer solstice sunrise. But the Moon which signals the rise of the tide here, signals its retreat in other regions far from this quarter of the heavens. And there is more: those who live north of me on the same coastline usually receive and give back each tide sooner than I do, and those /371/ to the south much later. In every region, the Moon holds to whatever rule of fellowship with the sea it received in the beginning. Hence we have ascertained that the *malina* often begins about 5 days before the new or full Moon, and the *ledones* the same number of days before the half Moon. Near the two equinoxes, the tide rises higher than usual, and is much lower at the winter and summer solstices. The Moon always *exerts less force when she is in the north and at her highest point away from the Earth*[45] *than when she ventures into the south and treads closer.* Natural reason [*naturalis ratio*] convinces us that the tides flow according to the patter of the Moon's cycle of 19 years; *hence the course of the sea returns to the beginning of its movements and to equal increments.*[46]

---

[44] Philippus Presbyter, *Commentarii in librum Job* 38(PL 26.752 C).

[45] By "Earth", Bede means "horizon"; *cf.* Chap. 26.

[46] Pliny, *HN* 2.99.215–216. (2.344). However, Pliny proposes an 8-year cycle for the tides. Bede accepted this model in *The Nature of Things* 39, but modifies it here.

8.2.5   Chapter 30: Equinoxes and Solstices

On the subject of the equinoxes and solstices, the opinion of many learned men, both worldly [philosophers] and Christians, is straightforward: the equinoxes are to be observed on the eighth kalends of April [25 March] and the eighth kalends of October [24 September], the solstices on the eighth kalends of July [24 June] and the eighth kalends of January [25 December].[47] Thus Pliny, who was born an orator and a philosopher, says in the second book of his /372/ *Natural History: Now the Sun itself has four turning-points: twice when the night is equal to the day, in spring and autumn, when he arrives at the mid-point of Earth in the eighth degree of Aries or Libra, and twice when the proportions [of the day and night] are reversed: in winter, in the eighth degree of Capricorn, when the days grow longer, an in summer, in the eighth degree of Cancer when the nights grow longer. Since an equal part of the universe is above and below the earth at all times, it is the angle of the zodiac which is the cause of this inequality. The signs which are upright when they rise hold their light for a longer time; those which rise at an angle, pass on in a briefer period of time.*[48] Hippocrates the premier physician, writing to King Antigonus on how he should watch over himself during the course of the year in order to prevent illness says this: *Let us begin therefore with the solstice, that is, with the 8th kalends of January [25 December]. From this day forward up until the spring equinox, a period of 90 days, moisture [humor] increases in bodies. This season activates a man's phlegm, /373/ so that people frequently catch catarrh, [suffer from] dripping of the uvula, pain in the side, weakness of vision, and ringing in the ears, and cannot smell anything. At such a season, take high-quality food, hot and seasoned with assafoetida, with pepper and mustard. Wash your head seldom. However, purge without ceasing. Indulge in wine and do not stint on sex for the first forty days. Ninety days follow from the above-mentioned day until the vernal equinox. From the above-mentioned 8th kalends of April [25 March], until the 8th ides of May [8 May], there are 45 days. In these days the sweet humours (that is, blood) increase in man. Eat food that is fragrant and very pungent. Again, from the 8th ides of May [8 May] until the 8th kalends of July [24 June] are 45 days. In these days the bitter bile (that is, red choler) increases. Eat sweet foods, indulge in wine, abstain from sex, and fast very little. The summer season begins on the 8th kalends of July [24 June]. At that time, the increase of red bile falls off, and black bile waxes; it is held that this occurs until the autumn equinox, which is from the 8th kalends of October [24 September], until the 8th kalends of January [25 December], the bitterness of the black bile wanes and the density of the [phlegmatic] humour increases. Use all*

---

[47] *Cf.* Isidore, *DNR* 8 (205.1–5); *Etym.* 5.34; Bede himself held this opinion when he wrote *On Times* 7.

[48] Pliny, *HN* 2.17.81 (1.224). Because of variations in the angle formed by the intersection of the ecliptic and the horizon (see Commentary on Chap. 25), some zodiac signs will rise more or less "upright", *i.e.* perpendicular to the horizon, while others will rise at an angle to the horizon. The former will require less time to clear the horizon than the latter, and thus will "hold their light for a longer time".

such foods as are hot and very pungent, abstain from sex and wash little. From the above-mentioned [date] /374/ until the Vergiliae [=Pleiades] set (that is, until the sixth ides of February [8 February]) there are 96 days, and then the Pleiades set. From that hour blood increases in men. It behoves one, therefore, to eat food that is very pleasant, and to indulge in wine and sex. There are 97 days in winter.[49]

This is what some of the pagans say; and very many of the Church's teachers recount things which are not dissimilar[50] to these about time, saying that our Lord was conceived and suffered on the 8th kalends of April [25 March], at the spring equinox, and that he was born at the winter solstice on the eighth kalends of January [25 December]. And again, that the Lord's blessed precursor and Baptist was conceived at the autumn equinox on the eighth kalends of July [24 June]. To this they add the explanation that it was fitting that the Creator of eternal light should be conceived and born along with the increase of temporal light, and that the herald of penance, who must decrease, should be engendered and born at a time when the light is diminishing.[51] But because, as we have learned in connection with the calculation of Easter, the judgement of all the men of the East (and especially of the Egyptians, who, it is agreed, where the most skilled in calculation) is in particular agreement that the spring equinox is on the 12th kalends of April [21 March], we think that the three other turning-points of the season ought to be observed a little before [the date] given in the popular treatises.[52]

So let us speak briefly about the spring equinox, which is the chief of the four annual changes we mentioned, as the Creation of the world indicates. [53] The rule of the Church's observance, confirmed at the council of Nicaea, holds that Easter Day /375/ is to be sought between the 11 kalends of April [22 March] and the 7th kalends of May [25 April]. Again the rule of catholic teaching commands that Easter is not to be celebrated before the passing of the spring equinox.[54] Therefore he who thinks that the eighth kalends of April [25 March] is the equinox is obliged either to declare that it is licit to celebrate Easter before the equinox, or to deny that it is licit to celebrate Easter before the seventh kalends of April [26 March], and also to confirm that the Pasch which our Lord kept with his disciples on the night before he suffered was either not on the ninth kalends of April[24 March] or was before the equinox.[55] For a doctrine not only of our own day but also of the Mosaic law

---

[49] Ps.-Hippocrates, *Ad Antigonum regem* 8–9, ed. E. Liechtenhan, J. Kollesch and D. Nickel, in *Marcelli de medicamentis liber*, ed. M. Niedermann, 2nd ed. Corpus medicorum latinorum 5 (Berlin: Akademie-Verlag, 1968):1.18–25.

[50] For examples of such statements in the works of the Fathers, see Jones, *BOT* 366.

[51] See *BOT* 366; for John the Baptist's statement "he must increase, but I must decrease", see John 3.30.

[52] Bede is not interested in fixing the dates of the autumn equinox or the solstices, as they are of no computistical consequence. For a signal exception, see *Letter to Wicthed* 6.

[53] See above, Chap. 6.

[54] *Cf.* Dionysisus Exiguus, *Ep. ad Petronium* (65.13–27); *Prologus Cyrilli* 4 (339–340).

[55] The traditional "historic" date of the Passion was 25 March (see above); therefore Christ's Last Supper could not have been a Passover meal (as the Synoptics indicate) if the equinox were on the 25th, since Passover must follow the equinox. *Cf. Letter to Wicthed* 12.

decrees that the day of the Paschal feast cannot be celebrated before this equinox has passed. Anatolius declares: *It is plain that Philo and Josephus teach thus; thus did their elders Agathabolus, and Aristobolus of Panaeda his student, who was one of the seventy elders sent by the priests to King Ptolemy to translate the Hebrew books into the Greek tongue, and who answered many [questions] about the traditions of Moses which the king proposed and asked particularly. So these men, when they explained the questions about Exodus, said that the Pasch could not be sacrificed before the spring equinox had passed.*[56]

Hence it is necessary, in order to preserve the rule of truth, that we state clearly both that the Pasch is not to be sacrificed before the equinox and before the darkness has been overcome, and that this equinox is rightfully to be assigned to the 12th kalends of April [March 21], as we are instructed not only by the authority of the Fathers, but also by examining the sundial [*horologica consideratione*]. /376/ But the other three termini of the seasons are, but an analogous reasoning, to be observed the same number of days before the eighth kalends of the following [month].

8.3   Study Questions

QUES. 8.1 Describe the movement of the sun through the zodiac.

a) What, according to Bede, is the shape of the heavens? And how is the zodiac arranged upon the heavens?

b) How many constellations comprise the zodiac? And what is the width of each? How many days does the sun spend in front of each zodiacal sign? Why does this present a problem? And how is it solved?

c) Describe the motion of the sun across the zodiac during the course of the year. From a geocentric standpoint, does the sun move *faster*, *slower*, or at the *same* speed as the celestial sphere containing the zodiac?

QUES. 8.2 Describe the movement of the moon through the zodiac.

a) Does the moon travel along the same path through the sky as the sun? Does it travel at the same rate? If not, how is it different?

b) What is meant by the "age" of the moon? How many days old is a new moon? A full moon? Write down a mathematical formula for determining the position of the moon, with respect to the zodiac, as a function of its age.

c) How is the trajectory of a full moon across the night sky correlated with the trajectory of the sun across the sky during daylight? For instance, when the sun's trajectory is high, is the full moon's trajectory high or low? Why is this?

---

[56] Anatolius 2 as recorded in Rufinius' trans. of Eusebius, *HE* 7.32.16–17 (725.1–8)

QUES. 8.3 Does the phase of the moon affect the weather?

a) To what faculty does Bede immediately appeal in addressing this question? And to what authority does he then appeal when discussing the rotation of the moon's position?
b) What does Bede mean when he says that the sun "ascends from the southern clime to the northern regions, and daylight increases"? To what season of the year is he here referring?
c) During this season, does the waxing crescent moon set to the north, to the south, or at the same position on the horizon as the sun? What about during the other three season?
d) How is the trajectory of a new moon across the sky correlated with the trajectory of the sun across the sky? For instance, when the sun's trajectory is high, is the new moon's trajectory high or low? Why is this?
e) During which season are the horns of the waxing crescent moon most upward (away from the horizon) directed?

QUES. 8.4 Does the moon affect the oceans?

a) Do the tides come and go at the same time daily? Why is this important for Bede's argument?
b) Why does Bede say that during two lunar months (59 days) the moon goes around the earth 57 times? How many tides occur during this time interval?
c) Are the tides equal in strength on all days? If not, is there a pattern?
d) Do the tides occur at the same time at all locations on the earth? How does Bede explain this phenomenon? How do the tides behave during the equinoxes? The solstices?

QUES. 8.5 How is the date of Easter computed?

a.) What are the traditional dates of Jesus' conception, birth, and passion? Of John the Baptist's conception and birth? What allegorical significance, based on astronomical events, was attached to these traditions?
b.) When, according to the Council of Nicea, could Easter be observed? On what basis does Bede say that Easter must be celebrated after the vernal equinox?
c.) What does Bede conclude regarding the date which should be assigned to the equinox? Is he suggesting that the equinox, as measured by a sundial, be rejected? Or that the calendar should be adjusted?

8.4   Exercises

Ex. 8.1 (THE MOON AND THE ZODIAC). Suppose that (i) the sun is in the middle of the constellation Gemini, and (ii) that the moon is 20 days old. Where is the moon located with respect to the zodiac?

Ex. 8.2 (CRESCENT MOON). Suppose that the moon is waxing crescent. Does it set before or after sunset? During which season (spring, summer, autumn, winter) do its "horns" point most southward?

Ex. 8.3 (TIDES). What (if any) configuration of the sun, moon and earth gives rise to the highest possible tide? How often can this occur?

Ex. 8.4 (LUNAR OBSERVATIONS). This field exercise consists of daily observations over the course of two or three weeks. You will need to set aside 15 min each day *at the same time* (to within 5 min) to make observations of the moon. To this end, you should find a location you can stand where you have a clear view of the sky and the horizon—such as a large open space or atop a building or hill. Here, measure the position and angular diameter of the moon using a device such as a cross-staff.[57] In addition to the position, diameter, and local solar time of your observation, make a sketch of the appearance of the moon in your astronomical notebook. What is the phase of the moon? What is the orientation of the moon relative to the horizon? In what constellation is the moon located? On a star chart, make a sketch of the position and phase of the moon each night as it moves through the zodiac.[58] By the end of your observation period, you should be able to answer the following questions. In which direction is the moon moving? How quickly? How many days comprise a lunar month? Is the sidereal month of the same length as the synodic month? What is the difference? Be sure to compare your measurements with accepted values.

Ex. 8.5 (CALENDAR CONSTRUCTION). If the synodic month were only 13 days, how might you construct a lunar calendar? How many days would be in each month? Would there be the need for a leap year? If so, how often?

8.5   Vocabulary

1. Zodiac
2. Milky Way
3. Puncti
4. Tropic of Cancer
5. Tropic of Capricorn
6. Equinox
7. Solstice
8. Brevity
9. Equinoctial hour
10. Waxing
11. Waning
12. Licit
13. Pasch
14. Kalends
15. Termini

---

[57] The construction and use of a cross-staff is described in Chap. 7 of the present volume.

[58] A very nice star chart can be found in Sinnott, R. W., *Sky and Telescope's Pocket Sky Atlas*, New Track Media, LLC, Cambridge, Massachusetts, 2006.

Chapter 9
From Astronomy to Cartography

Since thy Majesty is sacred throughout the vast world,
Maximilian Caesar, in the farthest lands,
Where the sun raises its golden head from the eastern waves
And seeks the straits known by Hercules' name,
Where the midday glows under its burning rays,
Where the Great Bear freezes the surface of the sea;
And since thou, mightiest of mighty kings, dost order
That mild laws should prevail according to thy will;
Therefore to thee in a spirit of loyalty this world map has been
dedicated
By him who has prepared it with wonderful skill.
—Martin Waldseemüller

9.1 Introduction

Martin Waldseemüller (*ca.* 1475–1522) was born in Wolfenweiler, Germany and matriculated from the University of Freiburg. He studied theology, was a cleric at the Diocese of Constance, and was canonized at Saint Dié. He was a highly influential geographer and cartographer whose famous twelve-panel map of the world, shown in Fig. 9.1, gave the name *America* to the new world after the journeys of the Italian explorer Amerigo Vespucci. The map, which measures 4 ft×8 ft when fully assembled, was the first to depict the entire Western Hemisphere as a separate continent between two vast bodies of water, the Atlantic and Pacific oceans. Interestingly, Waldseemüller's later maps omitted the name "America", simply referring to the new continent as "Terra Incognita." Nonetheless, other famous mapmakers had already adopted and popularized Waldseemüller's original terminology.[1] The text that follows is from Waldseemüller's *Cosmographiae Introductio*, which was published alongside his famous 1507 map. This *Introduction to Cosmography* makes extensive use of the meticulous geometrical and astronomical works of Ptolemy.

---

[1] For a wonderful account of the history of cartography up to and including the map of Waldseemüller, see Lester, T., *The fourth part of the world: the race to the ends of the earth, and the epic story of the map that gave America its name*, Free Press, New York, NY, 2009.

K. Kuehn, *A Student's Guide Through the Great Physics Texts*,
Undergraduate Lecture Notes in Physics, DOI 10.1007/978-1-4939-1360-2_9,
© Springer Science+Business Media, LLC 2015

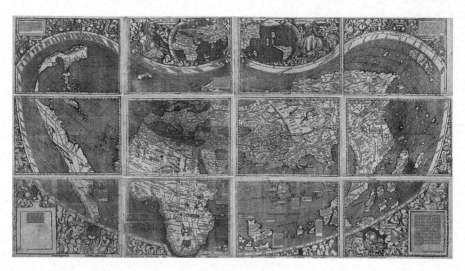

Fig. 9.1 Martin Waldseemüller's famous 1507 world map, purchased in 2001 from Prince Waldburg-Wolfegg by the Geography and Map Division of the United States Library of Congress. It is on permanent display in the Library's Thomas Jefferson Building

9.2   Reading: Waldseemuller, *Introduction to Cosmography*

Waldseemüller, *Introduction to Cosmography*, 31–81 pp., Ann Arbor: University Microfilms, Princeton, NJ, 1966.

9.2.1   Preface: To His Majesty Maximillian Cæsar Augustus Martinus Ilacomilus Wishes Good Fortune

If it is not only pleasant but also profitable in life to visit many lands and to see the most distant races (a fact that is made clear in Plato, Apollonius of Tyana, and many other philosophers, who went to the most remote regions for the purpose of exploration), who, I ask, most invincible Maximilian Cæ sar, will deny that it is pleasant and profitable to learn from books the location of lands and cities and of foreign peoples,

> Which Phœ bus sees when he buries his rays beneath the waves,
> Which he sees as he comes from the farthest east,
> Which the cold northern stars distress,
> Which the south wind parches with its torrid heat
> Baking again the burning sands?
>
> —Boethius

Who, I repeat, will deny that it is pleasant and profitable to learn from books the manners and customs of all these peoples? Surely—to express my own opinion—just as it is worthy of praise to travel far, so it can not be foolish for one who knows the world, even from maps alone, to repeat again and again that passage of the Odyssey which Homer, the most learned of poets, wrote about Ulysses:

Tell me, O Muse, of the man who after the capture of Troy
Saw the customs and the cities of many men.

Therefore, studying, to the best of my ability and with the aid of several persons, the books of Ptolemy from a Greek copy, and adding the relations of the four voyages of Amerigo Vespucci, I have prepared for the general use of scholars a map of the whole world—like an introduction, so to speak—both in the solid and projected onto the plane. This work I have determined to dedicate to your most sacred Majesty, since you are the lord of the world, feeling certain that I shall accomplish my end and shall be safe from the intrigues of my enemies under your protecting shield, as though under that of Achilles, if I know that I have satisfied, to some extent at least, your Majesty's keen judgment in such matters. Farewell, most illustrious Cæsar.

At St. Dié, in the year 1507 after the birth of our Saviour.

9.2.2 Order of Treatment

Since no one can obtain a thorough knowledge of Cosmography without some previous understanding of astronomy, nor even of astronomy itself without the principles of geometry, we shall in this brief outline say a few words:

1) Of the elements of geometry that will be helpful to a better understanding of the material sphere;
2) Of the meanings of *sphere, axis, poles, etc.*;
3) Of the circles of the heavens;
4) Of a certain theory, which we shall propose, of the sphere itself according to the system of degrees;
5) Of the five celestial zones, and the application of these and of the degrees of the heavens to the earth;
6) Of parallels;
7) Of the climates[2] of the earth;
8) Of the winds, with a general diagram of these and other things;
9) Of the divisions of the earth, of the various seas, of islands, and of the distances of places from one to another. There will be added also a quadrant useful to the cosmographer.

Lastly, we shall add the four voyages of Amerigo Vespucci. Thus we shall describe the cosmography, both in the solid and projected onto the plane.

---

[2] The word *climate* is here used in its ancient sense of a zone of the earth's surface comprised between two specified parallels of latitude.

9.2.3   Chapter I: Of the Principles of Geometry Necessary to an Understanding of the Sphere

Since in the following pages frequent mention will be made of the circle, the circumference, the center, the diameter, and other similar terms, we ought first of all briefly to discuss these terms one by one.

1) A circle is a plane figure bounded by a line drawn around, and in the middle there is a point, all straight lines drawn from which to the surrounding line are equal to one another.
2) A plane figure is a figure, no point of which rises above or falls below the lines that bound it.
3) The circumference is the line that so bounds the circle that all straight lines drawn from the center to the circumference are equal to one another. The circumference is also called in Latin *ambitus*, *circuitus*, *curvatura*, *circulus*, and in Greek *periphereia*.
4) The center of a circle is a point so situated that all straight lines drawn from it to the line bounding the circle are equal to one another.
5) A semicircle is a plane figure bounded by the diameter of the circle and one half of the circumference.
6) The diameter of a circle is any straight line passing through the center of the circle and extending in both directions to the circumference.
7) A straight line is the shortest distance between two points.
8) An angle is the mutual coming together of two lines. It is the portion of a figure increasing in width from the point of intersection.
9) A right angle is an angle formed by one line falling upon another line and making the two angles on either side equal to each other. If a right angle is bounded by straight lines, it is called plane; if bounded by curved lines it is called curved or spherical.
10) An obtuse angle is an angle that is greater than a right angle.
11) An acute angle is less than a right angle.
12) A solid is a body measured by length, breadth, and height.
13) Height, thickness, and depth are the same.
14) A degree is a whole thing or part of a thing which is not the result of a division into sixtieths.
15) A minute is the 60th part of a degree.
16) A second is the 60th part of a minute.
17) A third is the 60th part of a second, and so on.

9.2.4   Chapter II: Sphere, Axis, Poles, etc., Accurately Defined

Before any one can obtain a knowledge of cosmography, it is necessary that he should have an understanding of the material sphere. After that he will more easily comprehend the description of the entire world which was first handed down by Ptolemy and others and afterward enlarged by later scholars, and on which further light has recently been thrown by Amerigo Vespucci.

A sphere, as Theodosius defines it in his book on spheres, is a solid and material figure bounded by a convex surface in the center of which there is a point, all straight lines drawn from which to the circumference are equal to one another. And while, according to modern writers, there are ten celestial spheres, there is a material sphere like the eighth (which is called the fixed sphere because it carries the fixed stars), composed of circles joined together ideally by a line and axis crossing the center, that is, the earth.

The axis of a sphere is a line passing through the center and touching with its extremities the circumference of the sphere on both sides. About this axis the sphere whirls and turns like the wheel of a wagon about its axle, which is a smoothly rounded pole, the axis being the diameter of the circle itself. Of this Manilius speaks as follows:

> Through the cold air a slender line is drawn,
> Round which the starry world revolves.

The poles, which are also called *cardines* (hinges) and *verticies* (tops), are the points of the heavens terminating the axis, so fixed that they never move, but always remain in the same place. What is said here about the axis and the poles is to be referred to the eighth sphere, since for the present we have undertaken the limitation of the material sphere, which, as we have said, resembles the eighth sphere. There are accordingly two principal poles, one the northern, also called *Arcticus* (arctic) and *Borealis* (of Boreas), the other the southern, also called *Antarcticus* (antarctic). Of these Vergil says:

> The one pole is always above us, but the other
> The black Styx and the deep shades see 'neath our feet.

We who live in Europe and Asia see the arctic pole always. It is so called from *Arctus*, or *Arcturus*, the Great Bear, which is also named *Calisto*, *Helice*, and *Septentrionalis*, from the seven stars of the Wain, which are called *Triones*; there are seven stars also in the Lesser Bear, sometimes called *Cyanosura*. Wherefore Baptista Mantuanus says:

> Under thy guidance, Helice, under thine Cynosura,
> We set sail over the deep, *etc.*

Likewise, the wind coming from that part of the world is called *Borealis* and *Aquilonicus* (northern). Sailors are accustomed to call *Cyanosura* the star of the sea.

Opposite to the arctic pole is the antarctic, which it derives its name, for ʼαντί in Greek is the equivalent of *contra* in Latin. This pole is also called *Noticus* and *Austronoticus* (southern). It can not be seen by us on account of the curvature of the earth, which slopes downward, but is visible from the antipodes (the existence of which has been established). It should be remarked in passing that the downward slope of a spherical object means its swelling or belly; that convexity is the contrary of it and denotes concavity.

There are, besides, two other poles of the zodiac itself, describing two circles in the heavens, the arctic and antarctic. Since we have made mention of the zodiac, the arctic, and the antarctic (which are circles in the heavens), we shall treat of circles in the following chapter.

9.2.5   Chapter III: Of the Circles of the Heavens

There are two kinds of circles, called also *segmina* by authors, on the sphere and in the heavens, not really existing, but imaginary; namely, great and small circles.

A great circle is one which, described on the convex surface of the sphere, divides it into two equal parts. There are six great circles: the equator, the zodiac, the equinoctial colure, the solstitial colure, the meridian, the horizon.

A small circle on the sphere is one which, described on the same surface of the sphere, divides it into two unequal parts. There are four small circles: the arctic, the circle of Cancer, the circle of Capricorn, the antarctic. Thus there are in all ten, of which we shall speak in order, first of the great circles.

The equator, which is also called the girdle of the *primum mobile* and the equinoctial, is a great circle dividing the sphere into two equal parts. Any point of the equator is equally distant from both poles. It is so called because when the sun crosses it (which happens twice a year, at the first point of Aries, in the month of March, and at the first point of Libra, in the month of September), it is the equinox throughout the world and the day and night are equal. The equinox of March or of Aries is the vernal equinox, the equinox of September or of Libra the autumnal.

The zodiac is a great circle intersecting the equator at two points, which are the first points of Aries and Libra. One half of it inclines to the north, the other to the south. It is so called either from ζῴδιον, meaning an *animal*, because it has 12 animals on it, or from ζωή, meaning *life*, because it is understood that the lives of all the lower animals are governed by the movements of the planets. The Latins call it *signifer* (sign-bearing), because it has twelve signs on it, and the oblique circle. Therefore Vergil says:

Where the series of the signs might revolve obliquely.

In the middle of the width of the zodiac there is a circular line dividing it into two equal parts and leaving 6° of latitude on either side. This line is called the ecliptic, because no eclipse of the sun or moon ever takes place unless both of them pass under that line in the same or in opposite degrees,—in the same if it is to be an eclipse of the sun; in opposite, if it is to be an eclipse of the moon. The sun always passes with its center under that line and never deviates from it. The moon and the rest of the planets wander at one time under the line, at another on one side or the other.

There are two colures on the sphere, which are distinguished as solstitial and equinoctial. They are so called from the Greek κῶλον, which means a *member* and the Latin *uri boves* (wild oxen), which Cæsar says, in the fourth[3] book of his "Commentaries", are found in the Hercynian forest and are of the size of elephants, because, just as the tail of an ox when raised makes a semicircular and incomplete member, so the colure always appears to us incomplete, for one half is visible, while the other half is concealed.

The solstitial colure, which is also called the circle of declinations, is a great circle passing through the first points of Cancer and Capricorn, as well as through the poles of the ecliptic and the poles of the world.

The equinoctial colure, in like manner, is a great circle passing through the first points of Aries and Libra and the poles of the world.

The meridian is a great circle passing through the point vertically overhead and the poles of the world. These circles we have drawn 10° apart in our world map in the solid and projected on the plane. There is a point in the heavens directly over any object, which is called the zenith.

The horizon, also called *finitor* (limiting line) is a great circle of the sphere dividing the upper hemisphere (that is, the half of a sphere) from the lower. It is the circle at which the vision of those who stand under the open sky and cast their eyes about seems to end. It appears to separate the part of the heavens that is seen from the part that is not seen. The horizon of different places varies, and the point vertically overhead of every horizon is equally distant in all directions from the *finitor* or the horizon itself.

Having thus considered the great circles, let us now proceed to the small circles.

The arctic circle is a small circle which the other pole of the zodiac makes and describes about the antarctic pole of the world. We mean by the pole of the zodiac (of which we spoke also in the preceding chapter), the point that is equally distant from any point on the ecliptic, for the poles of the zodiac are the extremities of the axis of the ecliptic. The distance of the pole of the zodiac from the pole of the world is equal to the greatest declination of the sun (of which we shall say more presently.)

The tropic of Cancer is a small circle which the sun, when at the first point of Cancer, describes by the motion of the *primum mobile*. This point is also called the summer solstice.

The tropic of Capricorn is a small circle which the sun, when at the first point of Capricorn, describes by the motion of the *primium mobile*. This circle is also called the circle of the winter solstice.

Since we have mentioned declination, it should be remarked that declination occurs when the sun descends from the equinoctial to the tropic of Cancer, or from us to the tropic of Capricorn; that ascension, on the contrary, occurs when the sun approaches the equator from the tropics. It is, however, improperly said by some that the sun ascends when it approaches us and descends when it goes away from us.

Thus far we have spoken of circles. Let us now proceed to the theory of the sphere and a fuller consideration of the degrees by which such circles are distant from one another.

---

[3] The passage referred to is in the sixth book, chapter xxviii, of the *Commentaries*.

9.2.6 Chapter IV: Of a Certain Theory of the Sphere According to the System of Degrees

The celestial sphere is surrounded by five principal circles, one great and four small—the arctic, the circle of Cancer, the equator, the circle of Capricorn, and the antarctic. Of these the equator is a great circle, the other four are small circles. These circles, or rather the spaces that are between them, authors are wont to call zones. Thus Vergil, in the Georgics, says:

> Five zones the heavens contain; whereof is one
> Aye red with flashing sunlight, fervent aye
> From fire; on either side to left and right
> Are traced the utmost twain, stiff with blue ice,
> And black with scowling storm-clouds, and betwixt
> These and the midmost, other twain there lie,
> By the gods' grace to heart-sick mortals given,
> And a path cleft between them, where might wheel
> On sloping plane the system of the signs.

Of the nature of the zones more will be said in the following pages. Inasmuch as we have mentioned above the pole of the zodiac that describes the arctic circle, therefore in place of further consideration this must be understood to mean the upper pole of the zodiac (situated at an elevation of 66°9′, and distant from the arctic pole 24°51′[4]). It must be recalled also that a degree is the 30th part of a sign, that a sign is the 12th part of a circle, and that 30 multiplied by 12 gives 360. So it becomes clear that a degree can be defined as the 360th part of a circle.

The lower pole of the zodiac describes the antarctic circle, which is situated in the same degree of declination and is at the same distance from the antarctic pole as the upper pole of the zodiac is from the arctic. The inclination of the ecliptic, or the greatest declination of the sun toward the north (which is situated 33°51′[5] from the equinoctial), describes the tropic of Capricorn.

The distance between the tropic of Cancer and the arctic circle is 42°18′. The distance between the tropic of Capricorn and the antarctic circle is the same.

The middle of the heavens, being equally distant from the poles of the world, makes the equator.

Hitherto we have spoken of the five zones and of their distance from one another. We shall now briefly discuss the remaining circles.

The circle of the zodiac is determined by the poles of the zodiac. From the poles to the tropics (that is, to the greatest declinations of the sun or the solstices), the distance is 42°18′. The width of the zodiac from the ecliptic toward either of the tropics is 6°, or in all 12°.

The solstices and equinoxes mark the colures of declination and ascension. These intersect under the poles of the world along the axis of the heavens at spherical right

---

[4] Error for 23°51′.

[5] Error from 23°51′.

angles; likewise along the equator. But the equinoctial colures going along the zodiac make oblique angles, while they make right angles along the zodiac of the solstices. The meridional circle, which is movable, is contained by the same axis under the poles themselves.

The circle of the horizon is determined by the zenith, for, as its upper pole, the zenith is everywhere equally distant from it. The circle of the horizon also divides our hemisphere from the other from east to west, but for those who are beneath the equinoctial, through the two poles of the world. The zenith of every horizon is always distant 90°, which is the fourth part of a circle, from the circumference of the horizon, while the circumference of the horizon is four times as great as the distance between the zenith and the horizon.

It is worthy of notice that the axis of the world in the material sphere passes diametrically from the poles through the center of the world, which is the earth.

The axis of the zodiac, however, is not apparent in the sphere, but has to be conceived. This intersects the middle of the axis of the world, making unequal or oblique angles at the center.

In this way, in the very creation of the world there seems to be a wonderful order and arrangement. The old astronomers, in describing the form of the world, followed, as far as possible, in the footsteps of the Creator Himself, who made all things according to number, weight, and dimensions. We, too, while treating of this subject inasmuch as we are so hampered by the conditions of our space that our system of minutes can be perceived only with difficulty, or not at all, and, if perceived, would beget even annoyance as well as error, shall infer the positions of circles from the markings of degrees in full. For there is not much difference between 51′ and a full degree, which contains 60′, as we have said before, and in the book on the sphere and elsewhere it is indicated in exactly this way by specialists on this subject. Therefore in the diagram which we shall here insert for the better understanding of these matters (Fig. 9.2), the tropics of Cancer and Capricorn and the greatest declinations of the sun will be distant 24° from the equinoctial, the same as the distance of the poles of the zodiac or the arctic and antarctic circles from the poles of the world, situated at an elevation of over 66°.

9.2.7 Chapter V: Of the Five Celestial Zones and the Application of These and of the Degrees of the Heavens to the Earth

Up to this point we have spoken very briefly of several geometrical principles, of the sphere, the poles, the five zones the circles of the world, and of a certain theory in regard to these matters. Now, in regular order, if I am not mistaken, we come to the consideration of the application of these circles and degrees to the earth. It should therefore be known that on the earth there are five regions corresponding to the above-mentioned zones. Wherefore Ovid in the *Metamorphoses* says:

> And as two zones the northern heaven restrain,
> The southern two, and one the hotter midst,

Fig. 9.2 Angular orientation
of the axis of the zodiac with
respect to the axis of the
World. —[*K.K.*]

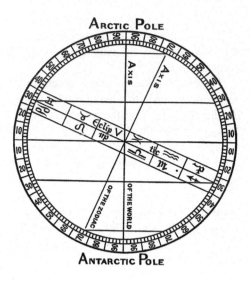

With five the Godhead girt th' inclosed earth,
And climates five upon its face imprest.
The midst from heat inhabitable: snows
Eternal cover two: 'twixt these extremes
Two temperate regions lie, where heat and cold
Meet in due mixture.

Metamorphonses, i, 45–51, translated by Howard.

In order to make the matter clearer, let us state that the four small circles, the arctic, the circle of Cancer, the circle of Capricorn, and the antarctic, divide and separate the five zones of the heavens.

In the following diagram (Fig. 9.3), let *a* represent the arctic pole of the world, *bc* the arctic circle, *de* the circle of Cancer, *fg* the circle of Capricorn, *hk* the antarctic circle, and *l* the south pole.

The first zone, or the arctic, is all the space included between *bac*. This zone, being frozen stiff with perpetual cold, is uninhabited.

The second zone is all the space included between *bc* and *de*. This is a temperate zone and is habitable.

The third zone is all the space included between *de* and *fg*. This zone, on account of its heat, is scarcely habitable; for the sun, describing circles there with a constant whirling motion along the line *fe* (which for us marks the ecliptic), by reason of its heat makes the zone torrid and uninhabited.

The fourth zone is all the space included between *fg* and *hk*. This is a temperate zone and is habitable, if the immense areas of water and the changed conditions of the atmosphere permit it.

The fifth zone is all the space included between *hkl*. This zone is always stiff with cold and uninhabited.

Fig. 9.3 The five celestial zones, mapped onto the surface of the Earth.—[*K.K.*]

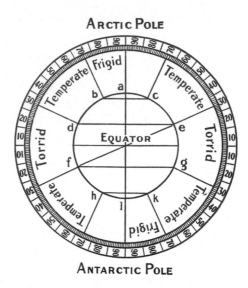

When we say that any zone of the heavens is either inhabited or uninhabited, we wish it to be understood that this applies to the corresponding zone lying beneath that celestial zone. When we say that any zone is inhabited or inhabitable, we mean that is it easily inhabitable. Likewise, when we say that any zone is uninhabited or uninhabitable, we understand that it is habitable with difficulty. For there are many people who now inhabit the dried-up torrid zone, such as the inhabitants of the Golden Chersonese,[6] the Taprobanenses,[7] the Ethiopians, and a very large part of the earth which had always been unknown, but which has recently been discovered by Amerigo Vespucci. In this connection we may state that we shall add the four voyages of Vespucci, translated from the Italian language into French and from French into Latin.

It must be understood, as the following diagram shows, that the first zone, which is nearest to the arctic pole, is 23°51′ in extent; the second, which is the antarctic, is equal to the arctic, and is therefore the same in extent; the third, a temperate zone, is 42°18′; the fourth, which is equal to it, is also 42°18′; the fifth, which is the torrid and is in the middle, is 47°42′.

Let us here insert the diagram (see Fig. 9.3).

---

[6] The peninsula of Malacca in India is probably meant.

[7] The people of what is now the island of Ceylon.

9.2.8 Chapter VI: Of Parallels

Parallels, which are also called Almuncantars, are circles or lines equidistant in every direction and at every point, and never running together even if extended to infinity. They bear the same relation to one another as the equator does to the four small circles on the sphere, not that the second is as distant from the third as the first is from the second, for this is false, as is clear from the preceding pages, but that any two circles joined together by a perpendicular are equally distant from each other throughout their extent. For the equator is neither nearer to nor more distant from one of the tropics at any one point than at any other, since it is everywhere distant 23°51′ from the tropics, as we have said before. The same must be said of the distance from the tropics to the two extreme circles, either of which is distant 42°44′[8] from the nearer tropic at all points.

Although parallels can be drawn at any distance apart, yet, to make the reckoning easier, it has seemed to us most convenient, as it seemed to Ptolemy also, in our representation of universal cosmography, both in the solid and projected on the plane, to separate the parallels by as many degrees from one another as the following table shows (see Table 9.1). To this table a diagram also will be subjoined, in which we shall extend the parallels through the earth on both sides to the celestial sphere (see Fig. 9.4).

9.3 Study Questions

QUES. 9.1 Which of the liberal arts forms the basis of cosmography?

a) What, according to Waldseemüller, are the benefits of foreign travel? Who are Plato, Boethius, and Achilles? And why do you think he mentions these?
b) Upon what source(s) does Waldseemüller draw in preparing his map? To whom does he dedicate it? Why?
c) According to Waldseemüller, with which of the liberal arts must one become familiar before obtaining a "thorough knowledge" cosmography?
d) What does Waldseemüller mean by the "material sphere"? And upon whose work does Waldseemüller base his conception of the world?
e) What definition of "sphere" does he use? How many celestial spheres are there? How are they arranged with respect to the earth? And what is special about the eighth sphere?

QUES. 9.2 Make a labeled sketch of the celestial sphere which includes the six great circles and four small circles described by Waldseemüller.

a) Where is the celestial equator, and why is it given this name?

---

[8] Error for 42°18′.

Table 9.1 This diagram shows by its numbers the climates, the degrees of the parallels, and the hours

Parallels from the equator	Degrees of the heavens	Greatest number of hours in a day	Number of miles in 1°
21 Of Thule 8	63	20	$28\frac{1}{2}$
20	61	19	
19	58	18	$32\frac{1}{2}$
18	56	17	$\frac{1}{2}$ (*sic?*)
17	54	17	$37\frac{1}{2}$
16 Of the Rhiphae an Mts. 7	$51\frac{1}{2}$	$16\frac{1}{2}$	$40\frac{1}{2}$
15 Of the Borysthenes (Dnieper) 6	$48\frac{1}{2}$	16	$42\frac{1}{2}$
14	45	$15\frac{1}{2}$	44
13	$43\frac{1}{12}$	$15\frac{1}{4}$	45
12 Of Rome 5	$40\frac{11}{12}$	15	47
11	$38\frac{7}{12}$	$14\frac{3}{4}$	$48\frac{1}{2}$
10 of Rhodes 4	36	$14\frac{1}{2}$	50
9	$33\frac{1}{3}$	$14\frac{1}{4}$	
8 of Alexandria 3	$30\frac{1}{3}$	14	54
7	$27\frac{2}{3}$	$13\frac{3}{4}$	
6 of Syene 2	$23\frac{5}{6}$	$13\frac{1}{2}$	57
5	$20\frac{1}{4}$	$13\frac{1}{4}$	
4 of Meroe 1	$16\frac{5}{12}$	13	
3	$12\frac{1}{2}$	$12\frac{3}{4}$	
2	$8\frac{5}{12}$	$12\frac{1}{2}$	
1	$4\frac{1}{4}$	$12\frac{1}{4}$	59
Equator equidistant from the poles		12 always	60
1	$4\frac{1}{4}$	$12\frac{1}{4}$	59
2	$8\frac{5}{12}$	$12\frac{1}{2}$	
3	$12\frac{1}{2}$	$12\frac{3}{4}$	
4 Anti-climate of Meroe	$16\frac{5}{12}$	13	
5	$20\frac{1}{4}$	$13\frac{1}{4}$	
6 Anti-Climate of Syene	$23\frac{5}{6}$	$13\frac{1}{2}$	52
7	$27\frac{2}{3}$	$13\frac{3}{4}$	

And so on toward the Antarctic Pole, as the following diagram shows.

b) What is the location and orientation of the zodiac? What is the significance of its name?

c) What is the ecliptic, and why is it given this name?

d) Where are the solstitial and equinoctial colures?

e) What are the meridian and horizon? Do they have unique locations?

f) What are the angular positions, and significance of, the four small circles?

g) What unit of angular measurement does Waldseemüller use, and why? How big is the uncertainty he claims in such measurements?

h) What concepts, according to Waldseemüller, did the Creator use when designing the world?

QUES. 9.3 Into how many zones does Waldseemüller divide the earth? Why does he do so? And how does he enact this division? What is the relationship between these zones and the climates of the earth?

QUES. 9.4 Is there a location on the earth at which the number of hours of day and of night are always equal?

Fig. 9.4 Earth's parallels
from the southern hemisphere
to the island of
Thule.—[*K.K.*]

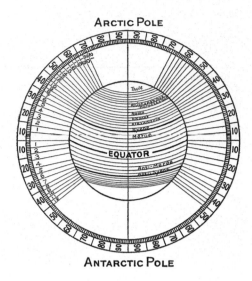

a) What are parallels? How many parallels does Waldseemüller draw? Why? What is the separation between parallels?
b) How long is the longest day in Rome? On the island of Thule? If one were to travel directly eastward from Rome, how many miles would one need to travel before returning to Rome?

9.4   Exercises

Ex. 9.1 (LATITUDES). According to the numbers reported in Table 9.1:

a) In which city will the sun be most nearly directly overhead at noon on the summer solstice?
b) On the island of Thule, what will be the angle of the sun above the horizon at noon on the summer solstice?
c) Which of the cities lie in the temperate zone of the northern hemisphere?
d) How many hours of daylight occur on the longest day in the city of Rome?
e) If one were to travel, by boat and by land, directly eastward from Rome, how many miles would one have to travel before returning to Rome?

Ex. 9.2 (WALDSEEMÜLLER'S WORLDVIEW). Clearly and unambiguously label the items listed below in Fig. 9.5. (a) the earth, (b) polaris, (c) the moon (where it is located today), (d) the celestial equator, (e) the ecliptic, (f) the axis of the zodiac, (g) the solstitial colure, (h) the equinoctial colure, (i) the tropic of Cancer, (j) the antarctic circle, (k) the constellation Taurus and (l) the constellation Libra.

Ex. 9.3 (LATITUDE AND LONGITUDE OBSERVATIONS). In the absence of notable geographic features, how can one determine his or her location on the surface of the

Fig. 9.5 The celestial sphere

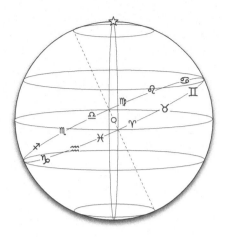

earth? The solution to this problem, employed by many wise seafarers, lies in carefully measuring the locations of celestial objects. In this two-part exercise, you will use the north star and the sun to determine your latitude and longitude, respectively.

First, let us measure your latitude. Go outside on a clear evening and measure the altitude of *Polaris*. An object's altitude is just its angle (in degrees) above the north point on the horizon. Reasonably accurate angular measurements can be done using a cross-staff, which can be constructed quite inexpensively in a couple of hours.[9] Alternatively, you can just hold your hand at arm's length and use the angular width of your hand (or fingers) to crudely measure the angular separation of two objects. To estimate the angular width of your hand (or fingers), when held at arm's length, you can just count how many hands (or fingers) can be fit, side by side, in the space between the horizon and the zenith. Now, from your measurement of Polaris' altitude, determine your latitude. Explain your calculation. Look up the latitude of your present location and compare it to your measured value. By how much were you off in degrees? In miles? Suppose that you were a seafarer sailing around the horn of Africa, whence you were unable to view *Polaris*. How, then, might you determine your latitude from celestial observation? Is it even possible? If so, how?

In order to measure the longitudinal separation of two locations, one can compare the local solar time measured at each location. This measurement requires that each of the two observers has a chronometer (a clock) which measures time independently of the sun's position. To think through this, consider the following questions. (i) Does the sun pass across the local meridian at the same instant for observers located in cities which are at the east and the west ends of your state? Explain. (ii) By how many degrees longitude are cities A and B separated if the sun crosses the local meridian one hour later at A than at B? Which city is farther east? (iii) What about if the sun crosses the local meridian twenty minutes earlier at A than at B?

---

[9] See the instructions for building a cross-staff in Ex. 7.5 of the present volume.

Now for the measurements. Acquire a *gnomon*: a vertical stick which casts a shadow. Stand it up on a flat surface before local solar noon. As time progresses, make marks on a sheet of paper on the ground which indicate the position and length of the gnomon's shadow. Measure the length of the shadow as a function of time on your chronometer. Then answer the following questions. (iv) At what time on your chronometer did the sun pass the local meridian? How do you know? (v) Are you in the eastern or western edge of your time zone? How many degrees east or west of the boundary of your time zone are you located? (vi) How far east or west are you of Greenwich, England? Look up the longitude of your present location and compare it to your measured value. How close are you? And what were your sources of error?

9.5   Vocabulary

1. Cosmography
2. Circumference
3. Right, obtuse, acute angle
4. Degree, minute, second, third
5. Convex, concave
6. Vertex
7. Arctus

8. Cynosura
9. Great circle
10. Solstice
11. Equinox
12. Declination
13. Primum mobile
14. Zenith

Chapter 10
Climates and Continents

> *The fourth part of the earth, which, because Amerigo discovered it, we may call Amerige, the land of Amerigo, so to speak, or America*
>
> —Martin Waldseemüller

10.1 Introduction

In the previous reading selection from the *Introduction to Cosmography*, Waldseemüller invoked the geometry of the heavens to mark off distinct zones on the surface of the earth. In the reading selection that follows, he now relates these zones to their respective climates. Perhaps most interestingly, he provides a rather detailed summary of the geography and the peoples of the known regions of Europe, Asia and Africa before mentioning the newly discovered "fourth part of the world."

10.2 Reading: Waldseemuller, *Introduction to Cosmography*

Waldseemüller, *Introduction to Cosmography*, pp. 31–81, Ann Arbor: University Microfilms, Princeton, NJ, 1966.

10.2.1 Chapter VII: Of Climates

Although the word *climate* properly means a region, it is hear used to mean a part of the earth between two equidistant parallels, in which from beginning to end of the climate there is a difference of a half-hour in the longest day. The number of any climate, reckoned from the equator, indicates the number of half-hours by which the longest day in that climate exceeds the day that is equal to the night. There are seven of these climates, although to the south the seventh has not yet been explored. But toward the north, Ptolemy discovered a country that was hospitable and habitable, at a distance represented by seven half-hours. These seven climates have obtained their names from some prominent city, river, or mountain.

K. Kuehn, *A Student's Guide Through the Great Physics Texts*,
Undergraduate Lecture Notes in Physics, DOI 10.1007/978-1-4939-1360-2_10,
© Springer Science+Business Media, LLC 2015

(1) The first climate is called Dia Meroes (of Meroe, modern Shendi), from διά, which in Greek means *through* and governs the genitive case, and Meroe, which is a city of Africa situated in the torrid zone of 16° on this side of the equator, in the same parallel in which the Nile is found. Our world map for the better understanding of which this is written, will clearly show you the beginning, the middle, and the end of this first climate and also of the rest, as well as the hours of the longest day in every one of them.

(2) Dia Sienes (of Syene, modern Assuan), from Syene, a city of Egypt, the beginning of the province of Thebias.

(3) Dia Alexandrias (of Alexandria), from Alexandria, a famous city of Africa, the chief city of Egypt, founded by Alexander the Great, of whom it has been said by the poet:

One world is not enough for the youth of Pella.

—Juvenal, x, 168.

(4) Dia Rhodon (of Rhodes,) from Rhodes, an island on the coast of Asia Minor, on which in our time there is situated a famous city of the same name, which bravely resisted the fierce and warlike attacks of the Turks and gloriously defeated them.

(5) Dia Rhomes (of Rome), from a well-known city of Europe, the most illustrious among the cities of Italy and at one time the famous conqueror of all nations and the capital of the world. It is now the abode of the great Father of Fathers.

(6) Dia Borysthenes (of Borysthenes, modern Dnieper), from a large river of the Scythians, the fourth from the Danube.

(7) Dia Rhipheon (of the Rhiphæan Mountains), from the Rhiphæan mountains, a prominent range in Sarmatian Europe, white with perpetual snow.

From these prominent places, through which approximately the median lines of the climates pass, the seven climates established by Ptolemy derive their names.

The eighth climate Ptolemy did not locate, because that part of the earth, whatever it is, was unknown to him, but was explored by later scholars. It is called Dia Tyles (of Thule, modern Iceland or Shetland), because the beginning of the climate, which is the 21st parallel from the equator, passes directly through Thule. Thule is an island in the north, of which our poet Vergil says:

The farthest Thule will serve.

—Georgics, i, 30.

So much for the climates north of the equator. In like manner we must speak of those which are south of the equator, six of which having corresponding names have been explored and may be called Antidia Meroes (Anti-climate of Meroe), Antidia Alexandrias, Antidia Rhodon, Antidia Rhomes, Antidia Borysthenes, from the Greek particle αντί, which means *opposite* or *against*. In the sixth climate toward the antarctic there are situated the farthest part of Africa, recently discovered, the islands of Zanzibar, the lesser Java, and Seula (Sumatra?), and the fourth part of the earth, which, because Amerigo discovered it, we may call Amerige, the land of

Amerigo, so to speak, or America. It is of these southern climates that these worlds of Pomponious Mela, the geographer, must be understood when he says:

> The habitable zones have the same seasons, but at different times of the year. The Antichthones inhabit the one, and we the other. The situation of the former zone being unknown to us on account of the heat of the intervening zone, I can speak only of the situation of the latter.
>
> —Perieg. i, I, 9.

Here it should be remarked that each one of the climates generally bears products different from any other, inasmuch as the climates are different in character and are controlled by different influences of the stars. Wherefore Vergil says:

> Nor can all climes all fruits of earth produce
> Here blithelier springs the corn, and here the grape,
> Their earth is green with tender growth of trees
> And grass unbidden. See how from Tmolus comes
> The saffron's fragrance, ivory from Ind,
> From Saba's weakling sons their frankincense,
> Iron from the naked Chalybs, castor rank
> From Pontus, from Epirus the prize-palms
> O' the mares of Elis.
>
> —Georgics, i, 54–59, translated by Rhoades.

10.2.2 Chapter VIII: Of the Winds

Since in the preceding pages we have mentioned the winds now and then (when we spoke of the north pole, the south pole, *etc.*), and as it is understood that a knowledge of winds is of some importance, or rather of great advantage to cosmography, we shall for these reasons say something in this chapter about winds, also called *spiritus* and *flatus* (breeze). A wind, therefore, as defined by the philosophers, is an exhalation, warm and dry, moving laterally around the earth, *etc.*

Now, inasmuch as the sun has a triple rising and setting, the summer rising and setting, the equinoctial rising and setting, and the winter rising and setting, according to its relation to the two tropics and the equator, and inasmuch as there are also two sides—to the north and to the south, all of which has winds peculiar to them; therefore it follows that there are twelve winds in all, three eastern, three western, three northern, and three southern. Of these the four which in the following diagram occupy the middle place are the principal winds; the others are secondary (see Table 10.1).

The poets, however, by poetic license, according to their custom, instead of the principal winds use their secondary winds, which are also called side winds. Thus Ovid says:

> Far to the east
> Where Persian mountains greet the rising sun
> Eurus withdrew. Where sinking Phœbus' rays
> Glow on the western shores mild Zephyr fled.

Table 10.1 Poetic classification of the twelve winds.—[*K.K.*]

		East	West
Side	Tropic of Cancer	Kaikias	Chorus
Principal	Equator	Subsolanus	Favonius or Zephyrus
Side	Tropic of Capricorn	Eurus or Vulturnus	Africus or Libs
		South	North
Side		Euronotus	Septentrio
Principal		Auster or Notus	Aquilo or Boreas
Side		Libonotus	Trachias or Circius

Terrific Boreas frozen Scythia seiz'd,
Beneath the icy bear. On southern climes
From constant clouds the showery Auster rains.

—Metamorphoses, i, 61–66, translated by Howard.

The east wind (Subsolanus), which is rendered by the sun purer and finer than the others, is very healthful.

The west wind (Zephyrus), having a mixture of heat and moisture, melts the snows. Whence Vergil's verse:

Melts from the mountain's hoar, and Zephyr's breath
Unbinds the crumbling clod.

—Georgics, i, 44, translated by Rhoades.

The south wind (Auster) frequently presages storms, hurricanes, and showers. Wherefore Ovid says:

Notus rushes forth
On pinions dripping rain.

—Metamorphoses, i, 264, translated by Howard.

The north wind (Aquilo), by reason of the severity of its cold, freezes the waters.

And frosty winter with his north the sea's face rough doth wear.

—Vergil, Æneid, iii, 285, translated by Morris.

In regard to these winds, I remember, our poet Gallinarius, a man of great learning, composed the following:

Eurus and Subsolanus blow from the east.
Zephyrus and Favonius fill the west with breezes.
Auster and Notus rage on Libya's farthest shores.
Boreas and Aquilo cloud-dispelling threaten from the north.

Although the north winds are naturally cold, they are softened because they pass through the torrid zone. This has been found to be true of the south wind, which passes through the torrid zone before it reaches us, as is shown in the following lines.

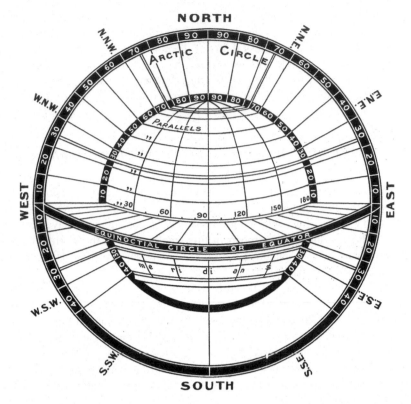

Fig. 10.1 On the back of this general map of the heavens and the earth (see Fig. 9.1), Walsdseemüller inserted the following text. "The purpose of this little book is to write a description of the world map, which we have designed, both as a globe and as a projection. The globe I have designed on a small scale, the map on a larger. As farmers usually mark off and divide their farms by boundary lines, so it has been our endeavor to mark the chief countries of the world by the emblems of their rulers. And (to begin with our own continent) in the middle of Europe we have placed the eagles of the Roman Empire (which rule the kings of Europe), and with the key (which is the symbol of the Holy Father) we have enclosed almost the whole of Europe, which acknowledges the Roman Church. The greater part of Africa and a part of Asia we have distinguished by crescents, which are the emblems of the supreme Sultan of Babylonia, the lord of all Egypt, and a part of Asia. The part of Asia called Asia Minor we have surrounded with a saffron-colored cross joined to a branding iron, which is the symbol of the Sultan of the Turks, who rules Scythia this side of the Imaus, the highest mountains of Asia and Sarmatian Scythia. Asiatic Scythia we have marked by anchors, which are the emblems of the great Tartar Khan. A red cross symbolizes Prester John (who rules both eastern and southern India and who resides in Biberith); and finally on the fourth division of the earth, discovered by the kings of Castile and Portugal, we have placed the emblems of those sovereigns. And what is to be borne in mind, we have marked with crosses shallow places in the sea where shipwreck may be feared. Herewith we close."

We have said enough about winds. We shall now insert a general map, indicating poles, the axes, the circles, great as well as small, the east, the west, the five zones, the degrees of longitude and latitude, both on the earth and in the heavens, the climates, the winds, *etc.* (see Fig. 10.1).

Wherever the cold south wind goes, it rages and binds the waters with tight fetters. But until with its blast it passes through the torrid regions, it comes welcome to our shores and hurls back the merciless shafts of the north wind. The latter wind, on the contrary, which deals harshly with us, slackening its flight, becomes in like manner gentler in the lowest part of the globe. The other winds, where they direct their various courses, soon change as the go, the natures which are proper to their homes.

10.2.3   Chapter IX: Of Certain Elements of Cosmography

It is clear from astronomical demonstrations that the whole earth is a point in comparison with the entire extent of the heavens; so that if the earth's circumference be compared to the size of the celestial globe, it may be considered to have absolutely no extent. There is about a fourth part of this small region in the world which was known to Ptolemy and is inhabited by living beings like ourselves. Hitherto it has been divided into three parts, Europe, Africa, and Asia.

Europe is bounded on the west by the Atlantic Ocean, on the north by the British Ocean, on the east by the river Tanais (modern Don), Lake Maeotis (modern Sea of Azov), and the Black Sea, and on the south by the Mediterranean Sea. It includes Spain, Gaul, Germany, Rætia, Italy, Greece, and Sarmatia. Europe is so called, after Europa, the 'daughter of King Agenor'. While with a girl's enthusiasm she was playing on the sea-shore accompanied by her Tyrian maidens and was gathering flowers in baskets, she is believed to have been carried off by Jupiter, who assumed the form of a snow-white bull, and after being brought over the seas to Crete seated upon his back to have given her name to the land lying opposite.

Africa is bounded on the west by the Atlantic Ocean, on the south by the Ethiopian Ocean, on the north by the Mediterranean Sea, and on the east by the river Nile. It embraces the Mauritanias, *viz.*, Tingitana (modern Tangiers) and Cæsarea, inland Libya, Numidia (also called Mapalia), lesser Africa (in which is Carthage, formerly the constant rival of the Roman empire), Cyrenaica, Marmarica (modern Barca), Libya (by which name also the whole of Africa is called, from Libs, a king of Mauritania), inland Ethiopia, Egypt, *etc.* It is called Africa because it is free from the severity of the cold. Asia, which far surpasses the other divisions in size and in resources, is separated from Europe by the river Tanais (Don) and from Africa by the Isthmus, which stretching southward divides the Arabian and the Egyptian seas. The principal countries of Asia are Bithynia, Galatia, Cappadocia, Pamphylia, Lydia, Cilicia, greater and lesser Armenia, Colchis, Hyrcania, Iberia, and Albania; besides many other countries which it would only delay us to enumerate one by one. Asia is so called after a queen of that name.

Now, these parts of the earth have been more extensively explored and a fourth part has been discovered by Amerigo Vespucci (as will be set fourth in what follows). Inasmuch as both Europe and Asia received their names from women, I see no reason why anyone should justly object to calling this part Amerige, *i.e.*, the land of Amerigo, or America, after Amerigo, its discoverer, a man of great ability. Its position and the customs of its inhabitants may be clearly understood from the four voyages of Amerigo, which are subjoined.

Thus the earth is now known to be divided into four parts. The first three parts are continents, while the fourth is an island, inasmuch as it is found to be surrounded on all sides by the ocean. Although there is only one ocean, just as there is only one earth, yet, being marked by many seas and filled with numberless islands it takes various names. These names may be found in the Cosmography, and Priscian in his translation of Dionysius enumerates them in the following lines:

> The vast abyss of the ocean, however, surrounds the earth on every side; but the ocean, although there is only one, takes many names. In the western countries it is called the Atlantic Ocean, but in the north, where the Arimaspi are ever warring, it is called the sluggish sea, the Saturnian Sea, and by others the Dead Sea,
>
> * * * * * * *
>
> Where, however, the sun rises with its first light, they call it the Eastern or the Indian Sea. But where the inclined pole receives the burning south wind, it is called the Ethiopian or the Red Sea,
>
> * * * * * * *
>
> Thus the great ocean, known under various names, encircles the whole world;
>
> * * * * * * *
>
> Of its arms the first that stretches out breaks through Spain with its waves, and extends from the shores of Libya to the coast of Pamphylia. This is smaller than the rest. A larger gulf is the one that enters into the Caspian land, which receives it from the vast waters of the north. The arm of the sea which Tethys (the ocean) rules as the Saturnian Sea is called the Caspian or the Hyrcanian. But of the two gulfs that come from the south sea, one, the Persian, running northward, forms a deep sea, lying opposite, the country where the Caspian waves roll; while the other rolls and beats the shores of Panchæa and extends to the south opposite to the Euxine Sea.
>
> * * * * * * *
>
> Let us begin in regular order with the waters of the Atlantic, which Cadiz makes famous by Hercules' gift of the pillar, where Atlas, standing on a mountain, holds up the columns that support the heavens. The first sea is the Iberian, which separates Europe from Libya, washing the shores of both. On either side are the pillars. Both face the shores, the one looking toward Libya, the other toward Europe. Then comes the Gallic Sea, which beats the Celtic shores. After this the sea, called by the name of the Ligurians, where the masters of the world grew up on Latin soil, extends from the north to Leucopetra; where the island of Sicily with its curving shore forms a strait. Cyrnos (modern Corsica) is washed by the waters that bear its name and flow between the Sardinian Sea and the Celtic. Then rolls the surging tide of the Tyrrhenian Sea, turning toward the south; it enters the sea of Sicily, which turns toward the east and spreading far from the shores of Pachynum extends to Crete, a steep rock, which stands out of the sea, where powerful Gortyna and Phæstum are situated in the midst of the fields. This rock, resembling with its peak the forehead of a ram, the Greeks have justly called $K\rho\iota o\hat{v}\ \mu\acute{\epsilon}\tau\omega\pi ov$ (ram's forehead). The sea of Sicily ends at Mt. Garganus on the coast of Apulia.
>
> Beginning there the vast Adriatic extends toward the northwest. There also is the Ionian Sea, famous throughout the world. It separates two shores, which, however, meet in one point. On the right fertile Illyria extends, and next to this the land of the warlike Dalmatians. But its left is bounded by the Ausonian peninsula, whose curving shores the three seas, the Tyrrhenian, the Sicilian, and the vast Adriatic, encircle on all sides. Each of these seas within its limits has a wind peculiar to itself. The west wind lashes the Tyrrhenian, the south wind the Sicilian, while the east wind breaks the waters of the Adriatic which roll beneath its blasts.

Leaving Sicily the sea spreads its deep expanse to the greater Syrtis which the coast of Libya encircles. After the greater Syrtis passes into the lesser, the two seas beat far and wide upon the re-echoing shores. From Sicily the Cretan Sea stretches out toward the east as far as Salmon is, which is said to be the eastern end of Crete.

Next come two vast seas with dark waves, lashed by the north wind coming from Ismarus, which rushes straight down from the regions of the north. The first, called the Pharian Sea, washes the base of a steep mountain. The second is the Sidonian Sea, which turns toward the north, where the gulf of Issus joins it. This sea does not continue far in a straight line; for it is broken by the, shores of Cilicia. Then bending westward it winds like a dragon because, forcing its way through the mountains, it devastates the hills and worries the forests. Its end hounds Pamphylia and surrounds the Chelidonian rocks. Far off to the west it ends near the heights of Patara.

Next look again toward the north and behold the Ægean Sea, whose waves exceed those of all other seas, and whose vast waters surround the scattered Cyclades. It ends near Imbros and Tenedos, near the narrow strait through which the waters of the Propontis issue, beyond which Asia with its great peoples extends to the south, where the wide peninsula stretches out. Then comes the Thracian Bosporus, the mouth of the Black Sea. In the whole world they say there is no strait narrower than this. There are found the Symplegades, close together. There to the east the Black Sea spreads out, situated in a northeasterly direction. From either side a promontory stands out in the middle of the waters; one, coming from Asia on the south, is called Carambis; the other on the opposite side juts out from the confines of Europe and is, called $K\rho\iota o\hat{v}\ \mu\acute{e}\tau\omega\pi o\nu$ (ram's forehead.) They face each other, therefore, separated by a sea so wide that a ship can cross it only in three days. Thus you may see the Black Sea looking like a double sea, resembling the curve of a bow, which is bent when the string is drawn tight. The right side resembles the string, for it forms a straight line, outside of which line is found Carambis only, which projects toward the north. But the coast that encloses the sea on the left side, making two turns, describes the arc of the bow. Into this sea toward the north Lake Mæotis (modern Sea of Azov) enters, enclosed on all sides by the land of the Scythians, who call Lake Mæotis the mother of the Black Sea. Indeed, here the violent sea bursts forth in a great stream, rushing across the Cimmerian Bosporus (modern Crimea), in those cold regions where the Cimmerians dwell at the foot of Taurus. Such is the picture of the ocean; such the glittering appearance of the deep.[1]

The sea, as we have said before, is full of islands, of which the largest and the most important, according to Ptolemy, are the following:

Taprobane (modern Ceylon), in the Indian Ocean under the equator; Albion, also called Britain and England; Sardinia, in the Mediterranean Sea; Candia, also called Crete, in the Ægean Sea; Selandia; Sicily, in the Mediterranean Sea; Corsica; Cyprus.

Unknown to Ptolemy: Madagascar, in the Prasodes Sea; Zanzibar; Java, in the East Indian Ocean; Angama; Peuta, in the Indian Ocean; Seula; Zipangri (Japan), in the Western Ocean. Of these Priscian says:

These are the large islands which the waters of the ocean surround. There are many other smaller islands, scattered about in different parts of the world, that are unknown, and that are either difficult of access to hardy sailors or suitable for harbors. Their names I cannot easily express in verse.[2]

---

[1] Priscian, *Periegesis*, 37, foll., ed. of Krehl.

[2] *Periegesis*, 609–613.

Table 10.2 Distance between places on Earth's surface, as it varies with latitude.—[*K.K.*]

	Degrees	Degrees	Italian miles	German miles
Equator	1 up to	12 cont'ng	60	15
	12	15	59	$14\frac{3}{4}$
Tropic	25	30	54	$13\frac{1}{2}$
	30	37	50	$12\frac{1}{2}$
	37	41	47	$11\frac{1}{4}$[a]
	41	51	40	10
	51	57	32	8
	57	63	28	7
	63	66	26	$6\frac{1}{2}$
Arctic circle	66	70	21	$5\frac{1}{4}$
	70	80	6	$1\frac{1}{2}$
Arcticpole	80	90		0

[a] Error for $11\frac{3}{4}$

In order to be able to find out the distance between one place and another, the elevation of the pole must first be considered. It should therefore be briefly remarked that, as is clear from what precedes, both poles are on the horizon for those who live on the parallel of the equator. But as one goes toward the north, the elevation of the pole increases the farther one goes away from the equator. This elevation of the pole indicates the distance of places from the equator. For the distance of any place from the equator varies as the elevation of the pole at that place. From this the number of miles is easily ascertained, if you will multiply the number of degrees of elevation of the pole. But according to Ptolemy, from the equator to the arctic pole miles are not equal in all parts of the world. For any one of the degrees from the first degree of the equator up to the 12th contains 60 Italian miles, which are equivalent to 15 German miles, four Italian miles being generally reckoned equal to one German mile. Any degree from the 12 degree up to the 25th contains 59 miles, or fourteen and three-quarter German miles.

In order to make the matter clearer, we shall insert the following table (see Table 10.2).

In like manner from the equator to either arctic or antarctic pole the number of miles in a degree of latitude varies. If you wish to find out the number of miles between one place and another, examine carefully in what degree of latitude the two places are and how many degrees there are between them; then find out from the above table how many miles there are in a degree of that kind, and multiply this number by the number of degrees between the places. The result will be the number of miles between them. Since these will be Italian miles, divide by four and you will have German miles.

All that has been said by way of introduction to the Cosmography will be sufficient, if we merely advise you that in designing the sheets of our world-map we have not followed Ptolemy in every respect, particularly as regards the new lands, where on the marine charts we observe that the equator is placed otherwise than Ptolemy represented it. Therefore those who notice this ought not to find fault with us, for we have done so purposely, because in this we have followed Ptolemy, and elsewhere

Fig. 10.2 Waldseemüller's
quadrant.—[*K.K.*]

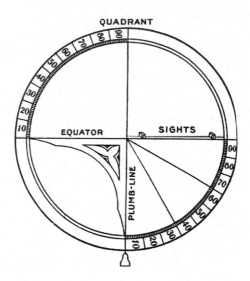

the marine charts. Ptolemy himself, in the fifth chapter of his first book, says that he was not acquainted with all parts of the continent on account of its great size, that the position of some parts on account of the carelessness of travelers was not correctly handed down to him, and that there are other parts which happen at different times to have undergone variations on account of the cataclysms or changes in consequence of which they are known to have been partly broken up. It has been necessary therefore, as he himself says he also had to do, to pay more attention to the information gathered in our own times. We have therefore arranged matters so that in the plane projection we have followed Ptolemy as regards the new lands and some other things, while on the globe, which accompanies the plane, we have followed the description of Amerigo that we subjoin.

10.2.4   Appendix

Before closing, we shall add to the forgoing, as an appendix or corollary, a quadrant, by which may be determined the elevation of the pole, the zenith, the center of the horizon, and the climates; although, if rightly considered, this quadrant, of which we shall speak, has a bearing on this subject. For a cosmographer ought to know especially the elevation of the pole, the zenith, and the climates of the earth. This quadrant, then, is constructed in the following way (Fig. 10.2). Divide any circle into four parts in such a way that the two diameters intersect the center at right angles.

One of these, which has sights at either end, will represent the axis of the poles of the world, the other the equator. Then divide that part of the circle which is between the semi-axis that has the sights and the other semi-diameter into ninety parts and the opposite part also into the same number, fix a plumb-line to the center, and your quadrant will be ready. The quadrant is used as follows: turn it so that you will see the pole directly through the openings in the sights and then toward the climate and the degree to which the plumb-line will fall. Your region, as well as your zenith and the center of your horizon, lies in that climate and at that degree of elevation.

Having now finished the chapters that we proposed to take up, we shall here include the distant voyages of Vespucci, setting forth the consequences of the several facts as they bear upon our plan.

The end of the outlines.

10.3 Study Questions

QUES. 10.1 What is the relationship between climate zones and the length of days? What are the most distant climates known by Waldseemüller? And what is the relationship between climate and natural resources?

QUES. 10.2 What is the relationship between the heavens and the winds? In particular, how does Waldseemüller define a wind? How many winds are there? And why this particular division?

QUES. 10.3 What are the four parts of the earth? And why are they named as they are?

a) How much of the world is inhabited? Into how many continents does he divide the world? And what is the etymology of the terms Europe? Africa? Asia? America?
b) How, exactly, can one calculate his latitude from observing the stars? What device does he recommend for so doing, and how is it used? And how can the distance between two cities be calculated from their difference in latitude?

10.4 Exercises

Ex. 10.1 (LATITUDE AND THE SHAPE OF THE EARTH). Suppose that you and your friend are located in distant cities. Speaking on the telephone just after noon, you discover that these cities lie along the same meridian, since you both observed the sun's culmination at the same instant. That same evening, after nightfall, you each use a quadrant to measure the altitude of Polaris above the local horizon; you measure 42°, your friend measures 44°. Which of you is farther north? By how many (Italian) miles? If instead you and your friend measure the altitude of Polaris to be 13 and 15°, respectively, then would the cities be the same distance apart as in the previous question? What does this imply about the shape of the earth?

10.5   Vocabulary

1. Temperate   4. Parallels
2. Torrid   5. Quadrant
3. Frigid   6. Plumb-line

Chapter 11
Heliocentrism: Hypothesis or Truth?

> *Among the many and varied literary and artistic studies upon which the natural talents of man are nourished, I think that those above all should be embraced and pursued with the most loving care which have to do with things that are very beautiful and very worthy of knowledge.*
>
> —Nicholaus Copernicus

11.1   Introduction

Nicholaus Copernicus (1473–1543) was born in the city of Torun in what is today Poland. He studied mathematics and Ptolemaic astronomy at the University of Cracow. Apart from his revolutionary work on astronomy, Copernicus engaged in a number of other pursuits: he was a practicing physician, trained at the university of Padua; he served as a close advisor in matters of government and diplomacy to the Bishop of Ermland, a small ecclesiastical principality; he wrote a short treatise on sound currency and the evils of inflation while serving as advisor to the Prussian Diet and King Sigismund I of Poland; and, owing to his reputation as a great astronomer, was invited to advise the Lateran Council on the pressing issue of calendrical reform.

In 1543 Copernicus published *On the Revolutions of the Heavenly Spheres*.[1] In this treatise, dedicated to Pope Paul III, Copernicus presented his heliocentric model of the world. To be sure, Copernicus was not the first to argue that the earth was itself moving in a circle around the center of the world; the Pythagoreans had famously done so many years before Ptolemy and Aristotle offered their strident defense of geocentrism.[2] So what set Copernicus apart from the Pythagoreans?

The reading selections in the next few chapters of this volume are from an English translation of Copernicus' latin text, *De Revolutionibus Orbium Cœlestium*, by Charles Glenn Wallis. In the first selection, contained in the present chapter, Copernicus begins by presenting the overall thesis of the work. He then proceeds to make

---

[1] For a fascinating account of Copernicus' famous book, see Gingerich, O., *The Book Nobody Read: Chasing the Revolutions of Nicholaus Copernicus*, Penguin Books, 2004.

[2] See Aristotle's criticism of the Pythagoreans in Book II, Chap. 13 of his *On the Heavens*, included in Chap. 7 of the present volume. Also, see Book I, Chap. 7 of Ptolemy's *Almagest*, included in Chap. 5 of the present volume.

K. Kuehn, *A Student's Guide Through the Great Physics Texts*,
Undergraduate Lecture Notes in Physics, DOI 10.1007/978-1-4939-1360-2_11,
© Springer Science+Business Media, LLC 2015

assertions regarding the nature of astronomy as a discipline. This introduction is quite controversial (as mentioned in a footnote below) since it was probably not written by Copernicus, and since it propounds a philosophy of science which differs quite remarkably from that of Copernicus himself, judging from subsequent chapters of the text.

11.2   Reading: Copernicus, *On the Revolutions of the Heavenly Spheres*

Copernicus, N., *On the Revolutions of the Heavenly Spheres*, Great Minds, Prometheus Books, Amherst, NY, 1995.

11.2.1   *Introduction: To the Reader Concerning the Hypotheses of this Work*[3]

/i [a]/[4] Since the newness of the hypotheses of this work—which sets the earth in motion and puts an immovable sun at the centre of the universe—has already received a great deal of publicity, I have no doubt that certain of the savants have taken grave offense and think it wrong to raise any disturbance among liberal disciplines which have had the right set-up for a long time now. If, however, they are willing to weigh the matter scrupulously, they will find that the author of this work has done nothing which merits blame. For it is the job of the astronomer to use painstaking and skilled observation in gathering together the history of the celestial movements, and then—since he cannot by any line of reasoning reach the true causes of these movements—to think up or construct whatever causes or hypotheses he pleases such that, by the assumption of these causes, those same movements can be calculated from the principles of geometry for the past and for the future too. This artist is markedly outstanding in both of these respects: for it is not necessary that these hypotheses should be true, or even probably; but it is enough if they provide a calculus which fits the observations—unless by some chance there is anyone so ignorant of geometry and optics as to hold the epicycle of Venus as probable and to believe this to be a cause why Venus alternately precedes and follows the sun at an angular distance of up to 40° or more. For who does not see that it necessarily follows from this assumption that the diameter of the planet in its perigee should appear more than four times greater, and the body of the planet more than 16 times

---

[3] This foreword, at first ascribed to Copernicus, is held to have been written by Andrew Osiander, a Lutheran theologian and friend of Copernicus, who saw the *De Revolutionibus* through the press.

[4] The numbers within the brackets refer to the pages of the first edition, published in 1543 at Nuremberg.

greater, than in its apogee? Nevertheless the experience of all the ages is opposed to that[5].

There are also other things in this discipline which are just as absurd, but it is not necessary to examine them right now. For it is sufficiently clear that this art is absolutely and profoundly ignorant of the causes of the apparent irregular movements. And if it constructs and thinks up causes—and it has certainly thought up a good many—nevertheless it does not think them up in order to persuade anyone of their truth but only in order that they may provide a correct basis for calculation. But since for one and the same movement varying hypotheses are proposed from time to time, as eccentricity or epicycle for the movement of the sun the astronomer much prefers to take the one which is easiest to /*ii* ᵃ/ grasp. Maybe the philosopher demands probability instead; but neither of them will grasp anything certain or hand it on, unless it has been divinely revealed to him. Therefore let us permit these new hypotheses to make a public appearance among old ones which are themselves no more probable, especially since they are wonderful and easy and bring with them a vast storehouse of learned observations. And as far as hypotheses go, let no one expect anything in the way of certainty from astronomy, since astronomy can offer us nothing certain, lest, if anyone take as true that which has been constructed for another use, he go away from this discipline a bigger fool than when he came to it. Farewell.

11.2.2   Preface and Dedication: To Pope Paul III

/*ii*ᵇ/ I can reckon easily enough, Most Holy Father, that as soon as certain people learn that in these books of mine which I have written about the revolutions of the

---

[5] Ptolemy makes Venus move on an epicycle the ratio of whose radius to the radius of the eccentric circle carrying the epicycle itself is nearly three to four. Hence the apparent magnitude of the planet would be expected to vary with the varying distance of the planet from the Earth, in the ratios stated by Osiander. Moreover, it was found that, whenever the planet happened to be on the epicycle, the mean position of the sun appeared in line with EPA. And so, granted the ratios of epicycle and eccentric, Venus would never appear from the Earth to be at an angular distance of much more than 40° from the centre of her epicycle. That is to say, from the mean position of the sun, as it turned out by observation.

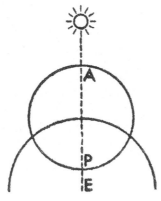

spheres of the world I attribute certain motions to the terrestrial globe, they will immediately shout to have me and my opinion hooted off the stage. For my own works do not please me so much that I do not weigh what judgments others will pronounce concerning them. And although I realize that the conceptions of a philosopher are placed beyond the judgment of the crowd, because it is his loving duty to seek the truth in all things, in so far as God has granted that to human reason; nevertheless I think we should avoid opinions utterly foreign to rightness. And when I considered how absurd this "lecture" would be held by those who know that the opinion that the Earth rests immovable in the middle of the heavens as if their centre had been confirmed by the judgments of many ages—if I were to assert to the contrary that the Earth moves; for a long time I was in great difficulty as to whether I should bring to light my commentaries written to demonstrate the Earth's movement, or whether it would not be better to follow the example of the Pythagoreans and certain others who used to hand down the mysteries of their philosophy not in writing but by word of mouth and only to their relatives and friends—witness the letter of Lysis to Hipparchus. They however seem to me to have done that not, as some judge, out of a jealous unwillingness to communicate their doctrines but in order that things of very great beauty which have been investigated by the loving care of great men should not be scorned by those who find it a bother to expend any great energy on letters—except on the money-making variety—or who are provoked by the exhortations and examples of others to the liberal study of philosophy but on account of their natural /*iii* ª/ stupidity hold the position among philosophers that drones hold among bees. Therefore, when I weighed these things in my mind, the scorn which I had to fear on account of the newness and absurdity of my opinion almost drove me to abandon a work already undertaken.

But my friends made me change my course in spite of my long-continued hesi-tation and even resistance. First among them was Nicholas Schonberg, Cardinal of Capua, a man distinguished in all branches of learning; next to him was my devoted friend Tiedeman Giese, Bishop of Culm, a man filled with the greatest zeal for the divine and liberal arts: for he in particular urged me frequently and even spurred me on by added reproaches into publishing this book and letting come to light a work which I had kept hidden among my things for not merely 9 years, but for almost four times 9 years. Not a few other learned and distinguished men demanded the same thing of me, urging me to refuse no longer—on account of the fear which I felt—to contribute my work to the common utility of those who are really interested in mathematics: they said that the absurder my teaching about the movement of the Earth now seems to very many persons, the more wonder and thanksgiving will it be the object of, when after the publication of my commentaries those same persons see the fog of absurdity dissipated by my luminous demonstrations. Accordingly, I was led by such persuasion and by that hope finally to permit my friends to undertake the publication, of a work which they had long sought from me.

But perhaps Your Holiness will not be so much surprised at my giving the results of my nocturnal study to the light—after having taken such care in working them out that I did not hesitate to put in writing my conceptions as to the movement of the Earth—as you will be eager to hear from me what came into my mind that in opposition to the general opinion of mathematicians—and almost in opposition to common sense I should dare to imagine some movement of the Earth. And so I am unwilling to hide from Your Holiness that nothing except my knowledge that mathematicians have not agreed with one another in their researches moved me to think out a different scheme of drawing up the movements of the spheres of the world. For in the first place mathematicians are so uncertain about the movements of the sun and moon that they can neither demonstrate nor observe the unchanging magnitude of the /*iii* ᵇ/ revolving year. Then in setting up the solar and lunar movements and those of the other five wandering stars, they do not employ the same principles, assumptions, or demonstrations for the revolutions and apparent movements. For some make use of homocentric circles only, others of eccentric circles and epicycles, by means of which however they do not fully attain what they seek. For although those who have put their trust in homocentric circles have shown that various different movements can be composed of such circles, nevertheless they have not been able to establish anything for certain that would fully correspond to the phenomena. But even if those who have thought up eccentric circles seem to have been able for the most part to compute the apparent movements numerically by those means, they have in the meanwhile admitted a great deal which seems to contradict the first principles of regularity of movement. Moreover, they have not been able to discover or to infer the chief point of all, *i.e.*, the form of the world and the certain commensurability of its parts. But they are in exactly the same fix as someone taking from different places hands, feet, head, and the other limbs—shaped very beautifully but not with reference to one body and without correspondence to one another—so that such parts made up a monster rather than a man. And so, in the process of demonstration which they call "method," they are found either to have omitted something necessary or to have admitted something foreign which by no means pertains to the matter; and they would by no means have been in this fix, if they had followed sure principles. For if the hypotheses they assumed were not false, everything which followed from the hypotheses would have been verified without fail; and though what I am saying may be obscure right now, nevertheless it will become clearer in the proper place.

Accordingly, when I had meditated upon this lack of certitude in the traditional mathematics concerning the composition of movements of the spheres of the world, I began to be annoyed that the philosophers, who in other respects had made a very careful scrutiny of the least details of the world, had discovered no sure scheme for the movements of the machinery of the world, which has been built for us by the Best and Most Orderly Workman of all. Wherefore I took the trouble to reread all the books by philosophers which I could get hold of, to see if any of them even supposed that the movements of the spheres of the world /*iv* ᵃ/ were different from those laid down by those who taught mathematics in the schools. And as a matter of fact, I found first in Cicero that Nicetas thought that the Earth moved. And afterwards I found in Plutarch that there were some others of the same opinion: I shall copy out his words here, so that they may be known to all:

Some think that the Earth is at rest; but Philolaus the Pythagorean says that it moves around the fire with an obliquely circular motion, like the sun and moon. Herakleides of Pontus and Ekphantus the Pythagorean do not give the Earth any movement of locomotion, but rather a limited movement of rising and setting around its centre, like a wheel.[6]

Therefore I also, having found occasion, began to meditate upon the mobility of the Earth. And although the opinion seemed absurd, nevertheless because I knew that others before me had been granted the liberty of constructing whatever circles they pleased in order to demonstrate astral phenomena, I thought that I too would be readily permitted to test whether or not, by the laying down that the Earth had some movement, demonstrations less shaky than those of my predecessors could be found for the revolutions of the celestial spheres.

And so, having laid down the movements which I attribute to the Earth farther on in the work, I finally discovered by the help of long and numerous observations that if the movements of the other wandering stars are correlated with the circular movement of the Earth, and if the movements are computed in accordance with the revolution of each planet, not only do all their phenomena follow from that but also this correlation binds together so closely the order and magnitudes of all the planets and of their spheres or orbital circles and the heavens themselves that nothing can be shifted around in any part of them without disrupting the remaining parts and the universe as a whole.

Accordingly, in composing my work I adopted the following order: in the first book I describe all the locations of the spheres or orbital circles together with the movements which I attribute to the earth, so that this book contains as it were the general set-up of the universe. But afterwards in the remaining books I correlate all the movements of the other planets and their spheres or orbital circles with the mobility of the Earth, so that it can be gathered from that how far the apparent movements of the remaining planets and their orbital circles can be saved by being correlated with the movements of the Earth. And I have no doubt that talented and learned mathematicians will agree with me, if—as philosophy /iv ᵇ/ demands in the first place—they are willing to give not superficial but profound thought and effort to what I bring forward in this work in demonstrating these things. And in order that the unlearned as well as the learned might see that I was not seeking to flee from the judgment of any man, I preferred to dedicate these results of my nocturnal study to Your Holiness rather than to anyone else; because, even in this remote corner of the earth where I live, you are held to be most eminent both in the dignity of your order and in your love of letters and even of mathematics; hence, by the authority of your judgment you can easily provide a guard against the bites of slanderers, despite the proverb that there is no medicine for the bite of a sycophant.

But if perchance there are certain "idle talkers" who take it upon themselves to pronounce judgment, although wholly ignorant of mathematics, and if by shamelessly distorting the sense of some passage in Holy Writ to suit their purpose, they dare to reprehend and to attack my work; they worry me so little that I shall even scorn their judgments as foolhardy. For it is not unknown that Lactantius, otherwise

---

[6] *De placitis philosophorum*, III. 13.

a distinguished writer but hardly a mathematician, speaks in an utterly childish fashion concerning the shape of the Earth, when he laughs at those who have affirmed that the Earth has the form of a globe. And so the studious need not be surprised if people like that laugh at us. Mathematics is written for mathematicians; and among them, if I am not mistaken, my labours will be seen to contribute something to the ecclesiastical commonwealth, the principate of which Your Holiness now holds. For not many years ago under Leo X when the Lateran Council was considering the question of reforming the Ecclesiastical Calendar, no decision was reached, for the sole reason that the magnitude of the year and the months and the movements of the sun and moon had not yet been measured with sufficient accuracy. From that time on I gave attention to making more exact observations of these things and was encouraged to do so by that most distinguished man, Paul, Bishop of Fossombrone, who had been present at those deliberations. But what have I accomplished in this matter I leave to the judgment of Your Holiness in particular and to that of all other learned mathematicians. And so as not to appear to Your Holiness to make more promises concerning the utility of this book than I can fulfill, I now pass on to the body of the work.

11.2.3 Book I: Introduction[7]

Among the many and varied literary and artistic studies upon which the natural talents of man are nourished, I think that those above all should be embraced and pursued with the most loving care which have to do with things that are very beautiful and very worthy of knowledge. Such studies are those which deal with the godlike circular movements of the world and the course of the stars, their magnitudes, distances, risings and settings, and the causes of the other appearances in the heavens; and which finally explicate the whole form. For what could be more beautiful than the heavens which contain all beautiful things? Their very names make this clear: *Caelum* (heavens) by naming that which is beautifully carved; and *Mundus* (world), purity and elegance. Many philosophers have called the world a visible god on account of its extraordinary excellence. So if the worth of the arts were measured by the matter with which they deal; this art—which some call astronomy, others astrology, and many of the ancients the consummation of mathematics—would be by far the most outstanding. This art which is as it were the head of all the liberal arts and the one most worthy of a free man leans upon nearly all the other branches of mathematics. Arithmetic, geometry, optics, geodesy, mechanics, and whatever others, all offer themselves in its service. And since a property of all good arts is to draw the mind of man away from the vices and direct it to better things, these arts can do that more plentifully, over and above the unbelievable pleasure of mind (which they furnish). For who, after applying himself to things which he sees established in the best order

---

[7] The three introductory paragraphs are found in the Thorn centenary and Warsaw editions.

and directed by divine ruling, would not through diligent contemplation of them and through a certain habituation be awakened to that which is best and would not wonder at the Artificer of all things, in Whom is all happiness and every good? For the divine Psalmist surely did not say gratuitously that he took pleasure in the workings of God and rejoiced in the works of His hands, unless by means of these things as by some sort of vehicle we are transported to the contemplation of the highest Good.

Now as regards the utility and ornament which they confer upon a commonwealth—to pass over the innumerable advantages they give to private citizens—Plato makes an extremely good point, for in the seventh book of the Laws he says that this study should be pursued in especial, that through it the orderly arrangement of days into months and years and the determination of the times for solemnities and sacrifices should keep the state alive and watchful; and he says that if anyone denies that this study is necessary for a man who is going to take up any of the highest branches of learning, then such a person is thinking foolishly; and he thinks that it is impossible for anyone to become godlike or be called so who has no necessary knowledge of the sun, moon, and the other stars.

However, this more divine than human science, which inquires into the highest things, is not lacking in difficulties. And in particular we see that as regards its principles and assumptions, which the Greeks call "hypotheses," many of those who undertook to deal with them were not in accord and hence did not employ the same methods of calculation. In addition, the courses of the planets and the revolution of the stars cannot be determined by exact calculations and reduced to perfect knowledge unless, through the passage of time and with the help of many prior observations, they can, so to speak, be handed down to posterity. For even if Claud Ptolemy of Alexandria, who stands far in front of all the others on account of his wonderful care and industry, with the help of more than 40 years of observations brought this art to such a high point that there seemed to be nothing left which he had not touched upon; nevertheless we see that very many things are not in accord with the movements which should follow from his doctrine but rather with movements which were discovered later and were unknown to him. Whence even Plutarch in speaking of the revolving solar year says, "So far the movement of the stars has overcome the ingenuity of the mathematicians." Now to take the year itself as my example, I believe it is well known how many different opinions there are about it, so that many people have given up hope of making an exact determination of it. Similarly, in the case of the other planets I shall try—with the help of God, without Whom we can do nothing—to make a more detailed inquiry concerning them, since the greater the interval of time between us and the founders of this art—whose discoveries we can compare with the new ones made by us—the more means we have of supporting our own theory. Furthermore, I confess that, I shall expound many things differently from my predecessors—although with their aid, for it was they who first opened the road of inquiry into these things.

11.2.4   Chapter 1: The World is Spherical

/1ª/ In the beginning we should remark that the world is globe-shaped; whether because this figure is the most perfect of all, as it is an integral whole and needs no joints; or because this figure is the one having the greatest volume and thus is especially suitable for that which is going to comprehend and conserve all things; or even because the separate parts of the world *i.e.*, the sun, moon, and stars are viewed under such a form; or because everything in the world tends to be delimited by this form, as is apparent in the case of drops of water and other liquid bodies, when they become delimited of themselves. And so no one would hesitate to say that this form belongs to the heavenly bodies.

11.2.5   Chapter 2: The Earth is Spherical too

The Earth is globe-shaped too, since on every side it rests upon its centre. But it is not perceived straightway to be a perfect sphere, on account of the great height of its mountains and the lowness of its valleys, though they modify its universal roundness to only a very small extent.

That is made clear in this way. For when people journey northward from anywhere, the northern vertex of the axis of daily revolution gradually moves overhead, and the other moves downward to the same extent; and many stars situated to the north are seen not to set, and many to the south are seen not to rise any more. So Italy does not see Canopus, which is visible to Egypt. And Italy sees the last star of Fluvius, which is not visible to this region situated in a more frigid zone. Conversely, for people who travel southward, the second group of stars becomes higher in the sky; while those become lower which for us are high up.

Moreover, the inclinations of the poles have everywhere the same ratio with places at equal distances from the poles of the Earth and that /1 ᵇ/ happens in no other figure except the spherical. Whence it is manifest that the Earth itself is contained between the vertices and is therefore a globe.

Add to this the fact that the inhabitants of the East do not perceive the evening eclipses of the sun and moon; nor the inhabitants of the West, the morning eclipses; while of those who live in the middle region—some see them earlier and some later.

Furthermore, voyagers perceive that the waters too are fixed within this figure; for example, when land is not visible from the deck of a ship, it may be seen from the top of the mast, and conversely, if something shining is attached to the top of the mast, it appears to those remaining on the shore to come down gradually, as the ship moves from the land, until finally it becomes hidden, as if setting.

Moreover, it is admitted that water, which by its nature flows, always seeks lower places—the same way as earth—and does not climb up the shore any farther than the convexity of the shore allows. That is why the land is so much higher where it rises up from the ocean.

11.2.6   Chapter 3: How Land and Water Make up a Single Globe

And so the ocean encircling the land pours forth its waters everywhere and fills up the deeper hollows with them. Accordingly it was necessary for there to be less water than land, so as not to have the whole earth soaked with water—since both of them tend toward the same centre on account of their weight—and so as to leave some portions of land—such as the islands discernible here and there—for the preservation of living creatures. For what is the continent itself and the *orbis terrarum* except an island which is larger than the rest? We should not listen to certain Peripatetics who maintain that there is ten times more water than land and who arrive at that conclusion because in the transmutation of the elements the liquefaction of one part of earth results in ten parts of water. And they say that land has emerged for a certain distance because, having hollow spaces inside, it does not balance everywhere with respect to weight and so the centre of gravity is different from the centre of magnitude. But they fall into error through ignorance of geometry; for they do not know that there cannot be seven times more water than land and some part of the land still remain dry, unless the land abandon its centre of gravity and give place to the waters as being heavier. For spheres are to one another as the cubes of their diameters. If therefore there were seven parts of water and one part of land, /2 ª/ the diameter of the land could not be greater than the radius of the globe of the waters. So it is even less possible that the water should be ten times greater. It can be gathered that there is no difference between the centres of magnitude and of gravity of the Earth from the fact that the convexity of the land spreading out from the ocean does not swell continuously, for in that case it would repulse the sea-waters as much as possible and would not in any way allow interior seas and huge gulfs to break through. Moreover, from the seashore outward the depth of the abyss would not stop increasing, and so no island or reef or any spot of land would be met with by people voyaging out very far. Now it is well known that there is not quite the distance of 2 miles—at practically the centre of the *orbis terrarum*—between the Egyptian and the Red Sea. And on the contrary, Ptolemy in his *Cosmography* extends inhabitable lands as far as the median circle, and he leaves that part of the Earth as unknown, where the moderns have added Cathay and other vast regions as far as 60° longitude, so that inhabited land extends in longitude farther than the rest of the ocean does. And if you add to these the islands discovered in our time under the princes of Spain and Portugal and especially America—named after the ship's captain who discovered her—which they consider a second *orbis terrarum* on account of her so far unmeasured magnitude—besides many other islands heretofore unknown, we would not be greatly surprised if there were antiphodes or antichthones. For reasons of geometry compel us to believe that America is situated diametrically opposite to the India of the Ganges.

   And from all that I think it is manifest that the land and the water rest upon one centre of gravity; that this is the same as the centre of magnitude of the land, since land is the heavier; that parts of land which are as it were yawning are filled with water; and that accordingly there is little water in comparison with the land, even if more of the surface appears to be covered by water.

Now it is necessary that the land and the surrounding waters have the figure which the shadow of the Earth casts, for it eclipses the moon by projecting a perfect circle upon it. Therefore the Earth is not a plane, as Empedocles and Anaximenes opined; or a tympanoid, as Leucippus; or a scaphoid, as Heracleitus; or hollowed out in any other way, as Democritus; or again a cylinder, as Anaximander; and it is not infinite in its lower part, with the density increasing rootwards, as Xenophanes thought; but it is perfectly round, as the philosophers perceived.

11.2.7   Chapter 4: The Movement of the Celestial Bodies is Regular, Circular, and Everlasting—or Else Compounded of Circular Movements

/2 ᵇ/ After this we will recall that the movement of the celestial bodies is circular. For the motion of a sphere is to turn in a circle; by this very act expressing its form, in the most simple body, where beginning and end cannot be discovered or distinguished from one another, while it moves through the same parts in itself.

But there are many movements on account of the multitude of spheres or orbital circles.[8] The most obvious of all is the daily revolution—which the Greeks call νυχϑήμερον; i.e., having the temporal span of a day and a night. By means of this movement the whole world—with the exception of the Earth—is supposed to be borne from east to west. This movement is taken as the common measure of all movements, since we measure even time itself principally by the number of days.

Next, we see other as it were antagonistic revolutions; i.e., from west to east, on the part of the sun, moon, and the wandering stars. In this way the sun gives us the year, the moon the months—the most common periods of time; and each of the other five planets follows its own cycle. Nevertheless these movements are manifoldly different from the first movement. First, in that they do not revolve around the same poles as the first movement but follow the oblique ecliptic; next, in that they do not seem to move in their circuit regularly. For the sun and moon are caught moving at times more slowly and at times more quickly. And we perceive the five wandering stars sometimes even to retrograde and to come to a stop between these two movements. And though the sun always proceeds straight ahead along its route, they wander in various ways, straying sometimes towards the south, and at other times towards the north—whence they are called "planets." Add to this the fact that sometimes they are nearer the Earth—and are then said to be at their perigee—and at other times are farther away and are said to be at their apogee.

---

[8] The "orbital circle" (orbis) is the great circle whereon the planet moves in its sphere (sphaera). Copernicus uses the word orbis which designates a circle primarily rather than a sphere because, while the sphere may be necessary for the mechanical explanation of the movement, only the circle is necessary for the mathematical.

We must however confess that these movements are circular or are composed of many circular movements, in that they maintain these irregularities in accordance with a constant law and with fixed periodic returns: and that could not take place, if they were not circular. For it is only the circle which can bring back what is past and over with; and in this way, for example, the sun by a movement composed of circular movements brings back to us the inequality of days and nights and the four seasons of the year. /3 ª/ Many movements are recognized in that movement, since it is impossible that a simple heavenly body should be moved irregularly by a single sphere. For that would have to take place either on account of the inconstancy of the motor virtue—whether by reason of an extrinsic cause or its intrinsic nature—or on account of the inequality between it and the moved body. But since the mind shudders at either of these suppositions, and since it is quite unfitting to suppose that such a state of affairs exists among things which are established in the best system, it is agreed that their regular movements appear to us as irregular, whether on account of their circles having different poles or even because the earth is not at the centre of the circles in which they revolve. And so for us watching from the Earth, it happens that the transits of the planets, on account of being at unequal distances from the Earth, appear greater when they are nearer than when they are farther away, as has been shown in optics: thus in the case of equal arcs of an orbital circle which are seen at different distances there will appear to be unequal movements in equal times. For this reason I think it necessary above all that we should note carefully what the relation of the Earth to the heavens is, so as not—when we wish to scrutinize the highest things—to be ignorant of those which are nearest to us, and so as not—by the same error—to attribute to the celestial bodies what belongs to the Earth.

11.3   Study Questions

QUES. 11.1 Is the study of astronomy the search for truth?

a) What, according to Osiander, is the job of the astronomer? And what criteria does, or should, the astronomer use for selecting theories?
b) Does Osiander believe that truth is ever accessible by any means whatsoever? Do you agree with him?
c) Does Copernicus agree with Osiander's views on the nature of astronomy? If not, why might this preface have been attached to Copernicus' work?

QUES. 11.2 Is it the duty of the scholar to share his work? Or is it wise to hide knowledge from the crowds?

a) To whom does Copernicus dedicate his work? Why?
b) What motivation, according to Copernicus, distinguishes the philosopher from the crowd?
c) Why did the Pythagoreans teach by word of mouth, rather than by writing books?
d) In what way did previous astronomical theories resemble "a monster rather than a man"? Why did this annoy Copernicus?

e) What was the primary result of Copernicus' observations? Was he the first to make such an assertion?

f) Which group is more likely to reject scientific discovery, the clergy or the laity? Why? Is this the case today?

QUES. 11.3 Is astronomy a liberal art?

a) What is a *liberal* art? How, in particular, does it differ from a *servile* art? And why does Copernicus describe astronomy as one of the liberal arts?

b) What is the goal, or end, of the study of the liberal arts for the individual? For the commonwealth? Do you agree with Plato's judgment regarding the necessity of the study of astronomy?

c) How do contemporary understandings of the liberal arts differ from that expressed by Copernicus?

QUES. 11.4 What is the shape of the World? Of the Earth? How do you know?

QUES. 11.5 What is the composition of the Earth?

a) Which is more prevalent on the Earth, land or water? How does Copernicus' opinion rely upon Aristotelian physics?

b) In what sense is his claim counterintuitive? And how does he employ geometry, and cartography, to support his claim?

QUES. 11.6 Do the planets move at a constant speed?

a) What is the first, and most obvious, motion of the celestial bodies? What is the second motion? And in what way(s) do the movements of the planets differ from those of the other celestial bodies?

b) In what geometrical shape(s) do the planets move? For what reason does Copernicus reject the possibility of irregular motion of the planets? How, then, does he explain the *apparent* irregularity of planetary motion?

c) Would you describe Copernicus' reasons as "scientific"? In what sense (if any) does Copernicus rely upon Aristotelian ideas?

11.4   Exercises

Ex. 11.1 (VENUS PROBLEM). Suppose that an astronomer watches Venus when it is at perigee, and, a bit later, when it is at apogee. Assuming the orbit of Venus to be circular, and centered upon the sun.

a) Determine the (constant) speed of Venus as it orbits the sun. (Hint: you may need to look up the mean distance of Venus from the sun and also the time it takes to orbit.) (ANSWER: approximately 1800 km/min.)

b) When Venus is at perigee, how many degrees does it appear to move through in one hour, from the vantage point of the earth? What about when it is at apogee?

(Hint: you may need to determine the distance between Venus and the earth at perigee and also at apogee.)

Ex. 11.2 (SIMPLICITY OF NATURE ESSAY). Do you agree with Copernicus that nature "takes very great care not to have produced anything superfluous or useless"? To what extent does Copernicus rely upon this principle? Is it a necessary presupposition of his world-view?

Ex. 11.3 (GOAL OF ASTRONOMY ESSAY). Osiander claims that astronomers do not think up ideas "in order to persuade anyone of their truth but only in order that they may provide a correct basis for calculation." Do you agree with this *instrumentalist* view of science? More generally, do you think that a person's understanding of the goal of scientific study affects what type(s) of theories he or she recognizes?

11.5 Vocabulary

1. Savant	19. Geodesy
2. Perigee	20. Liberal arts
3. Apogee	21. Convex
4. Eccentricity	22. Orbis terrarum
5. Epicycle	23. Peripatetic
6. Terrestrial	24. Transmutation
7. Luminous	25. Liquefaction
8. Homocentric	26. Center of gravity
9. Commensurable	27. Center of magnitude
10. Certitude	28. Median circle
11. Oblique	29. Cathay
12. Superficial	30. Longitude
13. Profound	31. Antipode
14. Sycophant	32. Antichthone
15. Ecclesiastical	33. Ecliptic
16. Lateran Council	34. Retrograde
17. Explicate	35. Extrinsic
18. Consummation	36. Intrinsic

Chapter 12
Earth as a Wandering Star

Since nothing hinders the mobility of the Earth, I think we
should now see whether more than one movement belongs to it,
so that it can be regarded as one of the wandering stars.
—Nicholaus Copernicus

12.1   Introduction

In the first four chapters of Book I of *On the Revolutions of the Heavenly Spheres*,
Copernicus adhered very closely to the opinions which he inherited from the ancients regarding both the *shape* of the heavens and the earth and the *circular motion*
of the planets. For example in Chap. 4 of Book I, Copernicus states that the sun, the
moon and the planets must all be attached to vast heavenly spheres which are themselves rotating at a constant speed—just as Aristotle and Ptolemy had taught many
centuries before. For otherwise, how could they exhibit "fixed periodic returns" according to a "constant law"? As Copernicus sees it, there is no other solution—apart
from attributing to the planets some type of variable "motor virtue", a supposition
which causes the mind to shudder. The difficulty now facing Copernicus is how he
might arrange this system of heavenly spheres in a configuration which is somehow
superior to that laid out so convincingly by Ptolemy over 1000 years earlier in his
Almagest. So in Chap. 5 of Book I, Copernicus begins to reconsider whether the
earth might spin about its own axis. On this point, his views begin to diverge from
those of Ptolemy (and Aristotle). In arguing his case, does Copernicus reject Aristotle's physics altogether—along with Ptolemy's astronomy—or does he attempt to
reconcile Aristotle with his own heliocentric world-view?

12.2   Reading: Copernicus, *On the Revolutions*
of the Heavenly Spheres

Copernicus, N., *On the Revolutions of the Heavenly Spheres*, Great Minds,
Prometheus Books, Amherst, NY, 1995. Book I.

K. Kuehn, *A Student's Guide Through the Great Physics Texts*,
Undergraduate Lecture Notes in Physics, DOI 10.1007/978-1-4939-1360-2_12,
© Springer Science+Business Media, LLC 2015

12.2.1   Chapter 5: Does the Earth Have a Circular Movement? And of its Place

Now that it has been shown that the Earth too has the form of a globe, I think we must see whether or not a movement follows upon its form and what the place of the Earth is in the universe. For without doing that it will not be possible to find a sure reason for the movements appearing in the heavens. Although there are so many authorities for saying that the Earth rests in the centre of the world that people think the contrary supposition inopinable and even ridiculous; if however we consider the thing attentively, we will see that the question has not yet been decided and accordingly is by no means to be scorned. For every apparent change in place occurs on account of the movement either of the thing seen or of the spectator, or on account of the necessarily unequal movement of both. For no movement is perceptible relatively to things moved equally in the same directions—I mean relatively to the thing seen and the spectator. Now it is from the Earth that the celestial circuit is beheld and presented to our sight. Therefore, if some movement should belong to the Earth /3 ^{b}/ it will appear, in the parts of the universe which are outside, as the same movement but in the opposite direction, as though the things outside were passing over. And the daily revolution in especial is such a movement. For the daily revolution appears to carry the whole universe along, with the exception of the Earth and the things around it. And if you admit that the heavens possess none of this movement but that the Earth turns from west to east, you will find—if you make a serious examination—that as regards the apparent rising and setting of the sun, moon, and stars the case is so. And since it is the heavens which contain and embrace all things as the place common to the universe, it will not be clear at once why movement should not be assigned to the contained rather than to the container, to the thing placed rather than to the thing providing the place.

As a matter of fact, the Pythagoreans Herakleides and Ekphantus were of this opinion and so was Hicetas the Syracusan in Cicero; they made the Earth to revolve at the centre of the world. For they believed that the stars set by reason of the interposition of the Earth and that with cessation of that they rose again. Now upon this assumption there follow other things, and a no smaller problem concerning the place of the Earth, though it is taken for granted and believed by nearly all that the Earth is the centre of the world. For if anyone denies that the Earth occupies the midpoint or centre of the world yet does not admit that the distance (between the two) is great enough to be compared with (the distance to) the sphere of the fixed stars but is considerable and quite apparent in relation to the orbital circles of the sun and the planets; and if for that reason he thought that their movements appeared irregular because they are organized around a different centre from the centre of the Earth, he might perhaps be able to bring forward a perfectly sound reason for movement which appears irregular. For the fact that the wandering stars are seen to be sometimes nearer the Earth and at other times farther away necessarily argues that the centre of the Earth is not the centre of their circles. It is not yet clear whether the

Fig. 12.1 The bisection of the
ecliptic by the
horizon.—[*K.K.*]

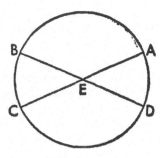

Earth draws near to them and moves away or they draw near to the Earth and move away.

And so it would not be very surprising if someone attributed some other movement to the earth in addition to the daily revolution. As a matter of fact, Philolaus the Pythagorean—no ordinary mathematician, whom Plato's biographers say Plato went to Italy for the sake of seeing—is supposed to have held that the Earth moved in a circle and wandered in some other movements and was one of the planets.

Many however have believed that they could show by geometrical reasoning that the Earth is in the middle of the world; that it has the proportionality of a point in relation to the immensity of the heavens, occupies the central position, and for this reason is immovable, because, when the universe moves, the centre /4 ª/ remains unmoved and the things which are closest to the centre are moved the most slowly.

12.2.2   Chapter 6: On the Immensity of the Heavens in Relation to the Magnitude of the Earth

It can be understood that this great mass which is the Earth is not comparable with the magnitude of the heavens, from the fact that the boundary circles—for that is the translation of the Greek ὁρίζοντες cut the whole celestial sphere into two halves; for that could not take place if the magnitude of the Earth in comparison with the heavens, or its distance from the centre of the world, were considerable. For the circle bisecting a sphere goes through the centre of the sphere, and is the greatest circle which it is possible to circumscribe.

Now let the horizon be the circle *ABCD*, and let the Earth, where our point of view is, be *E*, the centre of the horizon by which the visible stars are separated from those which are not visible (Fig. 12.1). Now with a dioptra or horoscope or level placed at *E*, the beginning of Cancer is seen to rise at point *C*; and at the same moment the beginning of Capricorn appears to set at *A*. Therefore, since *AEC* is in a straight line with the dioptra, it is clear that this line is a diameter of the ecliptic, because the six signs bound a semicircle, whose centre *E* is the same as that of the horizon. But when a revolution has taken place and the beginning of Capricorn arises at *B*, then the setting of Cancer will be visible at *D*, and *BED* will be a straight line and a diameter of the ecliptic. But it has already been seen that the line AEC is a diameter of the same circle; therefore, at their common section, point E will be their centre. So in this way the horizon always bisects the ecliptic, which is a great circle of the sphere. But on a sphere, if a circle bisects one of the great circles, then the circle bisecting is a great circle. Therefore the horizon is a great circle; and its centre is the same as that of the ecliptic, as far as appearance goes; although nevertheless the line passing through the centre of the Earth and the line touching to the surface are necessarily different; but on account of their immensity in comparison with the Earth they are like parallel lines, which on account of the great distance between the termini appear to be one line, when the space contained between them /4 ^{b}/ is in no perceptible ratio to their length, as has been shown in optics.

From this argument it is certainly clear enough that the heavens are immense in comparison with the Earth and present the aspect of an infinite magnitude, and that in the judgment of sense-perception the Earth is to the heavens as a point to a body and as a finite to an infinite magnitude. But we see that nothing more than that has been shown, and it does not follow that the Earth must rest at the centre of the world. And we should be even more surprised if such a vast world should wheel completely around during the space of 24 h rather than that its least part, the Earth, should. For saying that the centre is immovable and that those things which are closest to the centre are moved least does not argue that the Earth rests at the centre of the world. That is no different from saying that the heavens revolve but the poles are at rest and those things which are closest to the poles are moved least. In this way Cynosura [the pole star] is seen to move much more slowly than Aquila or Canicula because, being very near to the pole, it describes a smaller circle, since they are all on a single sphere, the movement of which stops at its axis and which does not allow any of its parts to have movements which are equal to one another. And nevertheless the revolution of the whole brings them round in equal times but not over equal spaces.

The argument which maintains that the Earth, as a part of the celestial sphere and as sharing in the same form and movement, moves very little because very near to its centre advances to the following position: therefore the Earth will move, as being a body and not a centre, and will describe in the same time arcs similar to, but smaller than, the arcs of the celestial circle. It is clearer than daylight how false that is; for there would necessarily always be noon at one place and midnight at another, and so the daily risings and settings could not take place, since the movement of the whole and the part would be one and inseparable.

But the ratio between things separated by diversity of nature is so entirely different that those which describe a smaller circle turn more quickly than those which describe a greater circle. In this way Saturn, the highest of the wandering stars, completes its revolution in thirty years, and the moon which is without doubt the closest to the Earth completes its circuit in a month, and finally the Earth itself will be considered to complete a circular movement in the space of a day and a night. So this same problem concerning the daily revolution comes up again. And also the question about the place of the Earth becomes even less certain on account of what was just said. For that demonstration proves nothing except that the heavens are of an indefinite magnitude with respect to the Earth. But it is not at all clear how far this immensity stretches out. On the contrary, since the minimal and indivisible corpuscles, which are called atoms, are not perceptible to sense, they do not, when taken in twos or in some small number, constitute a visible body; but they can be taken in such a large quantity that there will at last be enough to form a visible magnitude. So it is as regards the place of the earth; for although it is not at the centre of the world, nevertheless the distance is as nothing, particularly in comparison with the sphere of the fixed stars.

12.2.3   Chapter 7: Why the Ancients Thought the Earth Was at Rest at the Middle of the World as Its Centre

/5ᵃ/ Wherefore for other reasons the ancient philosophers have tried to affirm that the Earth is at rest at the middle of the world, and as principal cause they put forward heaviness and lightness. For Earth is the heaviest element; and all things of any weight are borne towards it and strive to move towards the very centre of it.

For since the Earth is a globe towards which from every direction heavy things by their own nature are borne at right angles to its surface, the heavy things would fall on one another at the centre if they were not held back at the surface; since a straight line making right angles with a plane surface where it touches a sphere leads to the centre. And those things which are borne toward the centre seem to follow along in order to be at rest at the centre. All the more then will the Earth be at rest at the centre; and, as being the receptacle for falling bodies, it will remain immovable because of its weight.

They strive similarly to prove this by reason of movement and its nature. For Aristotle says that the movement of a body which is one and simple is simple and the simple movements are the rectilinear and the circular. And of rectilinear movements, one is upward, and the other is downward. As a consequence, every simple movement is either toward the centre, *i.e.*, downward, or away from the centre, *i.e.*, upward, or around the centre, *i.e.*, circular. Now it belongs to earth and water, which are considered heavy, to be borne downward, *i.e.*, to seek the centre: for air and fire, which are endowed with lightness, move upward, *i.e.*, away from the centre. It seems fitting to grant rectilinear movement to these four elements and to give the heavenly bodies a circular movement around the centre. So Aristotle. Therefore, said Ptolemy of Alexandria, if the Earth moved, even if only by its daily rotation, the contrary of what was said above would necessarily take place. For this movement which would traverse the total circuit of the Earth in 24 hours would necessarily be very headlong and of an unsurpassable velocity. Now things which are suddenly and violently whirled around are seen to be utterly unfitted for reuniting, and the more unified are seen to become dispersed, unless some constant force constrains them to stick together. And a long time ago, he says, the scattered Earth would have passed beyond the heavens, as is certainly ridiculous; /5ᵇ/ and *a fortiori* so would all the living creatures and all the other separate masses which could by no means remain unshaken. Moreover, freely falling bodies would not arrive at the places appointed them, and certainly not along the perpendicular line which they assume so quickly.

And we would see clouds and other things floating in the air always borne toward the west.

12.2.4 Chapter 8: Answer to the Aforesaid Reasons and Their Inadequacy

For these and similar reasons they say that the Earth remains at rest at the middle of the world and that there is no doubt about this. But if someone opines that the Earth revolves he will also say that the movement is natural and not violent. Now things which are according to nature produce effects contrary to those which are violent. For things to which force or violence is applied get broken up and are unable to subsist for a long time. But things which are caused by nature are in a right condition and are kept in their best organization. Therefore Ptolemy had no reason to fear that the Earth and all things on the Earth would be scattered in a revolution caused by the efficacy of nature, which is greatly different from that of art or from that which can result from the genius of man. But why didn't he feel anxiety about the world instead, whose movement must necessarily be of greater velocity, the greater the heavens are than the Earth? Or have the heavens become so immense, because an unspeakably vehement motion has pulled them away from the centre, and because the heavens would fall if they came to rest anywhere else?

Surely if this reasoning were tenable, the magnitude of the heavens would extend infinitely. For the farther the movement is borne upward by the vehement force, the faster will the movement be, on account of the ever-increasing circumference which must be traversed every 24 h: and conversely, the immensity of the sky would increase with the increase in movement. In this way, the velocity would make the magnitude increase infinitely, and the magnitude the velocity. And in accordance with the axiom of physics that *that which is infinite cannot be traversed or moved in any way*, then the heavens will necessarily come to rest.

But they say that beyond the heavens there isn't any body or place or void or anything at all; and accordingly it is not possible for the heavens to move outward: in that case it is rather surprising that something can be held together by nothing. But if the heavens were infinite and were finite only with respect to a hollow space inside, then it will be said with more truth that there is nothing outside the heavens, since anything /6ª/ which occupied any space would be in them; but the heavens will remain immobile. For movement is the most powerful reason wherewith they try to conclude that the universe is finite.

But let us leave to the philosophers of nature the dispute as to whether the world is finite or infinite, and let us hold as certain that the Earth is held together between its two poles and terminates in a spherical surface. Why therefore should we hesitate any longer to grant to it the movement which accords naturally with its form, rather than put the whole world in a commotion—the world whose limits we do not and cannot know? And why not admit that the appearance of daily revolution belongs to the heavens but the reality belongs to the Earth? And things are as when Aeneas said in Virgil: "We sail out of the harbor, and the land and the cities move away." As a matter of fact, when a ship floats on over a tranquil sea, all the things outside seem to the voyagers to be moving in a movement which is the image of their own, and they think on the contrary that they themselves and all the things with them are at rest. So it can easily happen in the case of the movement of the Earth that the whole world should be believed to be moving in a circle. Then what would we say about the clouds and the other things floating in the air or falling or rising up, except that not only the Earth and the watery element with which it is conjoined are moved in this way but also no small part of the air and whatever other things have a similar kinship with the Earth? whether because the neighbouring air, which is mixed with earthly and watery matter, obeys the same nature as the Earth or because the movement of the air is an acquired one, in which it participates without resistance on account of the contiguity and perpetual rotation of the Earth. Conversely, it is no less astonishing for them to say that the highest region of the air follows the celestial movement, as is shown by those stars which appear suddenly—I mean those called "comets" or "bearded stars" by the Greeks. For that place is assigned for their generation; and like all the other stars they rise and set. We can say that that part of the air is deprived of terrestrial motion on account of its great distance from the Earth. Hence the air which is nearest to the Earth and the things floating in it will appear tranquil, unless they are driven to and fro by the wind or some other force, as happens. For how is the wind in the air different from a current in the sea?

But we must confess that in comparison with the world the movement of falling and of rising bodies is twofold and is in general compounded of the rectilinear and the circular. As regards things which move downward on account of their weight /6^{b}/ because they have very much earth in them doubtless their parts possess the same nature as the whole, and it is for the same reason that fiery bodies are drawn upward with force. For even this earthly fire feeds principally on earthly matter; and they define flame as glowing smoke. Now it is a property of fire to make that which it invades to expand; and it does this with such force that it can be stopped by no means or contrivance from breaking prison and completing its job. Now expanding movement moves away from the centre to the circumference; and so if some part of the Earth caught on fire, it would be borne away from the centre and upward. Accordingly, as they say, a simple body possesses a simple movement—this is first verified in the case of circular movement—as long as the simple body remain in its unity in its natural place. In this place, in fact, its movement is none other than the circular, which remains entirely in itself, as though at rest. Rectilinear movement, however, is added to those bodies which journey away from their natural place or are shoved out of it or are outside it somehow. But nothing is more repugnant to the order of the whole and to the form of the world than for anything to be outside of its place. Therefore rectilinear movement belongs only to bodies which are not in the right condition and are not perfectly conformed to their nature when they are separated from their whole and abandon its unity. Furthermore, bodies which are moved upward or downward do not possess a simple, uniform, and regular movement—even without taking into account circular movement. For they cannot be in equilibrium with their lightness or their force of weight. And those which fall downward possess a slow movement at the beginning but increase their velocity as they fall. And conversely we note that this earthly fire—and we have experience of no other—when carried high up immediately dies down, as if through the acknowledged agency of the violence of earthly matter.

Now circular movement always goes on regularly, for it has an unfailing cause; but (in rectilinear movement) the acceleration stops, because, when the bodies have reached their own place, they are no longer heavy or light, and so the movement ends. Therefore, since circular movement belongs to wholes and rectilinear to parts, we can say that the circular movement stands with the rectilinear, as does animal with sick. And the fact that Aristotle divided simple movement into three genera: away from the centre, toward the centre, and around the centre, will be considered merely as an act of reason, just as we distinguish between line, point, and surface, though none of them can subsist without the others or /7ᵃ/ without body.

In addition, there is the fact that the state of immobility is regarded as more noble and godlike than that of change and instability, which for that reason should belong to the Earth rather than to the world. I add that it seems rather absurd to ascribe movement to the container or to that which provides the place and not rather to that which is contained and has a place, *i.e.*, the Earth. And lastly, since it is clear that the wandering stars are sometimes nearer and sometimes farther away from the Earth, then the movement of one and the same body around the centre—and they mean the centre of the Earth—will be both away from the centre and toward the centre. Therefore it is necessary that movement around the centre should be taken more generally; and it should be enough if each movement is in accord with its own centre. You see therefore that for all these reasons it is more probably that the Earth moves than that it is at rest—especially in the case of the daily revolution, as it is the Earth's very own. And I think that is enough as regards the first part of the question.

12.2.5 Chapter 9: Whether Many Movements Can be Attributed to the Earth, and Concerning the Centre of the World

Therefore, since nothing hinders the mobility of the Earth, I think we should now see whether more than one movement belongs to it, so that it can be regarded as one of the wandering stars. For the apparent irregular movement of the planets and their variable distances from the Earth—which cannot be understood as occurring in circles homocentric with the Earth—make it clear that the Earth is not the centre of their circular movements. Therefore, since there are many centres, it is not foolhardy to doubt whether the centre of gravity of the Earth rather than some other is the centre of the world. I myself think that gravity or heaviness is nothing except a certain natural appetency implanted in the parts by the divine providence of the universal Artisan, in order that they should unite with one another in their oneness and wholeness and come together in the form of a globe. It is believable that this affect is present in the sun, moon, and the other bright planets and that through its efficacy they remain in the spherical figure in which they are visible, though they nevertheless accomplish their circular movements in many different ways. Therefore if the Earth too possesses movements different from the one around its centre, then they will necessarily be movements which similarly appear on the outside in the many bodies; and we find the yearly revolution among these movements. For if the annual revolution were changed from being solar to being terrestrial, and immobility were granted to the sun, /7ᵇ/ the risings and settings of the signs and of the fixed stars—whereby they become morning or evening stars—will appear in the same way; and it will be seen that the stoppings, retrogressions, and progressions of the wandering stars are not their own, but are a movement of the Earth and that they borrow the appearances of this movement. Lastly, the sun will be regarded as occupying the centre of the world. And the ratio of order in which these bodies succeed one another and the harmony of the whole world teaches us their truth, if only—as they say—we would look at the thing with both eyes.

12.3 Study Questions

QUES. 12.1 Does the earth spin about its axis?

a) What line of reasoning does Copernicus initially employ in order to call this view into question? Does his reasoning prove that the earth rotates?
b) What was the prevalent view, at Copernicus time, regarding the place, and motion, of the earth? Were there any notable dissenters?

QUES. 12.2 What is the location and size of the earth within the heavens?

a) Does the horizon bisect the zodiac? What does this imply about the location of the earth? Is the earth at the exact center of the heavens?
b) On what grounds does Copernicus criticize the view that the earth, as a body, rotates in unison with the heavenly sphere?
c) How does Copernicus argue, based on the notion of atoms, that the earth need not be at the exact center of the celestial sphere?

QUES. 12.3 Can Copernicus' views on the motion of the earth be reconciled with Aristotle's physics?

a) According to Aristotelian physics, what is the nature of the earthly element? What does this seem to imply about the position and motion of Earth itself?
b) What arguments were advanced by Ptolemy, based on Aristotelian physics, which suggest that the earth must be stationary?
c) According to Aristotelian physics, can motion which is produced by force or violence (as opposed to natural tendency) persist indefinitely?
d) What inconsistency does Copernicus find in Ptolemy's argument for a stationary earth and a rotating heaven?

e) How does he deploy Aristotle's axiom that "that which is infinite cannot be traversed or moved in any way"[1] against Ptolemy?

f) Does an object with a spherical form, such as the earth, have a natural motion? Where do you think Copernicus gets this idea?

g) How does Copernicus employ the poet Virgil in defense of his own world-view?

h) How does Copernicus address Ptolemy's concern regarding the effects that a hypothetically spinning earth would necessarily have on its atmosphere and on its inhabitants?

i) How does Copernicus account for the fact that objects dropped on a spinning earth would not follow a true rectilinear path when falling?

j) Is the earth noble and god-like? Does this imply a lack of motion?

QUES. 12.4 What is gravity?

a) Would Aristotle and Ptolemy agree with Copernicus' description of gravity as terrestrial heaviness?

b) What notable feature of nature does Copernicus explain using his understanding of gravity? Would Aristotle and Ptolemy agree with him?

c) Is Copernicus' view identical to the modern understanding of gravity? For example, in what way is Copernicus' view of gravity like, or unlike, that of Newton?

12.4   Exercises

Ex. 12.1 (COPERNICUS AND ARISTOTLE ESSAY). Would you consider Copernicus to be an Aristotelian? You might consider the following questions: Which Aristotelian doctrines did he uphold? Which (if any) did he reject? Why might he have been so reticent to break entirely from Aristotle's physics? Would you consider him to be a conservative or a radical?

12.5   Vocabulary

1. Inopinable	8. Tenable
2. Circumscribe	9. Vehement
3. Dioptra	10. Rectilinear
4. Horoscope	11. Velocity
5. Corpuscle	12. Acceleration
6. *A fortiori*	13. Homocentric
7. Efficacy	14. Angular elongation

---

[1] See Aristotle *On the Heavens*, Book I, Chaps. 5 and 6.

Chapter 13
Re-ordering the Heavenly Spheres

*We should rather follow the wisdom of nature, which, as it takes
very great care not to have produced anything superfluous or
useless, often prefers to endow one thing with many effects.*
—Nicholaus Copernicus

13.1 Introduction

In the preceding chapters from Book I of his *Revolutions*, Copernicus argued that
the *apparent* daily rotation of the heavens is due to an *actual* rotation of Earth
about its own axis. This idea wasn't entirely new, of course. Thirteen centuries
earlier Ptolemy had considered—and rejected—the same idea. Ptolemy's arguments
were largely based on common sense: do you *feel* like you are constantly spinning
around? Obviously not. Therefore you are not.[1] So Copernicus had the difficult
task of justifying a viewpoint which was opposed to both tradition and common
sense. What made his case even more difficult was the fact that both he and Ptolemy
accepted the basic tenets of Aristotelian physics—specifically that earth (the element)
exhibits a natural linear (not circular) motion. Why, then, would Earth (which is
made of earth) be spinning in a circle? Nonetheless, Copernicus finally concludes
at the end of Chap. 8 that "it is more probably that the Earth moves than that it
is at rest—especially in the case of the daily revolution." (Were you convinced by
Copernicus' arguments?) He then goes on to suggest that Earth may have other kinds
of movements; that Earth may in fact be orbiting the sun like all the other planets; and
that this orbital motion might explain the observed progressions and retrogressions
of the other planets through the zodiac.[2]

Now, in Chap. 10 of Book I, Copernicus begins to lay out detailed arguments in
defense of his heliocentric world-view. After summarizing the opinions of the ancients, he carefully reconsiders what the apparent motions, sizes and orbital periods
of the planets can tell us about their arrangement. Except where marked otherwise,
the generous footnotes which supplement this text were written by the translator,
Charles Glenn Wallis, to explain how Copernicus might have carried out some of his

---

[1] See Chap. 7 of Book I of the *Almagest*, which can be found in Chap. 5 of the present volume.

[2] For a discussion of the observed progression and retrogression of the planets through the zodiac,
see the introduction to Chap. 4 of the present volume. For Ptolemy's explanation of these motions,
see Chap. 7. To witness these motions on your own, do Ex. 3.2.

K. Kuehn, *A Student's Guide Through the Great Physics Texts,*
Undergraduate Lecture Notes in Physics, DOI 10.1007/978-1-4939-1360-2_13,
© Springer Science+Business Media, LLC 2015

calculations.[3] In addition to focusing on the technical aspects of this rather challenging text, you might consider: what role do the ideas of simplicity, beauty and dignity play in Copernicus' world-view? Do you think that these are appropriate criteria for establishing scientific theories? If not, then what are the most appropriate criteria?

13.2   Reading: Copernicus, *On the Revolutions of the Heavenly Spheres*

Copernicus, N., *On the Revolutions of the Heavenly Spheres*, Great Minds, Prometheus Books, Amherst, NY, 1995. Book I.

13.2.1   Chapter 10: On the Order of the Celestial Orbital Circles

I know of no one who doubts that the heavens of the fixed stars is the highest up of all visible things. We see that the ancient philosophers wished to take the order of the planets according to the magnitude of their revolutions, for the reason that among things which are moved with equal speed those which are the more distant seem to be borne along more slowly, as Euclid proves in his *Optics*. And so they think that the moon traverses its circle in the shortest period of time, because being next to the Earth, it revolves in the smallest circle. But they think that Saturn, which completes the longest circuit in the longest period of time, is the highest. Beneath Saturn, Jupiter. After Jupiter, Mars.

There are different opinions about Venus and Mercury, in that they do not have the full range of angular elongations from the sun that the others do.[4] Wherefore some place them above the sun, as Timaeus does in Plato; some, beneath the sun, as Ptolemy and a good many moderns. Alpetragius makes Venus higher than the sun and Mercury lower. Accordingly, as the followers of Plato suppose that all the planets—which are otherwise dark bodies—shine with light received from the sun, they think that if the planets were below the sun, they would on account of their slight distance from the sun be viewed as only half—or at any rate as only partly—spherical. For the light which they receive is reflected by them upward for the most part, *i.e.*, towards the sun, as we see in the case of the new moon or the old. Moreover, they say that necessarily the sun would sometimes be obscured through their interposition and that its light would be eclipsed in proportion to their magnitude; and as that has never

---

[3] Many of Copernicus' calculations are based on the observations contained in the first three chapters of Book IX of Ptolemy's *Almagest*.

[4] The greatest angular elongation of Venus from the sun is approximately 45°; that of Mercury, approximately 24°; while Saturn, Jupiter, and Mars have the full range of possible angular elongation, *i.e.*, up to 180°.

appeared to take place, they think that these planets cannot by any means be below the sun.[5]

On the contrary, those who place Venus and Mercury below the sun claim as a reason the amplitude of the space which they find between the sun and the moon. /8ª/ For they find that the greatest distance between the Earth and the moon, *i.e.*, $64\frac{1}{6}$ units, whereof the radius of the Earth is one, is contained almost 18 times in the least distance between the sun and the Earth. This distance is 1160 such units, and therefore, the distance between the sun and the moon is 1096 such units. And then, in order for such a vast space not to remain empty, they find that the intervals between the perigees and apogees—according to which they reason out the thickness of the spheres[6] —add up to approximately the same sum: in such fashion that the apogee of the moon may be succeeded by the perigee of Mercury, that the apogee of Mercury may be followed by the perigee of Venus, and that finally the apogee of Venus may nearly touch the perigee of the sun. In fact they calculate that the interval between the perigee and the apogee of Mercury contains approximately $177\frac{1}{2}$ of the aforesaid units and that the remaining space is nearly filled by the 910 units of the interval between the perigee and apogee of Venus.[7] Therefore they do not admit that these

---

[5] The transit of Venus across the face of the sun was first observed by means of a telescope—in 1639.

[6] That is to say, the thickness of the sphere would be measured by the ratio of the diameter of the epicycle to the diameter of the sphere, or, in the accompanying diagram, by the distance between the inmost and the outmost of the three homocentric circles.

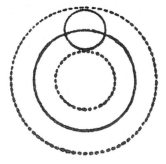

[7] The succession of the orbital circles according to their perigees and apogees may be represented in the following diagram, which has been drawn to scale.

planets have a certain opacity, like that of the moon; but that they shine either by their own proper light or because their entire bodies are impregnated with sunlight, and that accordingly they do not obscure the sun, because it is an extremely rare occurrence for them to be interposed between our sight and the sun, as they usually withdraw (from the sun) latitudinally. In addition, there is the fact that they are small bodies in comparison with the sun, since Venus even though larger than Mercury can cover scarcely one one-hundredth part of the sun, as al-Battani the Harranite maintains, who holds that the diameter of the sun is ten times greater, and therefore it would not be easy to see such a little speck in the midst of such beaming light. Averroes, however, in his paraphrase of Ptolemy records having seen something blackish, when he observed the conjunction of the sun and Mercury which he had computed. And so they judge that these two planets move below the solar circle.

But how uncertain and shaky this reasoning is, is clear from the fact that though the shortest distance of the moon is 38 units whereof the radius of the Earth is one unit—according to Ptolemy, but more than 49 such units by a truer evaluation, as will be shown below—nevertheless we do not know that this great space contains anything except air, or if you prefer, what they call the fiery element.

Moreover, there is the fact that the diameter of the epicycle of Venus—by reason of which Venus has an angular digression of approximately 45° on either side of the sun—would have to be six times greater than the distance from the centre of the Earth to its perigee, as will be shown in the proper place.[8] Then what will they say is contained in all this space which /**8^b**/ is so great as to take in the Earth, air, ether, moon and Mercury, and which moreover the vast epicycle of Venus would occupy if it revolved around an immobile Earth?

Furthermore, how unconvincing is Ptolemy's argument that the sun must occupy the middle position between those planets which have the full range of angular elongation from the sun and those which do not is clear from the fact that the moon's full range of angular elongation proves its falsity.

But what cause will those who place Venus below the sun, and Mercury next, or separate them in some other order—what cause will they allege why these planets do not also make longitudinal circuits separate and independent of the sun, like the

---

[8] According to Ptolemy, the ratio of the radius of Venus' epicycle to the radius of its eccentric is between 2 to 3 and 3 to 4, or approximately $43\frac{1}{6}$ to 60. Now since at perigee the epicycle subtracts from the mean distance, or radius of the eccentric circle, that which at apogee it adds to the mean distance, the ratio of Venus' distance at perigee to its distance at apogee is approximately 1 to 6. That is to say, in the passage from apogee to perigee, the ratio of increase in the apparent magnitude of the planet should be approximately 36 to 1, as the apparent magnitude varies inversely in the ratio of the square of the distance. But no such increase in the magnitude of the planet is apparent. This opposition between an appearance and the consequences of an hypothesis made to save another appearance is still present within Copernicus' own scheme.

other planets[9]—if indeed the ratio of speed or slowness does not falsify their order? Therefore it will be necessary either for the Earth not to be the centre to which the order of the planets and their orbital circles is referred, or for there to be no sure reason for their order and for it not to be apparent why the highest place is due to Saturn rather than to Jupiter or some other planet. Wherefore I judge that what Martianus Capella—who wrote the Encyclopedia—and some other Latins took to be the case is by no means to be despised. For they hold that Venus and Mercury circle around the sun as a centre; and they hold that for this reason Venus and Mercury do not have any farther elongation from the sun than the convexity of their orbital circles permits; for they do not make a circle around the earth as do the others, but have perigee and apogee interchangeable (in the sphere of the fixed stars). Now what do they mean except that the centre of their spheres is around the sun? Thus the orbital circle of Mercury will be enclosed within the orbital circle of Venus—which would have to be more than twice as large—and will find adequate room for itself within that amplitude.[10] Therefore if anyone should take this as an occasion to refer Saturn, Jupiter, and Mars also to this same centre, provided he understands the magnitude of those orbital circles to be such as to comprehend and encircle the Earth remaining

---

[9] Ptolemy makes the centres of the epicycles of Venus and Mercury travel around the Earth longitudinally at the same rate as the mean sun, and in such fashion that the mean sun is always on the straight line extending from the centre of the Earth through the centres of their epicycles, while the centres of the epicycles of the upper planets may be at any angular distance from the mean sun.

[10] As in the following diagram which has been drawn to scale.

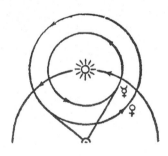

within them, he would not be in error, as the table of ratios of their movements makes clear.[11]

For it is manifest that the planets are always nearer the Earth at the time of their evening rising, *i.e.*, when they are opposite to the sun and the Earth is in the middle between them and the sun. But they are farthest away from the Earth at the time of

---

[11] Take the case of Mars. In Ptolemy, the ratio of its epicycle to its eccentric is $39\frac{1}{2}$ to 60, or approximately 2 to 3. Mars has 37 cycles of anomaly, or movement on the epicycle, and 42 cycles of longitude, or movement of the epicycle on the eccentric, in 79 solar years; or for the sake of easiness let us say that the ratio of the sun's movement to either of the planets' two movements is 2 to 1. Copernicus is here suggesting that if the centre of the planet's movement is placed around the moving sun then the Ptolemaic cycles of anomaly will represent the number of times the sun has overtaken the planet in longitude: thus the 37 cycles of anomaly plus the 42 cycles of longitude add up to the 79 solar revolutions. That is to say, the sun will now be traveling around the Earth on a circle which has the same relative magnitude as the Martian epicycle in Ptolemy and bears an epicycle having the same relative magnitude as Ptolemy's Martian eccentric circle, on which epicycle Mars travels in the opposite direction at half the speed of the sun. Under both hypotheses the appearances from the Earth will be the same as can be seen in the following diagrams.

PTOLEMAIC HYPOTHESIS

Movement of Sun=240°
Movement of Eccentric=120°
Movement of Epicycle=120°

For according to the Ptolemaic hypothesis, let the Earth be at the center of the approximately homocentric circles of the sun, Mars, and the ecliptic. Let the radius of the planet's epicycle be to the radius of the planet's eccentric as 2 to 3. Now, first, let the sun be viewed at the beginning of Leo, and let the planet at the perigee of its epicycle be viewed at the beginning of Aquarius, in opposition to the sun. Next, let the sun move 240° eastwards, to the beginning of Aries; and during the same interval let the epicycle move 120° eastwards to the beginning of Gemini, and the planet

their evening setting, *i.e.*, when they are occulted in the neighbourhood of the sun, namely, when we have the sun between them and the Earth. All that shows clearly enough that their centre is more directly related to the sun and is the same as that to which Venus and Mercury refer their revolutions.[12] But as they all have one common centre, it is necessary that the space left between the convex orbital circle of Venus and the concave orbital circle of Mars should be viewed as an orbital circle/9ª/or sphere homocentric with them in respect to both surfaces and that it should receive the Earth and its satellite the moon and whatever is contained beneath the lunar globe. For we can by no means separate the moon from the Earth, as the moon is incontestably very near to the Earth—especially since we find in this expanse a place for the moon which is proper enough and sufficiently large. Therefore we are not ashamed to maintain that this totality which the moon embraces—and the centre of the Earth too traverse that great orbital circle among the other wandering stars in an

---

120° eastwards on the epicycle. Now the planet will be found to appear in Taurus, about 36° west of the sun.

SEMI-COPERNICAN HYPOTHESIS

Movement of Sun=240°
Movement of Mars=120°

But if according to the semi—Copernican hypothesis, the sun is made to revolve around the Earth on a circle having the same relative magnitude as Mars' Ptolemaic epicycle, while Mars is placed on an epicycle which has the same relative magnitude as its Ptolemaic eccentric and has its centre at the sun; and if the apparent positions of Mars and the sun are first the same as before, and the sun moves 240° eastwards, bearing along the deferent of Mars, while Mars moves 120° westwards on its epicycle; then Mars will once more be found to appear in Taurus, approximately 36° west of the sun.

[12] Copernicus is asking what reason there is why the planets are always found to be at their apogees at the time of conjunction with the sun, and at their perigees at the time of opposition, since according to the Ptolemaic scheme the reverse is also possible—as is evident from the accompanying diagram.

annual revolution around the sun; and that the centre of the world is around the sun. I also say that the sun remains forever immobile and that whatever apparent movement belongs to it can be verified of the mobility of the Earth; that the magnitude of the world is such that, although the distance from the sun to the Earth in relation to whatsoever planetary sphere you please possesses magnitude which is sufficiently manifest in proportion to these dimensions, this distance, as compared with the sphere of the fixed stars, is imperceptible. I find it much more easy to grant that than to unhinge the understanding by an almost infinite multitude of spheres—as those who keep the earth at the centre of the world are forced to do. But we should rather follow the wisdom of nature, which, as it takes very great care not to have produced anything superfluous or useless, often prefers to endow one thing with many effects. And though all these things are difficult, almost inconceivable, and quite contrary to the opinion of the multitude, nevertheless in what follows we will with God's help make them clearer than day at least for those who are not ignorant of the art of mathematics (Fig. 13.1).

Therefore if the first law is still safe—for no one will bring forward a better one than that the magnitude of the orbital circles should be measured by the magnitude of time—then the order of the spheres will follow in this way beginning with the highest: the first and highest of all is the sphere of the fixed stars, which comprehends itself and all things, and is accordingly immovable. In fact it is the place of the universe, *i.e.*, it is that to which the movement and position of all the other stars are referred. For in the deduction of terrestrial movement, we will however give the cause why there are appearances such as to make people believe that even the sphere of the fixed stars somehow moves. Saturn, the first of the wandering stars follows; it completes

---

But if the sun and not the Earth is the centre of the planet's movements, the reason is obvious.

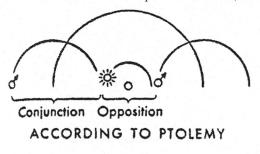

Fig. 13.1 The arrangement of
the sun, the planets and the
sphere of fixed stars based on
the planets' observed orbital
periods.—[*K.K.*]

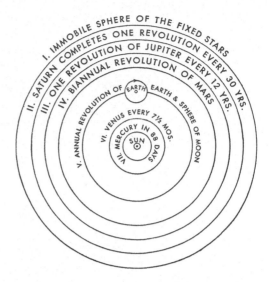

its circuit in 30 years. After it comes Jupiter moving in a 12-year period of revolution.
Then Mars, which completes a revolution every 2 years. The place fourth in order is
occupied by the annual revolution /9**ᵇ**/ in which we said the Earth together with the
orbital circle of the moon as an epicycle is comprehended. In the fifth place, Venus,
which completes its revolution in $7\frac{1}{2}$ months. The sixth and final place is occupied
by Mercury, which completes its revolution in a period of 88 days.[13]

In the center of all rests the sun. For who would place this lamp of a very beautiful
temple in another or better place than this wherefrom it can illuminate everything at
the same time? As a matter of fact, not unhappily do some call it the lantern; others,

---

[13] In order to see how Copernicus derived the length of his periods of revolution, consider the
following Ptolemaic ratios for the lower planets:

	Cycles of anomaly	Cycles of longitude	Solar years
Mercury	145	46+	46+
Venus	5	8–	8–

It is noteworthy that the number of cycles of longitude in one year is equal to the number of solar
cycles. Moreover, the two planets have a limited angular elongation from the sun. In order to explain
these two peculiar appearances Copernicus sets the Earth in motion on the circumference of a circle
which encloses the orbits of Venus and Mercury, with the sun at the centre of all three orbits. Thus
the planet's cycles of anomaly in so many years become the number of times the planet has overtaken
the Earth, as they revolve around the sun. That is to say, in so many solar years the planet will have
traveled around the sun a number of times which is equal to the sum of its cycles of anomaly and
its cycles in longitude. Thus, for example, Venus travels around the sun approximately 13 times
in 8 solar years; hence its period of revolution is approximately $7\frac{1}{2}$ months; and similarly, that of
Mercury is approximately 88 days—although for some obscure reason Copernicus actually writes
down 9 months for Venus (*nono mense reducitur*) and 80 days for Mercury (*octaginta dierum spatio*

circumcurrens). The reader may intuit from the following diagrams the equipollence, with respect to the appearances, of the Ptolemaic and the Copernican explanations of the movement of Venus.

COPERNICAN HYPOTHESIS

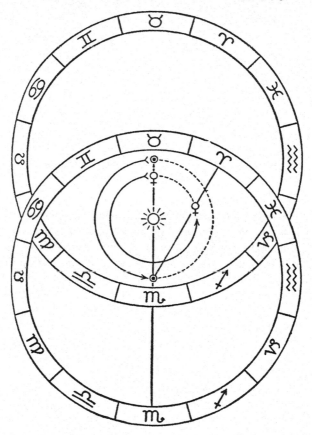

Movement of Earth=180°
Movement of Venus=292½°

Now, on Ptolemy's hypothesis, let the Earth be placed at the centre of the ecliptic, the solar circle, and the orbital circle of Venus, which carries the planetary epicycle. The radius of the epicycle is to that of the orbital circle approximately as 3:4. First let the sun be situated at the middle of Scorpio, and let Venus be in conjunction with the sun and at the perigee of its epicycle. Next let the sun move 180° eastwards to the middle of Taurus, and similarly the centre of the epicycle; during this same interval the planet will move 112½° eastwards on its epicycle and will be found to appear in the middle of Aries approximately, or 30° west of the sun. But according to the Copernican hypothesis, let us place the sun at the centre of the orbital circles of Venus and the Earth, which preserve the relative magnitudes of the Ptolemaic epicycle and orbital circle of Venus, but let us keep the Earth at the centre of the ecliptic, as far as appearances go, since the distance between the Earth and the sun is imperceptible in comparison with the magnitude of the sphere of the fixed stars. Now if the Earth is placed in the middle of Taurus, as viewed from the sun, and the planet at its perigee between the

the mind and still others, the pilot of the world. Trismegistus calls it a "visible god"; Sophocles' *Electra*, "that which gazes upon all things." And so the sun, as if resting on a kingly throne, governs the family of stars which wheel around. Moreover, the Earth is by no means cheated of the services of the moon; but, as Aristotle says in the *De Animalibus*, the earth has the closest kinship with the moon. The Earth moreover is fertilized by the sun and conceives offspring every year.

Earth and the sun, in such fashion that Venus and the sun would appear in the middle of Scorpio, while Venus moves eastwards $292\frac{1}{2}°$, then the sun will be found to appear in the middle of Taurus, and the planet itself in middle of Aries or 30° west of the sun.

But let us turn to the three upper planets.

	Cycles of anomaly	Cycles of longitude	Solar years
Mars	37	42+	79
Jupiter	65	6–	71–
Saturn	57	2+	59–

It is here noteworthy that according to the Ptolemaic hypothesis the sum of the revolutions of the eccentric circle and the revolutions in anomaly is equal to the number of solar cycles; and also that, the conjunctions with the sun take place at the planet's apogee, and the oppositions at its perigee. But according to Copernicus the Ptolemaic cycles of anomaly will now represent the number of times the Earth has overtaken the planet; and the period of revolution in longitude will stay the same. Thus, for example, Saturn will have two revolutions in longitude in 59 years, or one revolution around the sun in about 30 years. The planet will be revolving directly on its eccentric circle instead of on its Ptolemaic epicycle, and the Earth will now be revolving on an inner circle which has the same relative magnitude as the former epicycle. The two hypotheses, of course, are equipollent here too, with respect to appearances. In other words, in constructing a theory to account for four coincidences which were left unexplained by Ptolemy, namely, (1) the equality between the number of cycles in longitude and the solar cycles, in the two lower planets; (2) the equality between the solar cycles and the sum of the cycles of anomaly and longitude, in the upper planets; (3) the limited angular digressions of Mercury and Venus away from the sun; and (4) the apogeal conjunctions and perigeal oppositions of Saturn, Jupiter, and Mars; Copernicus has telescoped the eccentric circle of Venus and that of Mercury into one circle carrying the Earth; and he has furthermore collapsed the three epicycles of Saturn, Jupiter, and Mars into this same one circle. That is to say, one circle is now doing the work of five.

Therefore in this ordering we find /10ª/ that the world has a wonderful commensurability and that there is a sure bond of harmony for the movement and magnitude of the orbital circles such as cannot be found in any other way.[14]

For now the careful observer can note why progression and retrogradation appear greater in Jupiter than in Saturn and smaller than in Mars; and in turn greater in Venus than in Mercury.[15]

---

[14] Let us recall the Ptolemaic ratios between the radius of the epicycle and that of the eccentric circle, and also the eccentricity.

	Epicycle	Eccentric	Eccentricity
Mercury	$22\frac{1}{2}$	60	3
Venus	$43\frac{1}{6}$	60	$1\frac{1}{4}$
Mars	$39\frac{1}{2}$	60	6
Jupiter	$11\frac{1}{2}$	60	$2\frac{2}{5}$
Saturn	$6\frac{1}{2}$	60	$3\frac{1}{4}$

By the Ptolemaic scheme it is impossible to compute the magnitudes of the eccentric circles themselves relative to one another, as there is no common measure. But now that the eccentric circles of Mercury and Venus and the epicycles of Mars, Jupiter, and Saturn have all been reduced to the orbital circle of the Earth, it is easy to calculate the relative magnitudes of the orbital circles—heretofore the epicycles of the lower planets and the eccentric circles of the upper—since, by reason of the necessary commensurability between epicycle and eccentric, they are all commensurable with the orbital circle of the Earth. Thus, for example, if we take the distance from the Earth to the sun as 1, the planets will observe the following approximate distances from the sun.

Mercury	$\frac{1}{3}$	Earth	1	Jupiter	5
Venus	$\frac{3}{4}$	Mars	$1\frac{1}{2}$	Saturn	9

[15] In the three upper planets, the angles which measure the apparent progression and retrogradation have as their vertex the centre of the planet and as their sides the tangents drawn to the orbital circle of the Earth. In the two lower planets, however, the vertex of the angle is at the centre of the Earth and the sides are the tangents drawn to the orbital circle of the planet. It is easy to see that, on account of the relative magnitudes of the orbital circles, the arcs of progression and retrogradation will appear smaller in Saturn than in Jupiter, and smaller in Jupiter than in Mars, and greater in Venus than in Mercury.

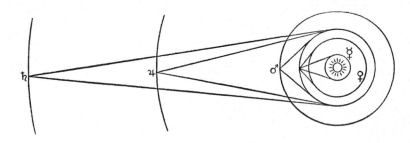

And why these reciprocal events appear more often in Saturn than in Jupiter, and even less often in Mars and Venus than in Mercury[16]. In addition, why when Saturn, Jupiter, and Mars are in opposition (to the mean position of the sun) they are nearer to the Earth than at the time of their occultation and their reappearance. And especially why at the times when Mars is in opposition to the sun, it seems to equal Jupiter in magnitude and to be distinguished from Jupiter only by a reddish color, but when discovered through careful observation by means of a sextant is found with difficulty among the stars of second magnitude?[17] All these things proceed from the same cause, which resides in the movement of the Earth.

But that there are no such appearances among the fixed stars argues that they are at an immense height away, which makes the circle of annual movement or its image disappear from before our eyes since every visible thing has a certain distance beyond which it is no longer seen, as is shown in optics. For the brilliance of their lights shows that there is a very great distance between Saturn the highest of the planets and the sphere of the fixed stars. It is by this mark in particular that they are distinguished from the planets, as it is proper to have the greatest difference between the moved and the unmoved. How exceedingly fine is the godlike work of the Best and Greatest Artist!

13.2.2 Chapter 11: A Demonstration of the Threefold Movement of the Earth

Therefore since so much and such great testimony on the part of the planets is consonant with the mobility of the Earth, we shall now give a summary of its movement, insofar as the appearances can be shown forth by its movement as by an hypothesis. We must allow a threefold movement altogether.

The first—which we said the Greeks called νυχθημέρινος—is the proper circuit of day and night, which goes around the axis of the earth from west to east—as the world is held to move in the opposite direction—and describes the equator or the equinoctial circle—which some, imitating the Greek expression /10ᵇ/ ἰσηέρινος, call the equidial.

---

[16] The interchanges of progression and retrogradation are proportional to the number of times the Earth overtakes the outer planets and the inner planets overtake the Earth. Now the Earth overtakes Saturn more often than Jupiter, Jupiter more often than Mars, Mars more often than overtaken by Venus, and overtaken less often by Venus than by Mercury. Hence the frequency of progression and retrogradation is in that order.

[17] According to the Ptolemaic scheme, it can be inferred only from the changes in magnitude of the planet Mars what its relative distances from the Earth are at perigee and apogee. But according to the Copernican scheme, it follows from the relative distances of the planet at perigee and at apogee—which are as one to five—that the apparent diameter of the planet should vary inversely in that ratio assuming that the planet could be seen when in conjunction with the sun.

Fig. 13.2 The tilt of Earth's
axis as it orbits the
sun.—[*K.K.*]

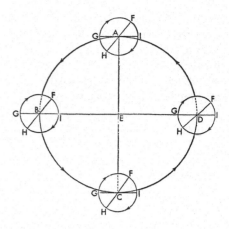

The second is the annual movement of the centre, which describes the circle of the (zodiacal) signs around the sun similarly from west to east, *i.e.*, towards the signs which follow (from Aries to Taurus) and moves along between Venus and Mars, as we said, together with the bodies accompanying it. So it happens that the sun itself seems to traverse the ecliptic with a similar movement. In this way, for example, when the centre of the Earth is traversing Capricorn, the sun seems to be crossing Cancer; and when Aquarius, Leo, and so on, as we were saying.

It has to be understood that the equator and the axis of the Earth have a variable inclination with the circle and the plane of the ecliptic. For if they remained fixed and only followed the movement of the centre simply, no inequality of days and nights would be apparent, but it would always be the summer solstice or the winter solstice or the equinox, or summer or winter, or some other season of the year always remaining the same. There follows then the third movement, which is the declination: it is also an annual revolution but one towards the signs which precede (from Aries to Pisces), or westwards, *i.e.*, turning back counter to the movement of the centre; and as a consequence of these two movements which are nearly equal to one another but in opposite directions, it follows that the axis of the Earth and the greatest of the parallel circles on it, the equator, always look towards approximately the same quarter of the world, just as if they remained immobile. The sun in the meanwhile is seen to move along the oblique ecliptic with that movement with which the centre of the earth moves, just as if the centre of the earth were the centre of the world—provided you remember that the distance between the sun and the earth in comparison with the sphere of the fixed stars is imperceptible to us.

Since these things are such that they need to be presented to sight rather than merely to be talked about, let us draw the circle *ABCD*, which will represent the annual circuit of the centre of the earth in the plane of the ecliptic, and let *E* be the sun around its centre (Fig. 13.2).

I will cut this circle into four equal parts by means of the diameters AEC and BED. Let the point A be the beginning of Cancer; B of Libra; C of Capricorn;[18] and D of Aries. Now let us put the centre of the earth first at A, around which we shall describe the terrestrial equator $FGHI$, but not in the same plane (as the ecliptic) except that the diameter GAI is the common section of the circles, i.e., of the equator and the ecliptic. Also let the diameter FAH be drawn at right angles to GAI; and let F be the limit of the greatest southward declination (of the equator), and H of the northward declination. With this set-up, the Earth-dweller will see the sun—which is at the centre E—at the point of the winter solstice in Capricorn —/11ª/ which is caused by the greatest northward declination at H being turned toward the sun; since the inclination of the equator with respect to line AE describes by means of the daily revolution the winter tropic, which is parallel to the equator at the distance comprehended by the angle of inclination EAH. Now let the centre of the Earth proceed from west to east; and let F, the limit of greatest declination, have just as great a movement from east to west, until at B both of them have traversed quadrants of circles. Meanwhile, on account of the equality of the revolutions, angle EAI will always remain equal to angle AEB; the diameters will always stay parallel to one another—FAH to FBH and GAI to GBI; and the equator will remain parallel to the equator. And by reason of the cause spoken of many times already, these lines will appear in the immensity of the sky as the same. Therefore from the point B the beginning of Libra, E will appear to be in Aries, and the common section of the two circles (of the ecliptic and the equator) will fall upon line $GBIE$, in respect to which the daily revolution has no declination; but every declination will be on one side or the other of this line. And so the sun will be soon in the spring equinox. Let the centre of the Earth advance under the same conditions; and when it has completed /11ᵇ/ a semicircle at C, the sun will appear to be entering Cancer. But since F the southward declination of the equator is now turned toward the sun, the result is that the sun is seen in the north, traversing the summer tropic in accordance with angle of inclination ECF. Again, when F moves on through the third quadrant of the circle, the common section GI will fall on line ED; whence the sun, seen in Libra, will appear to have reached the autumn equinox. But then as, in the same progressive movement, HF gradually turns in the direction of the sun, it will make the situation at the beginning return, which was our point of departure.

In another way: Again in the underlying plane let AEC be both the diameter (of the ecliptic) and its common section with the circle perpendicular to its plane (Fig. 13.3). In this circle let $DGFI$, the meridian passing through the poles of the Earth be described around A and C, in turn, i.e., in Cancer and in Capricorn. And let the axis of the Earth be DF, the north pole D, the south pole F, and GI the diameter of the equator. Therefore when F is turned in the direction of the sun, which is at E, and the inclination of the equator is northward in proportion to angle IAE, then the movement around the axis will describe—with the diameter KL and at the distance LI—parallel to the equator the southern circle, which appears with respect to the

---

[18] Corrected from E to C in English text.—[*K.K.*]

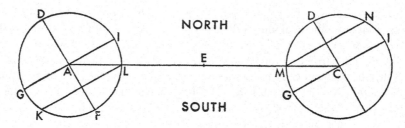

Fig. 13.3 The tilt of Earth's axis as viewed from the plane of the ecliptic.—[*K.K.*]

sun as the tropic of Capricorn. Or—to speak more correctly—this movement around the axis describes, in the direction of *AE*, a conic surface, which has the centre of the earth as its vertex and a circle parallel to the equator as its base.[19]

Moreover in the opposite sign, *C*, the same things take place but conversely. Therefore it is clear how the two mutually opposing movements, *i.e.*, that of the centre and that of the inclination, force the axis of the Earth to remain balanced in the same way and to keep a similar position, and how they make all things appear as if they were movements of the sun.

Now we said that the yearly revolutions of the centre and of the declination were approximately equal, because if they were exactly so, then the points of equinox and solstice and the obliquity of the ecliptic in relation to the sphere of the fixed stars could not change at all. But as the difference is very slight, /12ª/ it is not revealed except as it increases with time: as a matter of fact, from the time of Ptolemy to ours there has been a precession of the equinoxes and solstices of about 21°. For that reason some have believed that the sphere of the fixed stars was moving, and so they choose a ninth higher sphere. And when that was not enough, the moderns added a tenth, but without attaining the end which we hope we shall attain by means of the movement of the Earth. We shall use this movement as a principle and a hypothesis in demonstrating other things.

13.3 Study Questions

QUES. 13.1 What is the correct order of the celestial spheres? How do you know?

a) On what basis did the ancients order the spheres of the moon, sun and planets? Did they all agree?
b) Do planets shine with their own light? What does this have to do with the proposed arrangement of the planets?

---

[19] Or, in other words, the axis of the terrestrial equator describes around the axis of the terrestrial ecliptic a double conic surface having its vertices at the centre of the Earth, in a period of revolution equal approximately to that of the Earth's centre.

c) To what end did Ptolemy propose the existence of epicycles? How large were the proposed epicycles of Mercury and Venus? And what problem do their sizes present?

d) According to Ptolemy, where is the sun (in relation to the five planets)? Why did he place it here? And on what grounds does Copernicus criticize Ptolemy's argument?

e) What alternative was suggested by Martianus Capella? In what sense was Copernicus sympathetic with his view?

f) How can one ascertain when a planet, such as Mars, is at perigee? Also, when Mars is at perigee, are the sun and Mars in conjunction or opposition? How do these observations present a difficulty for the Ptolemaic world-view? Does Copernicus' world-view have the same difficulty?

g) Upon what law does Copernicus rely when deriving the order of the spheres?

h) What are the orbital times of each of the planets? How are these determined? And why do you suppose this is such a difficult problem?

i) In what sense is Copernicus' ordering of the planets simper than that of Ptolemy? Does this make it preferable? Does this make it right?

j) Which planet displays the larger angle of retrogression: Saturn or Jupiter? Jupiter or Mars? Mercury or Venus? How are these observations explained in Copernicus' world-view?

k) Which planet displays the more frequent retrogressions: Saturn or Jupiter? Jupiter or Mars? Mercury or Venus? Do these observations support Copernicus' world-view?

l) What about the variations in apparent size (magnitude) of the planets—does this support his world-view? Do the stars exhibit such a variation? What does this imply?

QUES. 13.2 If the earth really moves, then what type(s) of motion does it execute?

a) What observation suggests that the earth spins? In which direction does it spin?

b) What observation suggest that the earth orbits the sun? In which direction does it go around?

c) What feature of Copernicus' theory accounts for the variability of the length of the day during the year. Would you consider this to be a type of motion, as does Copernicus?

d) What is meant by the "precession of the equinoxes and solstices"? How does Copernicus account for this phenomenon?

13.4 Exercises

Ex. 13.1 (RACETRACK ANALOGY). Three runners race at constant but unequal speeds around a circular track. A circular stadium, filled with 12 equally spaced sections of cheering fans, completely surrounds the outermost lane of the race track (see Fig. 13.4). Runner A is in the innermost lane; she runs each lap in 1 min. Runner B

Fig. 13.4 A three-lane race track surrounded by twelve sections of cheering fans

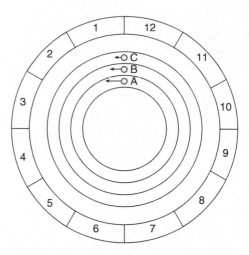

is in the middle lane; he completes each lap in 2 min. Runner C, in the outermost lane, completes each lap in 4 min.

a) From the perspective of runner B, when does runner A appear to be progressing (with respect to the cheering fans)? When does she appear to be retrogressing? And how often do A's retrogressions occur?

b) Again, from the perspective of runner B, when does runner C appear to be progressing? retrogressing? How often do C's retrogressions occur?

c) In this scientific parable, what most closely represents Venus? Earth? Mars? The zodiac?

Ex. 13.2 (PLANETARY DISCOVERY). How can one determine the orbital periods of the planets? This would be an easy task for an observer standing on the sun and watching them go around. But that is not the case for Copernicus who is standing here on Earth. So here is his method. By counting the number of cycles of anomaly (distinct periods of retrogression) that a planet undergoes during a certain number of cycles of longitude (progression through the entire zodiac), Copernicus was able to extract the orbital period of the planet. This makes sense if you grasp the racetrack analogy presented in Ex. 13.1. A cycle of anomaly occurs whenever the planet is lapping Earth (in the cases of Mercury and Venus), or when Earth is lapping the planet (in the cases of Mars, Jupiter and Saturn). The orbital period T (in Earth years) can then be computed from the number of cycles of anomaly, A, and the number of cycles of longitude, L:

$$T = \frac{L}{A + L} \qquad \text{for the inner planets} \qquad (13.1)$$

$$T = \frac{A + L}{L} \qquad \text{for the outer planets} \qquad (13.2)$$

Fig. 13.5 Copernicus'
worldview

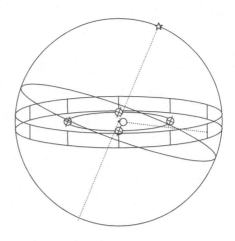

For example, Mars exhibits 37 cycles of anomaly during 42 cycles of longitude.[20]
This gives an orbital period $T = {}^{79}/_{42}$, or just under 2 Earth years. Now suppose
that a new planet, *Hera*, is found to exhibit a maximum angular elongation from
the sun of 180°, as viewed from the earth. Also, suppose that it exhibits five cycles
of anomaly during two cycles of elongation. Using a Copernican world view, what
is the orbital period of the planet Hera? And where does Hera's orbital circle lie in
relation to those of Mercury, Venus, Earth, Mars, Jupiter, and Saturn?

Ex. 13.3 (COPERNICAN WORLDVIEW). Identify the following objects in Fig. 13.5: (a)
the sun, (b) the north star, (c) the celestial equator, (d) Earth at the time of the vernal
equinox, (e) Earth today, and (f) the center of earth's orbit. When looking down
toward the earth from Polaris, does the earth spin clockwise or counterclockwise?
Does the earth orbit the sun clockwise or counterclockwise?

Ex. 13.4 (HELIOCENTRISM AND AESTHETICS ESSAY). Which of Copernicus' arguments
for the mobility of the earth would you consider most convincing? Why? What role
do the ideas of simplicity, beauty and dignity play in Copernicus' world-view? Are
these appropriate criteria for establishing scientific theories? If not, then what are?

Ex. 13.5 (ORBITAL PERIOD AND DISTANCE ESSAY). Speculating, do you think the
relationship between a planet's orbital period and its distance from the sun is a
matter of chance, or is there a necessary relationship between these two quantities?
Is there any other explanation?[21]

---

[20] See the table in the footnotes of the present chapter.

[21] Kepler will address the issue of the ordering of the planets using his *Doctrine of the Schemata*,
described in Part I of Book IV of his *Epitome of Copernican Astronomy*. See Chap. 15 of the present
volume, and especially Ex. 15.3.

13.5   Vocabulary

1. Interposition	13. Reciprocal
2. Perigee	14. Consonant
3. Apogee	15. Equinoctial
4. Impregnate	16. Equidial
5. Epicycle	17. Solstice
6. Manifest	18. Declination
7. Occult	19. Oblique
8. Convex	20. Imperceptible
9. Homocentric	21. Tropic
10. Superfluous	22. Quadrant
11. Comprehend	23. Precession
12. Commensurable	

Chapter 14
Celestial Physics

We knock at the doors of the science of the Magi ...
—Johannes Kepler

14.1 Introduction

Johannes Kepler (1571–1630) was born in the city of Weil der Stadt, Germany. He received his Batchelor of Arts and Master of Arts degrees at the University of Tübingen. After completing almost three additional years of study in theology, he became a teacher of mathematics in Graz, Styria. In 1601 he was named the successor of Tycho Brahe, who was Kepler's mentor, and the imperial mathematician to the Holy Roman Emperor Rudolf II in Prague. Shortly thereafter, he discovered the elliptical orbit of Mars, which provided the experimental foundation for what came to be known as Kepler's first law of planetary motion.[1]

Kepler was very prolific, publishing works on astronomy, optics and even the structure of the snowflake. His most influential work, the *Epitome Astronomiae Copernicanae* was written in Latin and published between 1618 and 1621. It was intended to serve as an astronomy textbook, and is hence organized (perhaps unsurprisingly considering Kepler's extensive theological training) like a catechism—that is—in the form of simple questions and direct answers. In his *Epitome of Copernican Astronomy*, Kepler attempts no less than to provide an alternative to the Aristotelian physics which underpinned astronomical reasoning up until his day. Even Copernicus' worldview, though certainly novel, was still built on Aristotle's notion of perfect spheres executing eternal uniform rotation—a view which had been called into serious question by Tycho Brahe's recent comet observations and Kepler's own discovery of the elliptical orbit of Mars.[2,3]

---

[1] For a more detailed biography of Kepler's life and work, see Voelkel, J. R., *Johannes Kepler and the New Astronomy*, Oxford Portraits in Science, Oxford University Press, New York, Oxford, 1999.

[2] Kepler's discovery of his first two laws of planetary motion were published in 1609. See Donahue, W. H. (Ed.), *Selections from Kepler's Astronomia Nova*, Green Lion Press, Santa Fe, NM, 2004.

[3] Read Brahe's comments on the comet of 1577 in the chapter entitled "Observational evidence against the Aristotelian Cosmology" in Dolling, L. M., Gianelli, A. F., and Statile, G. N. (Eds.), *The Tests of Time: Readings in the Development of Physical Theory*, Princeton University Press, Princeton, NJ, 2003.

K. Kuehn, *A Student's Guide Through the Great Physics Texts,*
Undergraduate Lecture Notes in Physics, DOI 10.1007/978-1-4939-1360-2_14,
© Springer Science+Business Media, LLC 2015

The reading selection in the present chapter contains the dedication and some introductory material which precedes Part I of Book IV of the *Epitome*. Kepler begins by explaining the motivation for his work and defending himself against the charge of "novelty-hunting"—an apparently serious charge by his opponents which was intended to discredit his reputation. In doing so, he outlines how his *Epitome* addresses many of the very same questions which had been raised centuries earlier by Aristotle.

14.2   Reading: Kepler, *Epitome of Copernican Astronomy*

Kepler, J., *Epitome of Copernican Astronomy & Harmonies of the World*, Great Minds, Prometheus Books, Amherst, NY, 1995. Book IV.

> Herein the natural and archetypal causes of *Celestial Physics* that is, of all the magnitudes, movements, and proportions in the heavens are explained and thus the Principle of the Doctrine on the Schemata are demonstrated.

> This book is designed to serve as a supplement to Aristotle's *On the Heavens*.

14.2.1   To the Reader

It has been 10 years since I published my *Commentaries on the Movements of the Planet Mars*. As only a few copies of the book were printed, and as it had so to speak hidden the teaching about celestial causes in thickets of calculations and the rest of the astronomical apparatus, and since the more delicate readers were frightened away by the price of the book too; it seemed to my friends that I should be doing right and fulfilling my responsibilities, if I should write an epitome, wherein a summary of both the physical and astronomical teaching concerning the heavens would be set forth in plain and simple speech and with the boredom of the demonstrations alleviated. I did that before many years had passed. But meanwhile various delays came between the book and publication: the little book itself was not up to date in spots, and, unless I am mistaken, it was also incomplete in the form in which it was given, and even the plan of publication began to totter. For in the "doctrine concerning the sphere"—published before three years were up—I seemed to certain people to be more diffuse in arguing about the diurnal movement or repose of the earth than befitted the form of an epitome. Accordingly I reflected that if the readers had not digested that part, which was however absent from no epitome of astronomy, all the more strange to them would be this Fourth Book, which airs so many new and unthought-of things concerning the whole nature of the heavens—so that you might doubt whether you were doing a part of physics or astronomy, unless you recognized that speculative astronomy is one whole part of physics.

On the other hand, I considered that this was a matter for the sake of my amplifying which and impressing it upon the public, *i.e.*, for the sake of my writing this little book, many men of letters had become my friends: that these speculations could not be omitted, unless I spent my devotion in giving attention to the darkness of a doctrine of schemata which was robbed of its proper principles. At least, necessity—how I wish she were sometimes less importunate!—cut short this disputation: for necessity makes that which cannot be done otherwise seem to be undertaken as if by design. The press groaned and the work on the doctrine of schemata was being struck off, when its lawful godfather, whom I mentioned in the foreword to the Spherical Doctrine, attained his former state, and was sleeping, or perhaps giving up the ghost, and as the liberality of this most eminent patron was paying for the parts of this book, it became necessary for me suddenly to set out and to break off the work. At that same time the printers had reached the end of the Fourth Book and the Frankfort market-day was at hand. I decided that it would be best if the Fourth Book, the subject-matter of which includes both physics and astronomy, were also published separately; whence, according to the choice of the astronomer buying, it could be passed over, or inserted into the rest of the epitome. Kind reader, you have the reasons for this publication, and I hope you will find them satisfactory.

But as regards this branch of philosophizing: it will not be out of keeping with the job at hand if I here set down in advance some things from the recent letter which I wrote to a man who is intimate with a great Prince and is himself also a great man. In this letter a comparison was undertaken between this book—or the related work *On the Harmonies*, published in the previous year—and Aristotle's books *On the Heavens* and *Metaphysics*; and this philosophy (*i.e.*, modern astronomy) was cleared of the worn-out charges of being esoteric and seeking after novelty.

Accordingly, these are the excerpts from the aforesaid letter which have to do with the present undertaking:

It seems to me I have nothing to worry about in the case of Aristotle: His Most Serene Highness is a Platonist in philosophy and a Christian in religion: His Most Serene Highness cannot dislike whatever is the more convincing, whether it be that the world was first made at a fixed beginning in time as was my work *On the Harmonies*, or will be destroyed at some time, or is merely liable to destruction, like the alterations of the ether and the celestial atmosphere; nor will he ever prefer the Master Aristotle to the truth of which Aristotle was ignorant.

But if His Most Serene Highness has a high opinion of Aristotle, wheresoever he reveals the mysteries of philosophy, if he makes any serious remark or any praiseworthy attempt; for indeed he is the man who in *On the Heavens* (Book II, Chap. 5) asks: "For what reason are there many movements?" So I ask: "What are the reasons for the number of the planets?" He asks in the following chapter: "For what reason are the heavens borne from east to west rather than from west to east?" So I ask: "Why is any planet moved with so much speed, no more, no less?" In Chap. 9 he asks: "Do the stars give forth sounds which are modulated [*contemperatos*] harmonically?" and answers no: I split up his judgment, for I grant that no sounds are given forth but I affirm and demonstrate that the movements are modulated according to harmonic proportions. In Chap. 10 he asks "about the order of the spheres, the intervals, and the ratio of the movements to the orbital circles"; but he merely asks and fails in the attempt. Not only do I answer these questions with most luminous demonstrations by means of the five regular solids, but also I add the number of the planets, which has been deduced from the Archetype, so that it may be clear that the world is created. In Chap. 12 he asks: "Why in the descent from the upper to the lower planets are not the movements of the single planets found to be more manifold?" and he pronounces a judgment most elegantly tempered by the modesty of confession and the wisdom of assertion. "Let us try," he says, "to say only that which appears as true; for we judge that the readiness" even to put forward what is probable "is worthy of being characterized as modesty rather than presumption, if anyone, in things concerning which there are very great difficulties, is content—in order to satisfy his thirst for philosophy—with even slight discussions such as these." But I myself, led on by this same praiseworthy thirst for philosophy, first wiped away from the eyes of astronomy those mists of the multiplicity of movements in the single planets: then I gave a demonstration of the following: that the movement of the planet is not uniform throughout its whole circuit—as Aristotle argued in Chaps. 6 and 7; but that in reality the movement is increased and decreased at places in its period which are fixed and are opposite to one another; and I explained the efficient or instrumental causes of this increase as the lessening of the interval between the planet and the sun, from which as from a source that movement arises. Then, as in each and every planet there is a very fast movement and a very slow movement and in a fixed proportion, I did not merely raise the question as to the reason for this proportion in the single planets separately and in all the planets in relation to one another; and why Saturn and Jupiter have middling eccentricities, Mars a great eccentricity, the Sun and Venus slight eccentricities, and Mercury a very great eccentricity; but I also brought forward a solution of this very great difficulty, and not a trifling discussion but one wholly legitimate; and I took my solution from the Archetype of the harmonic cosmos: whence it is established that this cosmos cannot be better than it is and that it is impossible that the world should not have been created at a fixed beginning in time.

This attempt of mine ought not to have been checked by shyness, but should have been brought forth into the light with strength of mind, namely, with the highest confidence in the visible works of God—if one has leisure for knowledge of them—or at the exhortation of Aristotle himself, who judged that in these questions you should not suppress or be silent about probabilities any more than about fully explored certainties. Then he is that same Aristotle who, in the *Metaphysics*, Book XII, Chap. 8, in which place he built up the most sublime part of his philosophy, the part concerning the gods and the number of them; who, I say, sends his students to the astronomers and who defers to the astronomers in respect to their authority and the weight of their testimony; indeed he would never have scorned Tycho Brahe or even myself, if that fatal necessity of the generations had made us contemporaries. For he orders his students "to read through both," that is to say, Eudoxus and Callippus, for the one had corrected the errors of the other; and today that would be to read both Ptolemy and Tycho: "but to follow" not, he says, the more ancient, but "the more accurate." And so, if Aristotle is dear to that most just Prince, I call Aristotle to witness that he has suffered no injury, if the astronomer, using the arguments which modern times have put forward concerning the heavens, has indicated that creatures arose in the heavens and will disappear once more—in opposition to the opinion of him who alleges experience, but experience not sufficiently long.

As regards the academies, they are established in order to regulate the studies of the pupils and are concerned not to have the program of teaching change very often: in such places, because it is a question of the progress of the students, it frequently happens that the things which have to be chosen are not those which are most true but those which are most easy. And by that division in things which makes different people form different judgements, it so happens that certain people are in error contrary to their own opinion. It seems to me that the truth concerning the mutable nature of the heavens can be taught conveniently; but someone else judges that students and teachers equally are thrown into confusion by this doctrine. But it is not without its use in explaining even those parts of the philosophy of Aristotle which are clearly false, as Book VIII of the *Physics* concerning celestial movement and Book II of *On the Heavens* concerning the eternity of the heavens—so that a comparison could be made between the philosophy of the gentiles and the truth of Christian dogma. Accordingly, if certain subtleties which are difficult to grasp should not be laid before beginners, or if they should not be preferred to the accepted and necessary teachings, it does not follow that therefore those things should neither be written nor read privately. You can count few academies in which it is a part of the program to explain the *Metaphysics* of Aristotle: yet Aristotle wrote the *Metaphysics* too, a very useful work in the judgement of the professors on all the faculties. Therefore, in order that no one should consider His Most Serene Highness blameworthy, if he observes the rules of the academies, and if he believes that the honour of the academics—even if they have sinned greatly in judgement—should be defended against presumptuous critics, against untimely quarrellers: so in turn I do not let myself be easily persuaded that this most wise Prince will seek to have all people remain publicly and privately inside the boundaries of academic philosophy; and to have no one labour privately in bringing forward these things, that is to say, in the manifestation of the works of God.

But His Most Supreme Highness will not pick a fight concerning the heavens; for he knows that the philosophers speak of the visible heavens; and Christ of the invisible heavens, or, as the schools say, of the empyrean, or, as the simple Christians take it, of the blessed seats, which no corruption will ever touch: since not Tycho, not I, but Christ Himself pronounces concerning this visible world: "Heaven and Earth shall pass away," and the Psalmist, "they shall grow old like a garment"; and Peter, "They shall be destroyed root and all, and be consumed by burning in the fire." And that will occur in order that the alterations in the heavens should not destroy their eternity, if there should be such an eternity, just as the terrestrial alterations, which are perrennial and return in a circle, destroy the Earth's eternity which was equally believed by Aristotle. But this kind of argument against Aristotle will perhaps seem too contentious. Therefore let us use his own testimony instead; for he is not everywhere consistent: in the *Metaphysics* he attributes movement to the celestial bodies for its own sake and teaches "that they are moved in order that they may be moved"; but in *On the Heavens*, being admonished by the things themselves, he attributes something or other like the terrestrial; something multiplex and turbulent to the stars or rather to their movers, who by means of these mechanisms and movements seek another end outside of the movement itself, and one mover attains this end with more difficulty than another: in this way, as a matter of fact, he adduces the fewness of movements in the moon as witness of the inferior condition of the moon and its closer kinship to the Earth.[4] For he means to say that the celestial bodies which cannot wholly attain the highest end by their own nature do not employ many motions; and that it would have been wholly useless for the Earth to have a movement to attain that end, but that the Earth is absolutely at rest there; that the moon progresses somewhere and stretches out towards that end; that the higher bodies attain the end, but by many movements; and the highest heaven, by the one simple movement. And so he compares the actions, the πράξεις of the moon—that is the word he uses—to the uniform life of plants, but the πράξεις of the higher bodies, to the more varied life of animals. Yet he makes all those bodies to be in need of these actions because they have their end and their blessedness outside of themselves. Accordingly, in the epilogue to the Fifth Book of the *Harmonies*, I wish for Aristotle as my reader and critic; as it is not right that I should wish to take up any more of the time of His Most Serene Highness, the highest judgement of the Prince. I am sure of one thing at least, that if he would direct the cultivated power of his mind toward those things which Aristotle wrote and toward my epilogue, everything would be agreed between us, and he would by his own judgement harmonize the discord which now, as you predict, he might feel between us.

In order to counter the envious charge of novelty-hunting, it would be first in my program, even though His Most Serene Highness can easily see all things for himself, to warn him fully of the distinction between the love—or thirst, to use the Aristotelian word—for the knowledge of natural things and the lust for contradicting and holding the opposite opinion. All philosophers, whether Greek or Latin, and all the poets too, recognize a divine ravishment in investigating the works of God: and not merely in investigating them privately but even in teaching them publicly: and it can be inferred that the false charge of esoteric novelty-hunting cannot cling to this ravishment.

> There is God in us, and our warmth comes from His movements: This Spirit has descended from the heavenly seats.

There is no need of this declamation before you, or before His Most Supreme Highness: only I must make some further mention of the boundary posts. For the boundary posts of investigation should not be set up in the narrow minds of a few men. "The world is a petty thing, unless everyone finds the whole world in that which he is seeking," as Seneca says. But the boundary posts of true speculation are the same as those of the fabric of the world; but the Christian religion has put up some fences around false speculation which is on the wrong track, in order that error may not rush headlong but may become in other respects harmless in itself. Antiquity teaches us by examples how vainly man sets up boundary posts where God has not set them up: how severely all the astronomers were blamed by the first Christians. Did not Eusebius write of an astronomer that he preferred to desert Christianity—I suppose

---

[4] See Chap. 12 of Book II of Aristotle's *On the Heavens*, included in Chap. 3 of the present volume—[K.K.]

because he was excommunicated—rather than his profession? Who today would opine that Eusebius is to be imitated? Did not those who taught that there were antipodes seem to Tertullian and to Augustine to be overwise? And, indeed, there was a Virgil Bishop of Salisbury who was removed from his office because he dared to assert this same fact. How many times were the Roman philosophers exiled from the city? And at that, under the ancient manners, wherewith the Roman State was established. Yet today we set up academies everywhere: we order that philosophy be taught, that astronomy be taught, that the antipodes be taught.

But I even in private free myself from the blame of seeking after novelty by suitable proofs: let my doctrines say whether there is love of truth in me or love of glory: for most of the ones I hold have been taken from other writers: I build my whole astronomy upon Copernicus' hypotheses concerning the world, upon the observations of Tycho Brahe, and lastly upon the Englishman, William Gilbert's philosophy of magnetism. If I rejoiced in novelty, I could have devised something like the Fracastorian or Patrician systems. Just as one who rejoices in occupations but rarely in companions, never of himself descends to dice or to a game of chess; similarly for me there is so much importance in the true doctrine of others or even in correcting the doctrines which are not in every respect well established, that my mind is never at leisure for the game of inventing new doctrines that are contrary to the true. Whatever I profess outwardly, that I believe inwardly: nothing is a worse cross for me than—I do not say, to speak what is contrary to my thought—to be unable to utter my inmost sentiments. I know that many innovators are produced by the same affect; but they are easily argued out of the error which seduces them. No one shows that I have committed an error. But because certain people cannot grasp the subtleties of things, they lay the charge of novelty-hunting upon me.

I now descend to the work itself, the *Harmonies*: I do not doubt that he who condemns the itch to devise new things and the presumption to profess new and grandiose things will find in the epilogue to the Fifth Book that which he will mark critically. For here the sun-spots and little flames are brought forward as evidence of there being exhalations from the sun which are analogous to exhalations from the Earth: here things corresponding to the generation of animals are established as occurring in the planets—here the confines of the mysteries of Christian religion are touched: we knock at the doors of the science of the Magi, of theurgy, of the idolatry of the Persians, and of those who worship the sun as god—as the interjection of frequent warnings does not dissimulate.

Accordingly, if what has been said so far concerning these esoteric things is not satisfactory: at any rate let this be impressed upon His Most Serene Highness: that this chapter contributes nothing in its own right except conjectures; and although it adds a good deal to the form of the work: because—as the opening of the chapter has it—reason itself leads "from the Muses to Apollo": nevertheless, since the other parts of the work are established by means of their proper demonstrations, the chapter, or epilogue, can be considered as cut off from the rest. For even without the epilogue, the following thesis is upheld by incontrovertible demonstrations: *that in the farthest movements of any two planets, the universe was stamped with the adornment of harmonic proportions; and, accordingly, in order that this adornment might be brought into concord with the movements, the eccentricities which fell to the lot of each planet had to be brought into concord.* The most wise Prince will easily reckon how great an addition this makes in illustrating the glory of the fabric of the world, and of God the Architect.

But if, however, even this inquiry is accused of being esoteric: I indeed confess that the head of astronomy is struck off. And since astronomy is studied either for its own sake as a philosophy or for the sake of making astronomical predictions; then, if I am to cast my ballot in the question of future contingencies, His Most Serene Highness repudiates any secondary end for this exact and subtle investigation of physical causes which does not offer itself for the uses of daily life: therefore the taking away from me of the primary end slays this whole subtle astronomy and plainly makes it useless.

Nevertheless, in order that I may arm myself against this eventuality also: I will grant that this work of mine, the *Harmonies*, is nothing except as it were a certain picture of the edifice of astronomy; and though it may be erased at the pleasure of him who spits upon it, nevertheless the house called astronomy stands by itself: and I know that astronomy is not condemned by His Most Serene Highness but is held of great value on account of its certitude in predicting movements: perhaps, therefore, he will judge its architect—who is almost the only renovator after the Master Tycho and who thought it worth while to devote his life to this work—to be not unworthy of his favour.

These extracts from the letter, most of which have to do with the investigation of very hidden causes which is to be viewed in this little book, should be spoken and understood. And now it is time for the reader to pass on to the little book.

14.2.2   First Book on the Doctrine of the Schemata: On the Position, Order, and Movement of the Parts of the World; or, on the System of the World

/433/ What is the subject of the doctrine of the schemata?

The proper movements of the planets; we call them the secondary movements; and the planets, the secondary movables.

Why do you call them the proper movements of the planets?

1. Because the apparent daily movement—with which the doctrine on the sphere is concerned—and which is common to both the planets and the fixed stars, and so to the whole world, is seen to travel from the east to the west; but the far slower single movements of the single planets travel in the opposite direction from west to east; and therefore it is certain that these movements cannot depend upon that common movement of the world—which we have discussed so far—but should be assigned to the planets themselves, and thus they are generically proper to the planets.

2. But even if in these proper movements of the single /434/ planets from west to east there is also present something common, not diurnal but annual, which is extrinsic and betrays that its cause lies in eyesight alone, outside the truth of the thing; and which meanwhile makes the planet in its proper movement have the appearance of retrograding, that is, from east to west, nevertheless because this common movement is so woven into the single periods of the single planets, and so variously transformed, that at first glance you cannot discern what is common to all the planets and what is proper to each: accordingly this whole composite movement of each planet, as it meets the eyes, is said to be proper to each planet specifically; especially since this movement which is common to many does not have its origin in that first common movement of the whole world, but in the proper movement of each planet.

How many parts are there to the doctrine of the schemata?

Above (in Book I, p. 15), the whole doctrine was divided into its three proper parts: the first, concerning the principles wherewith Copernicus demonstrates the secondary movements—the material of Book IV; the second, concerning the machinery whereby these movements are laid before the eyes, *viz.*, concerning the eccentric and similar circles—the material of Book V; and the third, concerning the apparent movements of the single planets and the common accidents of the planets taken together—the material of Book VI; and the fourth part, which is common to the doctrines on the sphere and on the schemata, concerns the apparent movement of the eighth sphere—the material of Book VII.

What are the hypotheses or principles wherewith Copernican astronomy saves the appearances in the proper movements of the planets?

They are principally: (1) that the sun is located at the centre of the sphere of the fixed stars—or approximately at the centre—and is immovable in place; (2) that the single planets move really around the sun in their single systems, which are compounded of many perfect circles /435/ revolved in an absolutely uniform movement; (3) that the Earth is one of the planets, so that by its mean annual movement around the sun it describes its orbital circle between the orbital circles of Mars and of Venus; (4) that the ratio of its orbital circle to the diameter of the sphere of the fixed stars is imperceptible to sense and therefore, as it were, exceeds measurements; (5) that the sphere of the moon is arranged around the Earth as its centre, so that the annual movement around the sun—and so the movement from place to place—is common to the whole sphere of the moon and to the Earth.

Do you judge that these principles should be held to in this Epitome?

Since astronomy has two ends, to save the appearances and to contemplate the true form of the edifice of the world—of which I have treated in Book I, folia 4 and 5—there is no need of all these principles in order to attain the first end: but some can be changed and others can be omitted; however, the second principle must necessarily be corrected: and even though most of these principles are necessary for the second end, nevertheless they are not yet sufficient.

Which of these principles can be changed or omitted and the appearances still be saved?

Tycho Brahe demonstrates the appearances with the first and third principles changed: for he, like the ancients, places the Earth immobile, at the centre of the world; but the sun—which even for him is the centre of the orbital circles of the five planets—and the system of all the spheres he makes to go around the Earth in the common annual movement, while at the same time in this common system any planet completes its proper movements. Moreover, he omits the fourth principle altogether and exhibits the sphere of the fixed stars as not much greater than the sphere of Saturn.

/436/ What in turn do you substitute for the second principle and what else do you add to the true form of the dwelling of the world or to what belongs to the nature of the heavens?

Even though the true movements are to be left singly to the single planets, nevertheless these movements do not move by themselves nor by the revolutions of spheres—for there are no solid spheres—but the sun in the centre of the world, revolving around the centre of its body and around its axis, by this revolution becomes the cause of the single planets going around.

Further, even though the planets are really eccentric to the centre of the sun: nevertheless there are no other smaller circles called epicycles, which by their revolution vary the intervals between the planet and the sun; but the bodies themselves of the planets, by an inborn force [*vi insite*], furnish the occasion for this variation.

What, then, will the material of Book IV be?

Book IV will contain celestial physics itself, or the form and proportions of the fabric of the world and the true causes of the movements. This will be the primary function of the astronomer—as we said in Book I, folium 5, namely, the demonstration of his hypotheses.

Review the principal parts of Book IV.

There will be three principal parts of Book IV.

The first is on the bodies themselves; the second, on the movements of those bodies; the third, on the real accidents of the movements.

For the first part will teach the conformation of the whole universe, its division into parts or principal regions; the place of the sun at its centre; the number, magnitude, and order or position of the planetary spheres; and lastly, the ratios of all the bodies of the world to one another.

The second part will teach the revolution of the sun around its axis, and its effect in making the planets revolve; the causes of the proportionality of the movements among themselves, *i.e.*, of the periodic /437/ times; the immobility of the centre of the sun and the annual movement of the centre of the Earth around the sun; the revolution of the Earth around its axis and its effect in making the moon revolve; the additional help in moving the moon given by the light of the sun; and what the causes of the proportions between the day, month, and year are.

The third part will disclose the causes of the threefold irregularity of the altitude, longitude, and latitude in the single planets—and how these irregularities are doubled in the moon by the force of the illumination from the sun.

14.3   Study Guide

QUES. 14.1 How does Kepler defend himself against the charge of novelty-hunting?

a) In what sense is Kepler's work similar to Aristotle's *On the Heavens* and his *Metaphysics*? In what notable ways does it deviate from (or perhaps supercede) Aristotle's? Does Kepler believe that Aristotle, given the opportunity, would have approved of his own doctrines?
b) Upon what types of doctrines do the academies tend to focus? Does Kepler himself believe that they should teach new, or perhaps difficult, doctrines? If not, then where should such doctrines be taught (if at all)?
c) What distinction does Kepler draw when speaking of "the heavens"? More specifically, does Kepler believe that the Christian scriptures should be employed against Aristotle's doctrines?
d) What, according to Kepler, motivates those who charge him with "novelty-hunting"? What more noble motivation does Kepler himself claim?
e) Are there any limitations which should be placed on the investigation of nature? If so, then by what (or who) should they be established? What is the track record, so to speak, of the establishment of boundaries on the investigation of nature?
f) Upon whose work does Kepler construct his astronomical theories? Why does he mention this?
g) What, according to Kepler, are the two purposes of studying astronomy? Which does he value more highly?

QUES. 14.2 What are the hypotheses or principles wherewith Copernican astronomy saves the appearances in the proper movements of the planets?

a) What type of movement is common to all planets? What type of movement is proper to each? And what does it mean for a hypothesis to *save the appearances*?
b) What are the five hypotheses of Copernicus? Which of these does Kepler accept? Which does he reject or modify? And with what does he replace these?

14.4   Exercises

EX. 14.1 (SPECULATIVE ASTRONOMY ESSAY). What is the significance of Kepler's claim that "speculative astronomy is one whole part of physics"? Do you believe that the discipline of physics (or natural science, for that matter) should have well-defined and articulated boundaries? If so, what should they be?

14.5   Vocabulary

1. Celestial	26. Presumptuous
2. Epitome	27. Terrestrial
3. Alleviate	28. Contentious
4. Diurnal	29. Adduce
5. Speculative	30. Epilogue
6. Importunate	31. Ravishment
7. Disputation	32. Opine
8. Eminent	33. Antipode
9. Serene	34. Overwise
10. Ether	35. Grandiose
11. Modulate	36. Dissimulate
12. Luminous	37. Incontrovertible
13. Deduce	38. Adornment
14. Archetype	39. Esoteric
15. Manifold	40. Contingency
16. Assertion	41. Repudiate
17. Multiplicity	42. Edifice
18. Eccentricity	43. Certitude
19. Trifling	44. Annual
20. Legitimate	45. Extrinsic
21. Cosmos	46. Retrograde
22. Exhortation	47. Accident
23. Sublime	48. Imperceptible
24. Mutable	49. Folia
25. Dogma	50. Epicycle

Chapter 15
Broken Spheres

> *For if the spheres were solid, the comets would not be seen to cross from one sphere into another.*
>
> —Johannes Kepler

15.1 Introduction

In the introduction to Book IV of his *Epitome of Copernican Astronomy*, Kepler explained the motivation of his work and how it may be charitably understood as a reaction to (and extension of) Aristotle's famous astronomy book, *On the Heavens*. In particular, Kepler defended his own work, saying that he was not motivated by a lust for glory and novelty—as some had charged—but rather by the love of knowledge and truth. He also outlined which aspects of Copernicus' heliocentric theory he accepted and which he rejected (or modified). Now, in Part I of Book IV, Kepler begins to articulate his understanding of the universe. Straightaway, he questions the Aristotelian division of the universe into two distinct regions—the *heavens* and the *earth*—which are marked off by a sphere which is centered on the earth and which extends out to the moon's orbit. The dissolution of this division is very significant in that it allows Kepler to appeal to *physics* in understanding *astronomy*. Prior to Kepler, physics was understood to be the study of the ever-changing properties of sub-lunary things, while astronomy (as a branch of mathematics) was understood to be the study of the immutable, ethereal and divine supra-lunary bodies.[1] In contrast, Kepler will here attempt to articulate the physical (rather than mathematical) principles which govern the motion of the heavenly bodies. Perhaps the most striking feature of the following reading selection is Kepler's allegorical use of Christian doctrine in understanding the form and the arrangement of the universe. While this approach may seem alien to many modern readers, one must keep in mind what Kepler is trying to do: he is not attempting to *reject* causal explanations of the observed order of the World altogether; rather, he is attempting to *replace* Aristotle's causal explanations—which he found wanting—with something more appropriate. So as you read the following text, you might simply ask yourself whether you find Kepler's explanations to be more (or less) reasonable than those of Aristotle. Is there perhaps another explanation which is better than either?

---

[1] See, for example, Ptolemy's Introduction to Book I of his *Almagest*, included in Chap. 5 of the present volume.

K. Kuehn, *A Student's Guide Through the Great Physics Texts*, Undergraduate Lecture Notes in Physics, DOI 10.1007/978-1-4939-1360-2_15, © Springer Science+Business Media, LLC 2015

Fig. 15.1 Kepler's principal
parts of the world: the center,
the surface and the
intermediate region.—[*K.K.*]

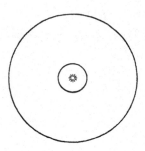

15.2   Reading: Kepler, *Epitome of Copernican Astronomy*

Kepler, J., *Epitome of Copernican Astronomy & Harmonies of the World*, Great
Minds, Prometheus Books, Amherst, NY, 1995. Book IV, Part I.

15.2.1   Chapter 1: On the Principal Parts of the World

/438/ What do you judge to be the lay-out of the principal parts of the world?

The Philosophy of Copernicus reckons up the principal parts of the world by dividing
the figure of the world into regions. For in the sphere, which is the image of God the
Creator and the Archetype of the world—as was proved in Book I—there are three
regions, symbols of the three persons of the Holy Trinity—the centre, a symbol of
the Father; the surface of the Son; and the intermediate space, of the Holy Ghost. So,
too, just as many principal parts of the world have been made—the different parts in
the different regions of the sphere (Fig. 15.1): the sun in the centre, the sphere of the
fixed stars on the surface, and lastly the planetary system in the region intermediate
between the sun and the fixed stars.

*I thought the principal parts of the world are reckoned to be the heavens and the
earth?*

Of course, our uncultivated eyesight from the Earth cannot show us any other more
notable parts—as was said in Book I, folia 8, 9, 10—since we tread upon the one with
our feet and are roofed over by the other, and since both parts seem to be commingled
and cemented together in the common limbo of the horizon—like a globe in which
stars, clouds, birds, man, and the various kinds of terrestrial animals are enclosed.

But we are practised in the discipline which discloses the causes of things, shakes
off the deceptions of eyesight, and carries the mind higher and farther, outside of
the boundaries of eyesight. Hence it should not be surprising to anyone that eyesight
should learn from reason, that the pupil should learn something new from his master
which he did not know before—namely, that the Earth, considered alone and by
itself, should not be reckoned among the primary parts of the great world but should
be added to one of the primary parts, *i.e.*, to the planetary region, the movable world,

and that the Earth has the proportionality of a beginning in that part; and that the sun in turn should be separated from the number of stars and set up as one of the principal parts of the whole universe. But I am speaking now of the Earth in so far as it is a part of the edifice of the world, and not of the dignity of the governing creatures which inhabit it.

By what properties do you distinguish these members of the great world from one another?

The perfection of the world consists in light, heat, movement, and the harmony of movements. These are analogous to the faculties of the soul: light, to the sensitive; heat, to the vital and the natural; movement, to the animal; harmony, to the rational. And indeed the adornment [*ornatus*] of the world consists in light; its life and growth, in heat; and, so to speak, its action, in movement; and its contemplation—wherein Aristotle places blessedness—in harmonies. Now since three things necessarily come together for every affection, namely, the cause *a qua*, the subject *in quo*, and the form *sub qua*—therefore, in respect to all the aforesaid affections of the world, the sun exercises the function of the efficient cause; the region of the fixed stars that of the thing forming, containing, and terminating; and the intermediate space, that of the subject—in accordance with the nature of each affection. Accordingly, in all these ways the sun is the principal body of the whole world.

For as regards light: since the sun is very beautiful with light and is as if the eye of the world, like a source of light or very brilliant torch, the sun illuminates, paints, and adorns the bodies of the rest of the world; the intermediate space is not itself light-giving, but light-filled and transparent and the channel through which light is conducted from its source, and there exist in this region the globes and the creatures upon which the light of the sun is poured and which make use of this light. The sphere of the fixed stars plays the role of the river-bed in which this river of light runs, and is as it were an opaque and illuminated wall, reflecting and doubling the light of the sun: you have very properly likened it to a lantern, which shuts out the winds.

Thus in animals the cerebrum, the seat of the sensitive faculty imparts to the whole animal all its senses, and by the act of common sense causes the presence of all those senses as if arousing them and ordering them to keep watch. And in another way, in this simile, the sun is the image of common sense; the globes in the intermediate space of /**440**/ the sense-organs; and the sphere of the fixed stars of the sensible objects.

As regards heat: the sun is the fireplace [*focus*] of the world; the globes in the intermediate space warm themselves at this fireplace, and the sphere of the fixed stars keeps the heat from flowing out, like a wall of the world, or a skin or garment—to use the metaphor of the Psalm of David. The sun is fire, as the Pythagoreans said, or a red-hot stone or mass, as Democritus said—and the sphere of the fixed stars is ice, or a crystalline sphere, comparatively speaking. But if there is a certain vegetative faculty not only in terrestrial creatures but also in the whole ether throughout the universal amplitude of the world—and both the manifest energy of the sun in warming and physical considerations concerning the origin of comets lead us to draw this inference—it is believable that this faculty is rooted in the sun as in the heart of the world, and that thence by the oarage of light and heat it spreads out into this most wide space of the world—in the way that in animals the seat of heat and of the vital faculty is in the heart and the seat of the vegetative faculty in the liver, whence these faculties by the intermingling of the spirits spread out into the remaining members of the body. The sphere of the fixed stars, situated diametrically opposite on every side, helps this vegetative faculty by concentrating heat, as they say; as it were a kind of skin of the world.

As regards movement: the sun is the first cause of the movement of the planets and the first mover of the universe, even by reason of its own body. In the intermediate space the movables, *i.e.*, the globes of the planets, are laid out. The region of the fixed stars supplies the movables with a place and a base upon which the movables are, as it were, supported; and movement is understood as taking place relative to its absolute immobility. So in animals the cerebellum is the seat of the motor faculty, and the body and its members are that which is moved. The Earth is the base of an animal body; the body, the base of the arm or head, and the arm, the base of the finger. And the movement of each part takes place upon this base as upon something immovable.

Finally, as regards the harmony of the movements: the sun occupies that place in which alone the movements of the planets /**441**/ give the appearance of magnitudes harmonically proportioned [*contemperatarum*]. The planets themselves, moving in the intermediate space, exhibit the subject or terms, wherein the harmonies are found; the sphere of the fixed stars, or the circle of the zodiac, exhibits the measures whereby the magnitude of the apparent movements is known. So too in man there is the intellect, which abstracts universals and forms numbers and proportions, as things which are not outside of intellect; but individuals [*individual*], received inwardly through the senses are the foundation of universals; and indivisible [*individuae*] and discrete unities, of numbers; and real terms of proportions. Finally, memory, divided as it were into compartments of quantities and times, like the sphere of the fixed stars, is the storehouse and repository of sensations. And further, there is never judgment of sensations except in the cerebrum; and the effect of joy never arises from a sense-perception except in the heart.

Accordingly, the aforesaid vegetating corresponds to the nutritive faculty of animals and plants; heating corresponds to the vital faculty; movement, to the animal faculty; light, to the sensitive; and harmony, to the rational. Wherefore most rightly is the sun held to be the heart of the world and the seat of reason and life, and the principal one among three primary members of the world; and these praises are true in the philosophic sense, since the poets honour the sun as the king of the stars, but the Sidonians, Chaldees, and Persians—by an idiom of language observed in German too—as the queen of the heavens, and the Platonists, as the king of intellectual fire.

These three members of the world do not seem to correspond with sufficient neatness to the three regions of a sphere: for the centre is a point, but the sun is a body; and the outer surface is understood to be continuous, yet the region of fixed stars does not shine as a totality, but is everywhere sown with shining points discrete from one another; and finally, the intermediate part in a sphere fills the whole expanse, but in the world the space between the sun /442/ and the fixed stars is not seen to be set in motion as a whole.

As a matter of fact, the question indicates the neatest answer concerning the three parts of the world. For since a point could not be clothed or expressed except by some body—and thus the body which is in the centre would fail of the indivisibility of the centre—it was proper that the sphere of the fixed stars should fail of the continuity of a spherical surface, and should burst open in the very minute points of the innumerable fixed stars; and that finally the middle space should not be wholly occupied by movement and the other affections, nor be completely transparent, but slightly more dense, since it could not be altogether empty but had to be filled by some body.

Are there solid spheres [orbes] *whereon the planets are carried? And are there empty spaces between the spheres?*

Tycho Brahe disproved the solidity of the spheres by three reasons: the first from the movement of comets; the second from the fact that light is not refracted; the third from the ratio of the spheres.

For if spheres were solid, the comets would not be seen to cross from one sphere into another, for they would be prevented by the solidity; but they cross from one sphere into another, as Brahe shows.

From light thus: since the spheres are eccentric, and since the Earth and its Surface—where the eye is—are not situated at the center of each sphere; therefore if the spheres were solid, that is to say far more dense than that very limpid ether, then the rays of the stars would be refracted before they reached our air, as optics teaches; and so the planet would appear irregularly and in places far different from those which could be predicted by the astronomer.

The third reason comes from the principles of Brahe himself; for they bear witness, as do the Copernican, that Mars is sometimes nearer the Earth than the sun is. But Brahe could not believe this interchange to be possible /**443**/ if the spheres were solid, since the sphere of Mars would have to intersect the sphere of the sun.

Then what is there in the planetary regions besides the planets?

Nothing except the ether which is common to the spheres and to the intervals: it is very limpid and yields to the movable bodies no less readily than it yields to the lights of the sun and stars, so that the lights can come down to us.

If it is ether, then it will be a material body having density. Therefore will not its matter resist the movable bodies somewhat?

On the contrary, the ether is more rarefied than our air, since it is very pure, being spread over a space which is practically immense.

How do you prove this?

In optics, by refractions. For our air, which is contiguous to the ether, causes a refraction of approximately 30′. But water contiguous to air causes a refraction of approximately 48°, whence the ratio of the density of water to air, and of air to ether is somehow established by taking the cubes of the numbers. For 30′ is contained approximately 100 times in 48°; and in squares, that is 10,000 times, and in cubes 1,000,000 times. Therefore air is that many times more rarefied than water, and ether than air.

Nevertheless the matter of the ether is not absolutely null: are the stars therefore still impeded by it?

We can without any inconvenience grant such a small impediment of movement and such a small resistance of the ether to the movable bodies, just as even before this it must be granted that they offer some resistance on account of the proper matter of their bodies, as will be made clear below. And what if no resistance should be granted to the ether, /**444**/ since it is fairly credible that the ether which surrounds the movable globe the most closely accompanies the globe on account of the very great limpidity [of the ether]?

15.2.2   Chapter 2: On the Place of the Sun at the Centre of the World

By what arguments do you affirm that the sun is situated at the centre of the world?

The very ancient Pythagoreans and the Italian philosophers supply us with some of those arguments in Aristotle (*On the Heavens*, Book II, Chap. 13); and these arguments are drawn from the dignity of the sun and that of the place, and from the sun's office of vivification and illumination in the world.

State the first argument from dignity.

This is the reasoning of the Pythagoreans according to Aristotle: the more worthy place is due to the most worthy and most precious body. Now the sun—for which they used the word "fire," as sects purposely hiding their teachings—is worthier than the Earth and is the most worthy and most precious body in the whole world, as was shown a little before. But the surface and centre, or midpoint, are the two extremities of a sphere. Therefore one of these places is due to the sun. But not the surface; for that which is the principal body in the whole world should watch over all the bodies; but the centre is suited for this function, and so they used to call it the Watchtower of Jupiter. And so it is not proper that the Earth should be in the middle. For this place belongs to the sun, while the Earth is borne around the centre of its yearly movement.

What answer does Aristotle make to this argument?

1. He says that they assume something which is not granted, namely, that the centre /**445**/ of magnitude, *i.e.*, of the sphere, and the centre of the things, *i.e.*, of the body of the world, and so of nature, *i.e.*, of informing or vivifying, are the same. But just as in animals the centre of vivification and the centre of the body are not the same—for the heart is inside but is not equally distant from the surface—we should think in the same way about the heavens, and we should not fear for the safety of the whole universe or place a guard at the centre; rather, we should ask what sort of body the heart of the world of the centre or vivification is and in what place in the world it is situated.

2. He tries to show the dissimilarity between the midpart of the nature and the midpart of place. For the midpart of nature, or the most worthy and precious body, has the proportionality of a beginning. But in the midpart of place is the last, in quantity considered metaphysically, rather than the first or the beginning. For that which is the midpart of quantity, *i.e.*, is the farthest in, is bounded or circumscribed. But the limits are that which bounds or circumscribes. Now that which goes around on the outside, and limits and encloses, is of greater excellence and worth than that which is on the inside and is bounded: for matter is among those things which are bounded, limited, and contained; but form, or the essence of any creature, is of the number of those things which limit, circumscribe, and comprehend. He thinks that he has proved in this way that not so much the midpart of the world as the extremity belongs to the sun, or as he understood it, to the fire of the Pythagoreans.

How do you rebut this refutation of Aristotle's?

1. Even if it be true that not in all creatures and least in animals is the principal part of the whole creature at the centre of the whole mass: however, since we are arguing about the world, nothing is more probable than this. For the figure of the world is spherical, and that of an animal is not. For animals need organs extending outside themselves, with which they stand upon the ground, and upon which they may move, and with which they may take within themselves the food, drink, /446/ forms of things, and sounds received from outside. The world on the contrary, is alone, having nothing outside, resting on itself immobile as a whole; and it alone is all things. And so there is no reason why the heart of the world should be elsewhere than in the centre in order that what it is, *viz.*, the heart, might be equally distant from all the farthest parts of the world, that is to say, by an interval everywhere equal.

2. Furthermore, as regards his telling us to ask what sort of body the principal part of the whole universe is: he is confused by that riddle of the Pythagoreans and believes that they claim that this element is principal. He is not wrong however in telling us to do that. And accordingly we, following the advice of Aristotle, have picked out the sun; and neither the Pythagoreans in their mystical sense nor Aristotle himself are against us. And when we ask in what place in the world the sun is situated, Copernicus, as being skilled in the knowledge of the heavens, shows us that the sun is in the midpart. The others who exhibit its place as elsewhere are not forced to do this by astronomical arguments but by certain others of a metaphysical character drawn from the consideration of the Earth and its place. Both we and they set a value upon these arguments; and they themselves too by means of these arguments do not show but seek the place of the sun. So if when seeking the place of the sun in the world, we find that it is the centre of the world; we are doing just as Aristotle; and his refutation does not apply to us.

3. As regards the fact that Aristotle, directly contradicting the Pythagoreans ascribes vileness to the centre, he does that contrary to the nature of figures and contrary to their geometrical or metaphysical consideration.

For above in Book I, the centre was absolutely not last in the sphere, but wholly its most regular beginning of generation in the mind, and it manifests the likeness of the Holy Trinity, in shadowing forth God the Father, who is the First Person.

4. Finally it can be seen by anyone that he who judges as a physicist of those things which are geometrical does not do rightly, unless what he questions concerning matter and form /**447**/ had been taken over by analogy from a consideration of geometrical figures. For indeed, in solid quantities the inward corporeality, everywhere spread out equally and not by itself partaking of any figure, is a true image of matter in physical things; but the outward figure of the corporeality, composed of fixed surfaces which bound the solidity, represent the form in physical things. And so this comparison is permitted to him simply; but it appears from that that he plays equivocally with "midpart" [*medium*]. For though the Pythagoreans spoke of the inmost point of the sphere [as the midpart or centre]; he understood the whole space within the surface as comprehended by the word "midpart." Accordingly we must grant him the victory as regards the space, but it is a useless victory; for the Pythagoreans and Copernicus win as regards the midpart of all this space. For even if the midpart as a space does not deserve the name of limit; nevertheless as centre it does deserve this name. And in this respect /**448**/ it must be added to the forms and boundaries: since above (in Book I) the centre was the origin of generation of the sphere, metaphysically considered.

Prove by means of the office of the sun that the centre is due to it.

That has already been partly done in rebutting the Aristotelian refutation. For (1) if the whole world, which is spherical, is equally in need of the light of the sun and its heat, then it would be best for the sun to be at the midpart, whence light and heat may be distributed to all the regions of the world. And that takes place more uniformly and rightly, with the sun resting at the centre than with the sun moving around the centre. For if the sun approached certain regions for the sake of warming them, it would draw away from the opposite regions and would cause alternations while it itself remained perfectly simple. And it is surprising that some people use jokingly the similitude of light at the centre of the lamp, as it is a very apt similitude, least fitted to satirize this opinion but suited rather to painting the power of this argument.

(2) But a special argument is woven together concerning light, which presupposes fitness, not necessity. Imagine the sphere of the fixed stars as a concave mirror: you know that the eye placed at the centre of such a mirror gazes upon itself everywhere: and if there is a light at the centre, it is everywhere reflected at right angles from the concave surface and the reflected rays come together again at the centre. And in fact that can occur at no other point in the concave mirror except at the centre. Therefore, since the sun is the source of light and eye of the world, the centre is due to it in order that the sun—as the Father in the divine symbolizing—may contemplate itself in the whole concave surface—which is the symbol of God the Son—and take pleasure in the image of itself, and illuminate itself by shining and inflame itself by warming. These melodious little verses apply to the sun:

Thou who dost gaze at thy face
and dost everywhere leap back
from the navel of the upper air
O gushing up of the gleams flowing
through the glass emptiness, Sun,
who dost again swallow thy reflections.

Nevertheless Copernicus did not place the sun exactly at the center of the world?

It was the intention of Copernicus to show that this node common to all the planetary systems—of which node we shall speak below—is as far distant from the centre of the sun as the ancients made the eccentricity of the sun to be. He established this node as the centre of the world, and was compelled to do so by no astronomical demonstration but on account of fitness alone, in order that this node and, as it were, the common centre of the mobile spheres would not differ from the very centre of the world. But if anyone else, in applying this same fitness, wished to contend that we should rather fear to make the sun differ from the centre of the world, and that it was sufficient that this node of the region of the moving planets should be situated very near, even if not exactly at the centre—anyone who wished to make this contention, I say, would have raised no disturbance in Copernican astronomy. So, firstly, the last arguments concerning the place of the sun at the centre are nevertheless unaffected by this opinion of Copernicus concerning the distance of this node from the sun. But secondly we must not agree to the opinion of Copernicus that this node is distant from the centre of the sun. For the common node of the region of the mobile planets is in the sun, as will be proved below; and so by some probable arguments either the one or the other point is set down at the centre of the sphere of the fixed stars, and by the same arguments the other point is brought to the same place, even with the approval of Copernicus.

15.2.3   Chapter 3: On the Order of the Movable Spheres

How are the planets divided among themselves?

Into the primary and the secondary. The primary planets are those whose bodies are borne around the sun, as will be shown below; the secondary planets are those whose own circles are arranged not around the sun but around one of the primary planets and who also share in the movement of the primary planet around the sun. Saturn is believed to have two such secondary planets and to draw them around with itself: they come into sight now and then with the help of a telescope. Jupiter has four such planets around itself: D, E, F, H. The Earth (B) has one (C) called the moon. It is not yet clear in the case of Mars, Venus, and Mercury whether they too have such a companion or satellite (Fig. 15.2).

Then how many planets are to be considered in the doctrine on schemata?

No more than seven: the six so-called primary planets: (1) Saturn, (2) Jupiter, (3) Mars, (4) the Earth—the sun to eyesight, (5) Venus, (6) Mercury, and (7) only one

Fig. 15.2 The sun, two
primary planets (Earth and
Jupiter), and their secondary
planets (moons).—[*K.K.*]

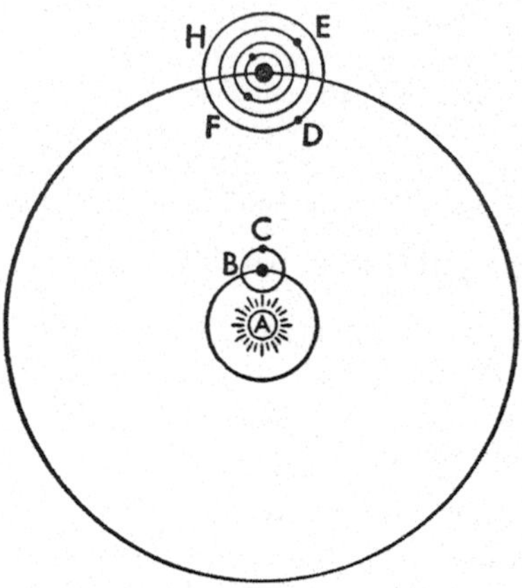

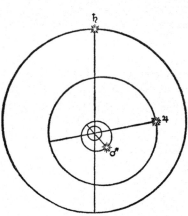

Fig. 15.3 Schema of Saturn,
Jupiter, Mars, and the Earth

of the secondary planets, the moon, because it alone revolves around our home, the
Earth; the other secondary planets do not concern us who inhabit the Earth, /**451**/
and we cannot behold them without excellent telescopes (see Figs. 15.3 and 15.4).

*In what order are the planets laid out: are they in the same heaven or in different
heavens?*

Eyesight places them all in that farthest and highest sphere of the fixed stars and
opines that they move among the fixed stars. But reason persuades men of all times
and of all sects that the case is different. For if the centres of all the planets were in the
same sphere and since we see that to sight they are fairly often in conjunction with
one another: accordingly one planet would impede the other, and their movements
could not be regular and perpetual.

But the reasoning of Copernicus and ancient Aristarchus, which relies upon observations, proves that the regions of the single planets are separated by very great
intervals from one another and from the fixed stars.

Fig. 15.4 Schema of the
Earth, Venus, and Mercury,
with orbit of the Earth
enlarged

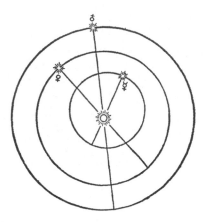

*What is the difference here between the reasoning of Copernicus and that of the
ancients?*

1. The reasoning of the ancients is merely probable, but the demonstration of
   Copernicus, arising from his principles, brings necessity.
2. They teach only that there is not more than one planet in any one sphere: Copernicus further adds how great a distance any planet must necessarily be above
   another.
3. Now the ancients built up one heaven upon another, like layers in a wall, or, to use
   a closer analogy, like onion skins: the inner supports the outer; for they thought
   that all intervals had to be filled by spheres and that the higher sphere must be set
   down as being only as great as the lower sphere of a known magnitude allows;
   and that is only a material conformation. Copernicus, having measured by his
   observation the intervals between the single spheres, showed that there is such a
   great distance between two planetary spheres, that it is unbelievable that it should
   be filled with spheres. And so this lay-out of his urges the speculative mind to
   spurn matter and the contiguity of spheres and to look towards the investigation of
   the formal lay-out or archetype, with reference to which the intervals were made.
4. The ancients, with their material structure, were forced to make the planetary or
   mobile world many parts greater than Copernicus was forced to do with his formal
   lay-out. But Copernicus, on the contrary, made the region of the mobile planets
   not very large, while he made the motionless sphere of the fixed stars immense.
   The ancients do not make it much greater than the sphere of Saturn.
5. The ancients do not explain and confirm as they desire the reason for their lay-out;
   Copernicus establishes his lay-out excellently by reasons.

*What do you mean by the reasons for the lay-out of the spheres, and how is Copernicus
outstanding in this respect?*

Aristotle teaches in *On the Heavens* (Book II, Chap. 10) that nothing is more consonant with reason than that the times of revolution of each planet should correspond to
the altitude or amplitude of its sphere. Now for the ancients, the highest planet was

the same as the slowest, namely, Saturn, because it takes 30 years. Jupiter follows it in place and in time, and takes 2 years; Mars, which takes less than 2 years, follows Jupiter. But for the ancients, this proportionality was changed in the remaining planets. For unless you grant to the Earth an annual movement around the sun, then the sun, Venus, and Mercury—three distinct planets—have the same time of revolution of a year, nevertheless they give them different spheres: the upper to the sun, the middle to Venus, and the third to Mercury. Finally they give the lowest place to the moon, as it takes the shortest time, namely a month.

But Copernicus, postulating that the Earth moves around the sun, keeps the same proportion of movement and time in all the planets. For him the sun is at the centre of the world and is thus the farthest in; it is without the revolution of the centre, that is to say, it is motionless with respect to the centre and the axis. But a few years after this, the body /**453**/ of the sun was perceived to move around its motionless axis more quickly than the space of one month. Mercury, the nearest, circles around the sun in the smallest sphere and completes its revolution in 3 months; around this sphere moves Venus in a larger sphere and in a longer period of time, *viz.*, $7^1/_2$ months. Around the heaven of Venus moves the Earth with its satellite the moon—for the moon is a secondary planet, whose proportionality is not counted among the primary planets—and it revolves in a period of 12 months. After, follow Mars, Jupiter, and Saturn, as with the ancients, each with its satellite. After Saturn, comes the sphere of the fixed stars—and it is distant by such an immense interval that it is absolutely at rest.

What measure does Copernicus use in measuring the intervals of the single planets?

We must use a measure so proportioned that the other spheres can be compared, a measure very closely related to us and thus somehow known to us: such is the amplitude of the sphere whereon the centre of the Earth and the little sphere of the moon revolve—or its semidiameter, the distance of the Earth from the sun. This distance, like a measuring rod, is suitable for the business. For the Earth is our home; and from it we measure the distances of the heavens; and it occupies the middle position among the planets and for many reasons—on which below—it obtains the proportionality of a beginning among them. /**455**/ But the sun, by the evidence and judgment of our sight, is the principal planet. But by the vote of reason cast above, the sun is the heart of the region of moving planets proposed for measurement. And so our measuring rod has two very signal termini, the Earth and the sun.

How great therefore are the intervals between the single spheres?

The Copernican demonstrations show that the distance of Saturn is a little less than ten times the Earth's from the sun; that of Jupiter, five times; that of Mars, one and one-half times; that of Venus, three-quarters; and that of Mercury, approximately one-third.

And so the diameter of the sphere of Saturn is less than twice the length of its ncighbour Jupiter's; the diameter of Jupiter is three times that of the lower planet Mars; the diameter of Mars is one and one-half times that of the terrestrial sphere placed around the sun; the diameter of the Earth's sphere is more than one and one-third that of Venus; and that of Venus is approximately five-thirds or eight-fifths that of Mercury. However, it should be noted that the ratios of the distances are different in other parts of the orbits, especially in the case of Mars and Mercury.

What is the cause of the planetary intervals upon which the times of the periods follow?

The archetypal cause of the intervals is the same as that of the number of the primary planets, being six.

I implore you, you do not hope to be able to give the reasons for the number of the planets, do you?

This worry has been resolved, with the help of God, not badly. Geometrical reasons are co-eternal with God—and in them there is first the difference between the curved and the straight line. Above (in Book I) it was said that the curved somehow bears a likeness to God; the straight line represents creatures. And first in the adornment of the world, the farthest region of the fixed stars has been made spherical, in that geometrical likeness of God, because as a corporeal God—worshipped by the gentiles under the name of Jupiter—it had to contain all the remaining things in itself. Accordingly, rectilinear **/456/** magnitudes pertained to the inmost contents of the farthest sphere; and the first and most beautiful magnitudes to the primary contents. But among rectilinear magnitudes the first, the most perfect, the most beautiful, and most simple are those which are called the five regular solids. **/457/** More than 2000 years ago Pythagoreans said that these five were the figures of the world, as they believed that the four elements and the heavens—the fifth essence—were conformed to the archetype for these five figures.

But the truer reason for these figures including one another mutually is in order that these five figures may conform to the intervals of the spheres. Therefore, if there are five spherical intervals, it is necessary that there be six spheres: just as with four linear intervals, there must necessarily be five digits.

What are these five regular figures?

The cube, tetrahedron, dodecahedron, icosahedron, and octahedron . . .

15.3   Study Questions

QUES.15.1 What, according to Kepler, is the layout of the world? Into how many parts does he divide it? To what does he liken each of these parts? And how does he justify his departure from the ancient understanding?

QUES.15.2 By what properties does Kepler distinguish the parts of the world from one another?

a) What are the three so-called "perfections" of the world? And how are these analogous to the so-called "faculties of the soul"?

b) What are the three things which come together in every "affection"? What role do the sun, the fixed stars, and space play?

c) How is the sun the "principle body", regarding its light, heat and movement? Do such arguments adequately support a heliocentric world-view? If not, then what would?

QUES. 15.3 Is interplanetary space empty? On what basis did Tycho Brahe reject the existence of solid celestial spheres? What problem arises from filling this space with "aether"? So how does Kepler address this issue?

QUES. 15.4 Is the sun at the center of the world?

a) How does Kepler make use of the concept of "dignity" in assigning the sun's location? Is he unique in using such reasoning?

b) On what biological grounds did Aristotle reject the opinion of the Pythagoreans regarding the position of the sun? How does Kepler rebut Aristotle's opinion?

c) In what way does Kepler employ "purpose" or "fitness" in justifying his opinion regarding the sun's location? Is he unique in using such reasoning?

d) Did Copernicus place the sun at the exact center of the spheres of the planets and the fixed stars? Does this pose a problem for the previous arguments provided by Kepler?

QUES. 15.5 What is the number and the spacing of the planets?

a) How many planets are there? Does Kepler employ the same definition of "planets" as the one to which you are accustomed?

b) Do the planets lie amongst the fixed stars? In answering this question, does Kepler appeal to eyesight or to reason?

c) What accounts for the spacing of the planets, according to the ancients? Did Copernicus agree with their reasoning? Why or why not? To whose opinion is Kepler sympathetic?

d) What unit, or standard, of measurement does Kepler employ in measuring astronomical distances? Why is this suitable?

e) What, then, is the spacing of the planets? What cause, or reason, for this spacing does Kepler offer? Is Kepler's approach scientific? Is it appropriate? Is it correct?

15.4   Exercises

EX. 15.1 (THE PLACE OF THE SUN). What do you think: is assigning a location for the sun an empirical problem or a metaphysical problem?

EX. 15.2 (SCIENCE AND METAPHOR). Is the use of metaphor to describe or understand nature commonly employed by scientists? Can you think of any examples? In what

Fig. 15.5 Kepler's Doctrine
of the Schemata from his
1619 publication *Harmonies
of the World*

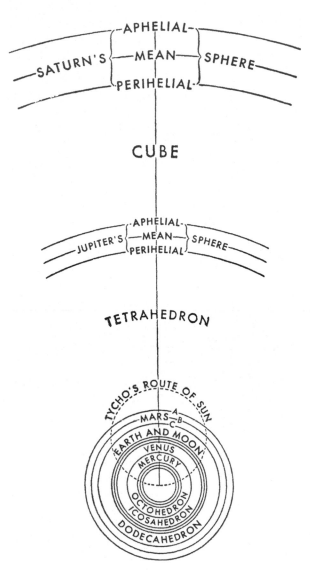

sense is Kepler's use of metaphor similar to, or different than, these? Is Kepler's use
of metaphor appropriate?

Ex. 15.3 (PLATONIC SOLIDS). At the conclusion of the previous reading selection
Kepler refers to his so-called *Doctrine of the Schemata*. This is his ingenious attempt
at explaining the cause of the spacing and the number of the planets. Noting that
the vast intervals separating the orbits of the six known planets could very nearly
accommodate the five Platonic solids, as shown in Fig. 15.5, Kepler reasoned that
God himself had employed these unique geometrical shapes as a plan, or scheme,

when constructing the World. In this problem, we will explore some of the properties of the Platonic solids, which are described mathematically in Book XIII of Euclid's *Elements* and which figure prominently in Plato's *Timaeus*.[2]

a) First, what are the platonic solids, and how many sides does each have?
b) How are the platonic solids defined? What makes these objects unique? Might there be more than five?
c) What is the ratio of the radii of the spheres which circumscribe and inscribe a cube?
d) Why do you think Kepler places each of the five platonic solids where he does? What empirical constraints did he need to consider?
e) Do you think Kepler's *Doctrine of the Schemata* is a valid scientific theory? If not, then what constitutes a valid scientific theory?

15.5   Vocabulary

1. Uncultivated
2. Commingle
3. Limbo
4. Terrestrial
5. Dignity
6. Cerebrum
7. Inference
8. Zodiac
9. Refract
10. Limpid
11. Contiguous
12. Impediment
13. Credible
14. Vivify
15. Circumscribe
16. Refutation
17. Metaphysics
18. Equivocate
19. Similitude
20. Apt
21. Presuppose
22. Concave
23. Melodious
24. Contention
25. Conjunction
26. Conformation
27. Spurn
28. Consonant
29. Amplitude
30. Archetype
31. Rectilinear
32. Tetrahedron
33. Octahedron
34. Dodecahedron
35. Icosahedron

---

[2] See Densmore, D. (Ed.), *Euclid's Elements*, 2nd ed., Green Lion Press, Santa Fe, NM, 2003.

Chapter 16
Kepler's Third Law

> *The ratio of the times is not equal to the ratio of the spheres, but greater than it, and in the primary planets exactly the ratio of the $3/2$th powers.*
>
> —Johannes Kepler

16.1 Introduction

In the previous reading selection from Part I of Book IV of the *Epitome of Copernican Astronomy*, Kepler argued that it would be most fitting for the sun to be located at the center of the World, from which point it can readily illuminate, warm, regulate and (perhaps) even move the planets which orbit around it at divinely ordained distances. Their motion, however, can not be caused by the motion of celestial spheres. Why? Because there are no such spheres, as Tycho Brahe has demonstrated. Then what is the cause of their motion? In Part II of Book IV he begins to address this question. After summarizing the observed motions of the planets, he carefully recounts the planetary models of Aristotle and Ptolemy, which he rejects. He then proceeds to reconsider the ancient idea that an intelligence, or *mind*, is required to govern the complex yet orderly motion of the planets. Such a solution might seem particularly tempting for Kepler since he has already explicitly rejected *matter*—in the form of vast rotating solid celestial spheres—as the cause of the orbital motion the planets. To what cause does Kepler finally attribute the motion of the planets? In the course of this text, he casually mentions the true trajectory of the planetary orbits and he also introduces what is now known as *Kepler's third law of planetary motion*. What does this law state? He also asserts that the planets themselves must must have *inertia*. Why is this significant? And in what way is this contradictory to Aristotelian physics?

16.2 Reading: Kepler, *Epitome of Copernican Astronomy*

Kepler, J., *Epitome of Copernican Astronomy & Harmonies of the World*, Great Minds, Prometheus Books, Amherst, NY, 1995. Book IV, Part II: On the Movement of the Bodies of the World.

K. Kuehn, *A Student's Guide Through the Great Physics Texts,*
Undergraduate Lecture Notes in Physics, DOI 10.1007/978-1-4939-1360-2_16,
© Springer Science+Business Media, LLC 2015

16.2.1   Chapter 1: How Many and of What Sort are the Movements?

/499/ *What was the opinion of Copernicus concerning the movement of bodies? For him, what was in motion and what was at rest?*

There are two species of local movement: for either the whole thing turns, while remaining in its place, but with its parts succeeding one another. This movement can be called δίνητις—lathe-movement, or cone-movement—from the resemblance; or rotation from a rotating pole. Or else the whole thing is borne from place to place circularly. The Greeks call this movement φορά, the Latins *circuitus*, or *circumlatio*, or *ambitus*. But they call both movements generally revolution.

Accordingly Copernicus lays down that the sun is situated at the centre of the world and is motionless as a whole, *viz.*, with respect to its centre and axis. Only a few years ago, however, we grasped by sense that the sun turns with respect to the parts of its body, *i.e.*, around its centre and axis—as reasons had led me to assert for a long time—and with such great speed that one rotation is completed in the space of 25 or 26 days.

Now according as each of the primary bodies is nearer the sun, so it is borne around the sun in a shorter period, under the same common circle of the zodiac, and all in the same direction in which the parts of the solar body precede them/500/—Mercury in the space of 3 months, Venus in $7\frac{1}{2}$ months, the Earth with the lunar heaven in 12 months, Mars in $22\frac{1}{2}$ months or less than 2 years, Jupiter in 12 years, Saturn in 30 years. But for Copernicus the sphere of the fixed stars is utterly immobile.

The Earth meanwhile revolves around its own axis too, and the moon around the Earth—still in the same direction (if you look towards the outer parts of the world) as all the primary bodies.

Now for Copernicus all these movements are direct and continuous, and there are absolutely no stations or retrogradations in the truth of the matter.

By what arguments is it proved that the sphere of the fixed stars does not move?

It was shown in Book I that the sphere of the fixed stars does not rotate around its centre and axis. For we attribute wholly to the Earth whatever appearance of this meets the eyes. Let the other arguments be sought there, in folium 104 *et seqq*. Let us repeat two things alone as proper to this place, one as regards the speed. For if the outmost sphere contains at least 4,000,000 diameters of the sun in its diameter; the circumference will be more than 12,566,370 solar diameters in length. And if all that revolves in 24 h, then in 1 h 523,600 diameters will revolve; in one minute 8727; in 1 s—which is approximately equal to the heart-beat of man—145 diameters of the sun, which is not less than 13,000 German miles; and so during the space of time during which the artery once dilates and again contracts, with a twin pulse-beat, around 7,500,000 (German) miles of the greatest circle would be revolved—and Saturn, in an orbit 2000 times narrower, would still traverse approximately 4000 miles.

The second argument destroys completely every movement of the sphere of the fixed stars. For it is not apparent for whose good, since nothing is outside of it, it changes its position and appearances by being moved to what place or from what place, and since it obtains by rest /501/ whatever it could acquire by any movement. For the movements of all bodies are understood from its rest; and unless it gives them a place, as it can do perfectly by being at rest, nothing can be moved.

How is the ratio of the periodic times, which you have assigned to the mobile bodies, related to the aforesaid ratio of the spheres wherein those bodies are borne?

The ratio of the times is not equal to the ratio of the spheres, but greater than it, and in the primary planets exactly the ratio of the $^3/_2$th powers. That is to say, if you take the cube roots of the 30 years of Saturn and the 12 years of Jupiter and square them, the true ratio of the spheres of Saturn and Jupiter will exist in these squares. This is the case even if you compare spheres which are not next to one another. For example, Saturn takes 30 years; the Earth takes 1 year. The cube root of 30 is approximately 3.11. But the cube root of 1 is 1. The squares of these roots are 9.672 and 1. Therefore the sphere of Saturn is to the sphere of the Earth as 9672:1000. And a more accurate number will be produced, if you take the times more accurately.

What is gathered from this?

Not all the planets are borne with the same speed, as Aristotle wished, otherwise their times would be as their spheres, and as their diameters; but according as each planet is higher and farther away from the sun, so it traverses less space in 1 h by its mean movement: Saturn—according to the magnitude of the solar sphere believed in by the ancients—traverses 240 German miles (in 1 h), Jupiter 320 German miles, Mars 600, the centre of the Earth 740, Venus 800, and Mercury 1200. And if this is to be according to the solar interval proved by me in the above, the number of miles must everywhere be tripled.

16.2.2   Chapter 2: Concerning the Causes of the Movement of the Planets

/502/ State the opinion of the ancient astronomers as to how the planets move.

The ancients, Eudoxus and Callippus, and their follower Ptolemy did not advance beyond circles, wherewith they were accustomed to demonstrate the phenomena—not worrying as to how the planets completed these circles: for in Book XIII of the *Almagest*, Chap. 2, Ptolemy writes as follows:

> But let no one judge that these interweavings of circles which we postulate are difficult, on the ground that he sees that for men the manual imitation of these interweavings is quite intricate. For it is not right for our human things to be compared on a basis of equality with the immortal gods, and for us to seek the evidence for very lofty things from examples of very unlike things.
> For is anything more unlike anything than those things which are always in the same state are unlike those things which never stay like themselves, and than those things which can

everywhere be impeded by all things are unlike those things which can be impeded not even by themselves? Indeed we must try hard to fit the most simple hypotheses to the celestial movements, in so far as that is possible; but if that is not successful, whatever sort of hypotheses can be used. For if only all things which appear in the heavens are given as a consequence of these hypotheses, then there is no reason for being surprised that interweavings of this sort can occur in the movements of the celestial bodies. For these [interwoven circles] do not have a nature which may impede their movement, but only a nature which has grown fitted to give way and to offer a place for the natural motions of each planet, even if the motions happen to be contrary to one another: so much so that all the circles, speaking absolutely, can interpenetrate all circles with no more difficulty than the movements can be perceived. And these movements occur with ease not only around the single circles, but also around the whole spheres, and around the axes of curved and closed surfaces. For even if the various interweavings of circles, on account of the different movements, and the engrafting of one circle in another are very difficult in the customary representations which are constructed by the human hand, and do not succeed so easily that the movements themselves are not at all impeded: nevertheless we see in the heavens that such a manifold concourse of movements by no means stands in the way of the single movements taking place. Indeed, we should not judge what is simple in celestial bodies by the examples of things which seem to us to be simple, since not even here does the same thing seem to be equally simple in all lands. For it will easily happen that he who wishes to judge celestial things in this way will not recognize as simple any of those movements which take place in the heavens, not even the invariable constancy of the first movement: because it is not only difficult but utterly impossible to find among men this thing (namely, something which stays in the same state perpetually). Therefore we must not form our judgement upon terrestrial things, but upon the natures of the things which are in the heavens and upon the unchanging steadfastness of their movements. So it comes about that in this way all the movements are seen to be simple, and much more simple than those movements which seem to us to be simple. For we are unable to suspect them of any labor or any difficulty in their revolutions.[1]

So Ptolemy.

What do you find lacking in this opinion of Ptolemy's?

Even if it is true for many reasons that we should not judge of the ease of celestial movements from the difficulty of the movements of the elements, nevertheless it does not follow that with respect to the celestial movements no terrestrial cases are akin; and Ptolemy seems to draw out this excuse to such lengths that he undermines the whole possibility [*universalem rationem*] of astronomy; and so the excuse satisfies neither the astronomers nor the philosophers, and cannot be tolerated in a Christian discipline.

For as regards astronomy, he brings all hypotheses under suspicion of falsity so long as he argues so strongly for the diversity of celestial and terrestrial things, so that even reason is put down as erring in its judgment /504/ of what is geometrically simple. For if that which to our reasoning concerning the heavens seems to be composite, because our reason compounds circles, is simple in the heavens themselves, therefore in the heavens circles are not compounded with one another in order to fashion one movement. Therefore the astronomer is making a false supposition, and,

---

[1] I have, for obvious reasons, translated Kepler's Latin rather than Ptolemy's Greek. [*C. G. Wallis*].

as is extremely astonishing, is eliciting the truth from things which are absolutely false. But that is to destroy the honor of astronomy, which Aristotle upholds in his books of *Metaphysics*, believing that "the astronomers should be listened to on the form, lay-out, and movements of the celestial bodies." But in truth Ptolemy reveals himself as regards what he desires: for he says to construct hypotheses which are as simple as possible, if that can be done. And so if anyone constructs simpler hypotheses than he—understanding simplicity geometrically—he on the contrary will not defend his composite hypotheses by this excuse but will say to prefer the hypotheses which seem simpler to us men of the earth, even if we employ terrestrial examples.

As regards philosophy: the philosophers will deny that it is sufficient that the matter of the celestial body should be liquid and permeable by the globes and so should not resist the motions of the globes through it. For they ask what this thing is which leads the globe around, especially if it is established that the matter of the globes resists the movers. They ask by what force the mover moves the body from place to place, as there is no immobile field remaining underneath, and since a round body does not possess the services of feet or wings, by the motion of which animals transport their bodies through the ether, or birds through the air by pressing upon and springing up from the air-current. They ask by what light of the mind, by what means the mover perceives or forms the centres of the circles and the encircling orbits. Finally, neither theology nor the nature of things can bear that Ptolemy, who is steeped in pagan superstition, should make the stars to be visible gods—namely, by inferring immortal life from their eternal motion—and should attribute more to them than belongs to God Himself the Founder—that is to say, that geometrical reasons which are really composite, and the understanding where of /**505**/ God wished man His image to have in common with Him, should be simple in the stars.

State Aristotle's opinion as to how the planets move in a circle.

Aristotle, believing that the heavens were joined together by solid spheres—though of an equivocal matter—and the later philosophers, whom the Arabs seem to have followed, and after them Peurbach the writer on the schemata—they, I say, at first believed astronomy as regards the number of circles necessary in order to demonstrate the appearances: so Aristotle believed Eudoxus and Callippus concerning the 25 spheres. He attributed to the spheres the same number of motor intelligences, who were to revolve in their mind the time of the period and the region of the world into which the motion was to proceed. But since it was probable that all the spheres should look to the same beginning, Aristotle judged that 24 other spheres should be placed between these 25 spheres, and he called them ἀνελίττοντες, or counter-turners: namely, in order that each lower sphere should be freed by the interposition of the counter-turner from the carrying off which it was going to suffer from the higher sphere on account of the contiguity of the surfaces. The counter-turners move in an equal time and in a direction opposite to that of the higher sphere, and by that resistance give an appearance of rest, wherein as in an immobile place the lower sphere is stayed and completes its own proper period.[2]

And so the mover of each sphere was appointed to give to his own sphere and to all the lower spheres which it embraced a most regular movement within the higher sphere which was placed in contiguity to this sphere. But since that philosopher had decided that movement was eternal, he appointed movers which were also eternal and immaterial, because material things could not have an infinite power. Therefore it followed that the movers were separate and immobile beginnings. But since this eternal duration of the celestial essence seemed to him to be the goodness and perfection of the whole world, as being opposed to destruction, which was something evil; he also gave to these beginnings the highest perfection and the understanding of this perfection, and from understanding the good the will /506/ to pursue it, lest [the intelligence] should not do well that which is good; in this way he introduced to us separate minds and finally gods, as the administrators of the everlasting movement of the heavens—just as Ptolemy did. As a matter of fact Scaliger, who professed Christianity, and other followers of Aristotle dispute as to whether this movement of the spheres is voluntary and as to whether the beginning of will in the movers is understanding and desire. And indeed, if the world were eternal as Aristotle contended, at any rate the fixed region in which the planet revolves would bear witness concerning the understanding. For we Christians cannot deny that the highest wisdom has presided over the instituting of the movements whereby the planet is made to run into its own region and is dispatched into its own spaces as if from the barriers; but Aristotle assigned this office to the movers themselves, as being eternal.

Furthermore, motor souls were added, tightly bound to the spheres and informing them, in order that they might assist the intelligences somewhat; or because it seemed necessary for the first mover and the movable to unite in some third thing; or because the power of movement was finite with respect to the space to be traversed and the movement was not of an infinite speed but was described in a time measured out according to space: and that argued that the ratio of the motor power to the movable body and to the spaces was fixed and measured.

And so by this solidity of the spheres and by the constant strength of the motor power absolutely all the movements or celestial appearances were so taken care of, that—given the beginning of movement—then indeed every variation in the movements would arise from the lay-out and plurality of the spheres without any labour or worry on the part of the intelligence; and the spheres moved around poles which were at rest—in approximately the way in which in Book I the terrestrial body was said to rotate around its axis and its poles. And by that movement every sphere—and certain people make them wholly of adamant, so that they by no means yield to any body—carried around its planet, which was bound to the sphere at a fixed place; and

---

[2] Kepler is here describing a mechanical model, adopted by Aristotle, in which a number of rotating (and counter-rotating) concentric spheres are connected in such a way as to govern the motion of all the celestial bodies. See the introductory notes in Chap. 4 of the present volume—[K.K.]

one sphere supported another /507/ as was said above: and there was no fear that the globes or spheres would fall, bound to one another in this way.

How do you feel about this philosophy?

Again, I do not raise as an objection to it so much the authority of the Christian discipline as the absurdity of the teaching which fashions gods whose functions are among the works of nature and which meanwhile ascribes to them from eternity such things as are necessarily started by one first beginning of all things at the commencement of time. And since this reasoning cannot do without its theology, the whole thing is overthrown by the denial of gods.

Further, solid spheres cannot be granted, as was proved above. But once more, this philosophy rests upon solid spheres, and it is overthrown by undermining them. For Aristotle will readily grant that a body cannot be transported by its soul from place to place, if the sphere lacks the organ which reaches out through the whole circuit to be traversed, and if there is no immobile body upon which the sphere may rest.

Moreover, even if we grant solid spheres, nevertheless there are vast intervals between the spheres. Either these intervals will be filled by useless spheres which contribute nothing to the state of movement; or else, if there are not solid spheres throughout these intervals, then the spheres will not touch one another or carry one another.

Finally this theory abandons itself, in seeing to it that one sphere rests upon another, but forgetting the lowest sphere. For if we are to grant that spheres are supported by spheres and that they are contiguous to one another, then what supports the lowest sphere of the moon or by what columns is it supported upon the Earth, which, as they suppose, is at rest? Since nowhere on the surface of the Earth is any solidity met with: the winds, clouds, and birds freely and easily come and go everywhere. Why doesn't the great weight of the heavens sink down upon us, especially when the denser parts of the spheres /508/ approach our zenith? Or if the heavens have no weight, what need do we have of spheres for carrying the planetary globes?

If there are no solid spheres, then there will seem to be all the more need of intelligences in order to regulate the movements of the heavens, although the intelligences are not gods. For they can be angels or some other rational creature, can they not?

There is no need of these intelligences, as will be proved; and it is not possible for the planetary globe to be carried around by an intelligence alone. For in the first place, mind is destitute of the animal power sufficient to cause movement, and it does not possess any motor force in its assent alone, and it cannot be heard or perceived by the irrational globe; and even if mind were perceived, the material globe would have no faculty of obeying or of moving itself. But before this, it has already been said that no animal force is sufficient for transporting the body from place to place, unless there are organs and some body which is at rest and on which the movement can take place. Therefore the question falls back to the above.

But on the contrary the natural powers which are implanted in the planetary bodies can enable the planet to be transported from place to place.

But let it be posited as sufficient for movement that the intelligence should will movement into this or that region: then the discovery of the figure whereon the line of movement is ordered will be irrational. For we are convinced by the astronomical observations which have been taken correctly that the route of a planet is approximately circular and as a matter of fact eccentric—that is, the centre [of the circle] is not at the centre of the world or of some body; and furthermore that during the succession of ages the planet crosses from place to place. Now as many arguments can be drawn up against the discovery of such an orbit as there are parts of it already described.

For firstly, the orbit of the planet is not a perfect circle. But if mind caused the orbit, it would lay out the orbit in a perfect circle, /509/ which has beauty and perfection to the mind. On the contrary, the elliptic figure of the route of the planet and the laws of the movements whereby such a figure is caused smell of the nature of the balance or of material necessity rather than of the conception and determination of the mind, as will be shown below.

Finally, in order that we may grant that a different idea from that of a circle shines in the mind of the mover: it is asked by what means the mind can apply this or that [idea] to the regions of the world. Now the circle is described around some one fixed centre, but the ellipse, which is the figure of the planetary orbits, is described around two centres.

Then what seat will you give to mind, so that it may measure out a circle or an elliptic orbit on the liquid plains of the ether? You do not place the mind at the centre, do you? For then you are placing it in the ether, which is not different from all the remaining space of the world, because the orbit of the planet is eccentric to the solar body. But this is exceedingly absurd, since elsewhere the beginning of individuation of souls is assigned to the matter and to the body, to which the soul is added, and this matter differs in place and time and in many other marks from the remaining matter of the world. Surely no other position belongs to the soul and to the mind than that which comes through its body, which the soul informs. And by what force will mind be moved from place to place in a small circle around the centre of the world, so that it may be at the centres of the planetary orbits in the succession of ages, if the mind is without a body and is no more able to be moved than to be given a position in space? By what means will mind view its position or its distance from the centre of the world?

But let it be granted that the mind has a view from its seat at the centre: then how will it cause the planet, which is very distant, to trace its orbit around this centre? If the mind had the planet tied by a rope, perhaps the planet would fly around, being tied to the centre. Perhaps the mind, looking out from the centre, could perceive— especially if it were endowed with bodily eyes—whether the planet were moving in a circle, if the planet were always viewed making an equal angle; but if it should go outside of its circle, in what way would the mind lead it back, if it did not see the orbit by itself? /510/ But how does the mind understand the orbit, which is not stamped on the body as its special property? For here there is no question of the intellectual idea of a circle, wherein there is no distinction of great and small, but of the real route of the planet, which has a fixed magnitude in addition to the idea.

But if you place the motor mind outside of the centre of its orbit, its condition will
be worse. For either it will be in the body which is at the centre of the world; and
thus all the minds will be in the same body, and the above difficulties with respect
to keeping the planet in its orbit and with respect to the discovery of the orbit will
remain. Or else the mind will be in the globe of the planet: then in both cases it is
asked by what means the mind knows where the centre is, around which the orbit
of the planet should be organized; and how great the distance of the mind and its
globe from that point is. For Avicenna rightly judged that if the mover of the planet
is a mind, it has need of knowledge of the centre and of its distance from the centre.
For the circle is defined and perfected by the same things, the centre and the equal
curvature around it, *viz.*, the distance of the circumference from the centre; and so,
however much you exalt the motor mind, nevertheless the circle is nothing else to
God except what has already been said. And this same thing should be understood
proportionally concerning the figure of the ellipse.

*Why do you say that a celestial body, which is unchanging with respect to its matter,
cannot be moved by assent alone? For if the celestial bodies are neither heavy nor
light, but most suited for circular movement, then do they resist the motor mind?*

Even if a celestial globe is not heavy in the way in which a stone on the earth is said
to be heavy, and is not light in the way in which among us fire is said to be light:
nevertheless by reason of its matter it has a natural ἀδυναμία or powerlessness of
crossing from place to place, and it has a natural inertia or rest whereby it rests /511/
in every place where it is placed alone. And hence in order that it may be moved
out of its position and its rest, it has need of some power which should be stronger
than its matter and its naked body, and which should overcome its natural inertia.
For such a faculty is above the capacity of nature and is a sprout of form, or a sign
of life.

*Whence do you prove that the matter of the celestial bodies resists its movers, and is
overcome by them, as in a balance the weights are overcome by the motor faculty?*

This is proved in the first place from the periodic times of the rotation of the single
globes around their axes, as the terrestrial time of 1 day and the solar time of approximately 25 days. For if there were no inertia in the matter of the celestial globe—and
this inertia is as it were a weight in the globe—there would be no need of a virtue
[*virtute*] in order to move the globe; and if the least virtue for moving the globe were
postulated, then there would be no reason why the globe should not revolve in an
instant. But the revolutions of the globes take place in a fixed time, which is longer
for one planet and shorter for another: hence it is apparent that the inertia of matter
is not to the motor virtue in the ratio in which nothing is to something. Therefore the
inertia is not *nil*, and thus there is some resistance of celestial matter.

    Secondly, this same thing is proved by the revolution of the globes around the
sun—considering them generally. For one mover by one revolution of its own globe
moves six globes, as we shall hear below. Wherefore if the globes did not have a
natural resistance of a fixed proportion, there would be no reason why they should not
follow exactly the whirling movement of their mover, and thus they would revolve

with it in one and the same time. Now indeed all the globes go in the same direction as the mover with its whirling movement, nevertheless no globe fully attains the speed of its mover, and one follows another more slowly. Therefore they mingle the inertia of matter with the speed of the mover in a fixed proportion.

The ratio of the periodic times seems to be the work of a mind and not of material necessity.

The most accurately harmonic attunement of the extreme movements—the slowest and the fastest movement in any given planet—is the work of the highest and most adored creator Mind or Wisdom. But if the lengths of the periodic times were the work of a mind, they would have something of beauty, like the rational ratios, duplicate, triplicate, and so on. But the ratios of the periodic times are irrational [*ineffabiles, irrationales vulgo*] and thus partake of infinity, wherein there is no beauty for the mind, as there is no definiteness [*finitio*].

  Secondly, these times cannot be the work of a mind—I am not speaking of the Creator but of the nature of the mover; because the unequal delays in different parts of the circle add up to the times of one period. But the unequal delays arise from material necessity, as will be said below, and as if by reason of the balance [*ex ratione staterae*].

Therefore by what force do you suspend your material globes and the Earth in especial, so that each remains within the boundaries of its region, though it is destitute of the bonds of the solid spheres?

Since it is certain that there are no solid spheres, it is necessary that we should take refuge in this inertia of matter, whereby any globe, placed in any place on the world beyond the motor virtues, naturally rests in that place, because matter, as such, has no faculty of transporting its body from place to place.

Then what is it which makes the planets move around the sun, each planet within the boundaries of its own region, if there are not any solid spheres, and if the globes themselves cannot be fastened to anything else and made to stick there, and if without solid /513/ spheres they cannot be moved from place to place by any soul?

Even if things are very far removed from us and which are without a real exemplification are difficult to explain and give rise to quite uncertain judgements, as Ptolemy truly warns; nevertheless if we follow probability [*verisimilitudinem*] and take care not to postulate anything which is contrary to us, it will of necessity be clear that no mind is to be introduced which should turn the planets by the dictation of reason and so to speak by a nod, and that no soul is to be put in charge of this revolution, in order that it should impress something into the globes by the balanced contest of the forces, as takes place in the revolution around the axis; but that there is one only solar body, which is situated at the centre of the whole universe, and to which this movement of the primary planets around the body of the sun can be ascribed.

16.2.3   Chapter 3: On the Revolution of the Solar Body Around its Axis and its Effect in the Movement of the Planets

By what reasons are you led to make the sun the moving cause or the source of movement for the planets?

1. Because it is apparent that in so far as any planet is more distant from the sun than the rest, it moves the more slowly—so that the ratio of the periodic times is the ratio of the $3/2$th powers of the distances from the sun. Therefore we reason from this that the sun is the source of movement.
2. Below we shall hear the same thing come into use in the case of the single planets—so that the closer any one planet approaches the sun during any time, it is borne with an increase of velocity in exactly the ratio of the square.
3. /514/ Nor is the dignity or the fitness of the solar body opposed to this, because it is very beautiful and of a perfect roundness and is very great and is the source of light and heat, whence all life flows out into the vegetables: to such an extent that heat and light can be judged to be as it were certain instruments fitted to the sun for causing movement in the planets.
4. But in especial, all the estimates of probability are fulfilled by the Sun's rotation in its own space around its immobile axis, in the same direction in which all the planets proceed: and in a shorter period than Mercury, the nearest to the sun and fastest of all the planets.

For as regards the fact that it is disclosed by the telescope in our time and can be seen every day that the solar body is covered with spots, which cross the disk of the sun or its lower hemisphere within 12 or 13 or 14 days, slowly at the beginning and at the end, but rapidly in the middle, which argues that they are stuck to the surface of the sun and turn with it; I proved in my *Commentaries on Mars*, Chap. 34, by reasons drawn from the very movement of the planets, long before it was established by the sun-spots, that this movement necessarily had to take place.

16.3   Study Questions

QUES. 16.1 How fast do the planets move?

a) What are the two types of local movement that Kepler describes? Which of these two does the sun exhibit? The planets? The moon? The earth?
b) For what reason(s) does Kepler reject the idea that the stars are themselves moving?
c) What is the mathematical relationship between the times of orbit of the various planets? What does this relationship imply about their speed?

QUES. 16.2 What is the cause of motion of the planets?

a) What, according to Ptolemy, is the cause of motion of the planets? How does he address the difficulty of intersecting spheres?
b) Why does Kepler find Ptolemy's reasoning suspect from astronomical, philosophical and theological perspectives?
c) What was the opinion of Aristotle regarding the cause of the planets' motion? How many rotating (and counter-rotating) spheres did Aristotle invoke?
d) How does Kepler feel about Aristotle's opinion? What four counterarguments does he provide?
e) According to Kepler, is there a need for *intelligence* or *mind* to cause or regulate the movements of the heavenly bodies? Why? And what does this have to do with the shape of planetary orbits?
f) Do the planets have inertia? In other words, do they resist changes in their direction of motion? What does this imply about the cause of their motion?
g) What does the law of the periodic times, which is obeyed by the planets' orbits, suggest regarding the cause of their motion?
h) What holds the planets up, if not solid spheres? And what causes their motion?
i) Does the sun in fact rotate? How do you know?

16.4   Exercises

Ex. 16.1 (KEPLER'S THIRD LAW). Does Kepler's third law of planetary motion apply to man-made satellites or just to planets? To find out, look up the orbital time and altitude of the International Space Station (ISS). What is the ratio of the orbital times of the ISS and of the moon? What is the ratio of their orbital radii? Do these ratios obey Kepler's third law of planetary motion?

Ex. 16.2 (MIND IN NATURE). Is *intelligence* or *mind* an appropriate causal explanation for planetary motion? Is it an appropriate causal explanation for human or animal motion? If not, then what is the cause? If so, then where should one draw the line beyond which the concept of intelligence should certainly *not* be invoked as a causal explanation of motion? Justify your position.

Ex. 16.3 (SUNSPOTS LABORATORY). In this field exercise, you will use a solar telescope[3] to measure the period of rotation of the sun about its axis. You will need to set aside 15 min each day *at the same time* (to within 5 min) to make observations of the sun over the course of 1 or 2 weeks. On your first day of observation, begin by learning how to set up the sunspotter. Your first setup will likely take a bit longer than the others since you will be unfamiliar with the apparatus. Each day, you should use the solar telescope to project an image of the sun on a sheet of paper and mark

---

[3] An economical and easy-to-use solar telescope is the *Sunspotter*, manufactured by Science First Inc. A detailed description of this apparatus was written by Philip M. Sadler and Mary Lou West.

the locations of any observed sunspots. This is a bit trickier than it sounds, since the image of the sun continually drifts across the paper. It will help to have one person drawing features and another person gently moving the paper between each marking. Be sure to record the direction of the sun's motion after you have traced the features so that you know which direction is westward on your diagram. After several days of viewing, you can begin to address the question: is the sun rotating? And if so, at what rate?[4]

16.5   Vocabulary

1. Assert	10. Essence
2. Manifold	11. Ascribe
3. Steadfast	12. Commence
4. Akin	13. Zenith
5. Terrestrial	14. Succession
6. Elicit	15. Endow
7. Permeable	16. Inertia
8. Interpose	17. Attune
9. Contiguity	18. Dictation

---

[4] To assist in your calculations, you will need to determine the coordinates of some of your observed sunspots as a function of time. To do this most precisely, you might make use of a Stonyhurst grid to read off the heliographic coordinates of any observed sunspots. Stonyhurst grids are essentially sets of longitude and latitude lines overlaying a disc which represents the sun. Since the observed orientation of the axis of the sun varies from day to day over the course of the year, you will need to print on a sheet of paper the particular Stonyhurst grid which is appropriate for the date of observation.

Chapter 17
Kepler's First and Second Laws

*The routes or single circuits of the planets are not arranged
exactly in a perfect circle but are ellipses.*

—Johannes Kepler

*The planet is really slower at its greater distance from the sun,
and faster at its lesser.*

—Johannes Kepler

17.1 Introduction

In the previous reading selection from Part II of Book IV of his *Epitome*, Kepler
stated that the ratio of the orbital periods of any two planets is related in a particular
way to the ratio of the semi-major axes of the ellipses which describe their orbits.
This is *Kepler's third law of planetary motion.*

$$\frac{T_1}{T_2} = \left(\frac{a_1}{a_2}\right)^{3/2} \tag{17.1}$$

Kepler then argued that the sun itself is the cause of this motion.[1] For the sun, he
later claims, emits a certain immaterial power, or *virtue*, in outwardly directed lines
from its interior. The source of these lines of power is a *soul*, albeit not an intelligent
soul like that of a person or animal. As the body of the sun spins around its axis, these
immaterial threads of power also rotate, forming a sort of invisible whirlpool which
impels the planets in their orbits around the sun. By positing such an immaterial
power to move the planets, Kepler is walking a fine line between material causes
(rotating spheres) and intelligent causes (animism). Kepler's theory of planetary
motion is inspired, he says, by the recent work of William Gilbert, who claimed that
magnets have a native strength, or vigor, which allows them to act on nearby magnets
through the empty space separating them.[2] If magnets can act on nearby magnets,
Kepler reasoned, then why can't the sun also act on nearby planets?

---

[1] One is tempted here to assume that Kepler is proposing gravitational attraction between the
sun and the planets. This, however, is not the case. Kepler's lines of force are neither attractive
nor repulsive. Newton's theory of gravitational attraction comes several decades later when his
Principia is published. See Chap. 27 of volume II.

[2] See Chap. 4 of Book II of Gilbert, W., *De Magnete*, Dover Publications, New York, 1958.

K. Kuehn, *A Student's Guide Through the Great Physics Texts*,
Undergraduate Lecture Notes in Physics, DOI 10.1007/978-1-4939-1360-2_17,
© Springer Science+Business Media, LLC 2015

Our final reading selection from Kepler is taken from Part III of Book IV of his *Epitome*. It is subtitled *On the real and true irregularity of the planets and its causes*. By "irregularity," Kepler means the fact that the planets, when observed carefully, do not exhibit regular, uniform, circular motion, as had been maintained by Aristotle. This irregular motion was, of course, recognized by Ptolemy. But he attempted to rationalize it as only *apparently* irregular motion. He invoked an equant—a hypothetical point about which the planet moves uniformly. The equant point of each planet was itself offset from the center of the deferent circle about which the planet orbits Earth—all of this to maintain a nominal allegiance to perfect circular motion.[3] Kepler, on the other hand, describes this apparent irregular motion as a *true* irregular motion. As suggested by the subtitle of Part III, he not only defends a real and true irregularity of the planets' motion, he also offers a causal explanation as to why the planets might exhibit such irregular motion. What is it?

17.2   Reading: Kepler, *Epitome of Copernican Astronomy*

Kepler, J., *Epitome of Copernican Astronomy & Harmonies of the World*, Great Minds, Prometheus Books, Amherst, NY, 1995. Book IV, Part III: On the Real and True Irregularity of the Planets and its Causes.

17.2.1   Introduction

From what do the planets have that name which signifies wanderers *in this language?*

From the manifold variety of their proper movements. For if you follow the judgment of the eyes, that variety has no law, no determined circle, no definite time, if a comparison is made with the fixed stars.

In how many ways do the planets seem to wander?

In three ways: (1) In the longitude of the sphere of the fixed stars, which we said extends along the ecliptic. (2) In latitude /570/ or to the two sides of the ecliptic, towards its poles. (3) In altitude *i.e.*, in the straight line stretching from the centre of vision into the depth of the ether. Nevertheless this variety is not uncovered by the eyes alone; but reasoning from the diverse apparent magnitude of the bodies and the arcs assents to it.

What must be held concerning these wanderings of the planets? Do they really wander in so many various ways, or is sight merely deceived?

---

[3] See the introduction of Chap. 7 of the present volume. The equant is the center of the eccentric circle in Fig. 7.1.

Even though that movement is not wholly such as meets the eyes, it is present in the planetary bodies themselves; but much deception of sight winds its way in here. Nevertheless when these deceptions have been removed by the mind, some irregularity of movements still remains and is really present in all the planets.

Then what is the true movement of the planets through their surroundings?

It is constant with respect to the whole periods; and proceeds around the sun, the centre of the world, always eastward towards the signs which follow. It never sticks in one place, as though stationary, and much less does it ever retrograde. But nevertheless it is of irregular speed in its parts; and it makes the planet in one fixed part of its circuit digress rather far from the sun, and in the opposite part come very near to the sun: and so the farther it digresses, the slower it is; and the nearer it approaches, the faster it is. Finally, in one part of the circle it departs from the ecliptic to the north, and in the other, to the south. And so the planet is left with a real irregularity and one which is threefold, too: in longitude, in latitude and in altitude. The astronomers prove that by suitable evidence—concerning which in Book VI.

17.2.2   Chapter 1: The Causes of the True Irregularities

/571/ State what the ancients thought about the causes of this irregularity.

The ancients wished it to be the office of the astronomer to bring forward such causes of this apparent irregularity as would bear witness that the true movement of the planet or spheres is most regular, most equal, and most constant, and also of the most simple figure, that is, exactly circular. And they judged that you should not listen to him who laid down that there was actually any irregularity at all in the real movements of these bodies.

Do you judge that this axiom should be kept?

I make a threefold answer: I. That the movements of the planets are regular, that is, ordered and described according to a fixed and immutable law is beyond controversy. For if this were not the case, astronomy would not exist, nor could the celestial movements be predicted. II. Therefore it follows that there is some conformity between the whole periods. For that law, of which I have spoken, is one and everlasting; the circuits or traversings of the celestial course are numberless. But if they all have the same law and rule, then all the circuits are similar to one another and equal in the passage of time.

III. But it has not yet been granted that the movement is really regular even in the diverse parts of any given circuit. (1) For astronomy bears witness that, if with our mind we remove all deceptions of sight from that confused appearance of the planetary motion, the planet is left with such a circuit that in its different parts, which are really equal, the speed of the planet is irregular—just as there is apparent inequality in the angles at the sun which are equal with respect to time. And Ptolemy himself, by setting up different centres in accordance with the rule of movement of eccentrics and epicycles, makes those circles of his to move more swiftly at one time, and more slowly at another. (2) /572/ Finally, astronomy, if handled with the right subtlety, bears witness in this case that the routes or single circuits of the planets are not arranged exactly in a perfect circle but are ellipses.

But by what arguments did the ancients establish their opinion which is the opposite of yours?

By four arguments in especial: (1) from the nature of movable bodies, (2) from the nature of the motor virtue, (3) from the nature of the place in which the movement occurs, and (4) from the perfection of the circle.

Will you state their argument from the nature of bodies?

They reasoned that those bodies are not composed of the elements, and so neither generation or corruption nor alteration has any rights over them. The experience of all the ages bears witness to this: for the bodies are always viewed as the same and are not found to have changed at all in mass or in number or in form. But the movements of the bodies made up of elements are for this very reason various and inconstant, because the elements are variously mixed in the constitution of the bodies and are at war with one another within the mixed bodies. Therefore in the celestial bodies, where there is no such mixture and no war of the elements as in mixed bodies, there is also no place for turbulence, none for irregularity.

What answer do you judge should be made to this argument?

If the argument is speaking of disordered turbulence of movements, there is none such in the heavens: there are no celestial disturbances as in thunderstorms,

> Flame and drops of water at war with one another

because the composition of the bodies of the world is of a very different family. But if the argument is in opposition to every regular irregularity also; /573/ then not every irregularity, certainly not that regular intensification and remission of movements, comes from the war of the elements mixed together in the moved bodies, nor from the bodies being mutable. For some irregularity arises just because they are bodies, bodies which are moved and which give movement too, and because they are made up of their own matter, their own magnitude, and their own figure both inwardly and outwardly, and in accordance with their magnitudes and figures they are endowed with their natural power too. And in accordance with their natural power they are less movable at a distance than at near-by, where the faculties of mover and moved are in agreement rather than at war. Thus by one part of its body the loadstone attracts iron, and by the other repels iron; not in either case on account of any mixture of elements, but on account of the inward rectilinear configuration, in accordance with which the loadstone has an inborn virtue. Thus the same loadstone attracts more strongly iron when near-by than when farther away—not that when the loadstone is nearer, it has more of fire or Earth, but because its virtue is weakened with its distance. Nevertheless the celestial bodies—*i.e.*, the bodies of the world—remain everlasting and immutable as regards their total masses [*moles*]. For the changes which come about on their surfaces can bring on nothing sufficient to disturb [*nullum momentum ad turbandos*] the movements of the total masses [*molium*]. And upon this everlastingness of the whole globes and upon the fact that in the world there is nothing disordered which impedes their movements, there depend this regularity of the circlings, and the everlasting similarity, and the constant regularity, with respect to the whole cycles, of the irregularity in the single parts.

Will you review the second argument of the ancients taken from the mover cause?

They said, the motor virtues of the celestial bodies are of the most simple substance; they are minds divine and most pure, which do unceasingly what they do; they are everlastingly similar; they employ a most equal struggle of forces, and they are never tired, because they feel no labour. /574/ And so there is no reason why they should move their globes differently at different times. And accordingly even the figures of the movements, on account of the very nature of the minds, are most perfect circles.

What do you oppose to this?

Even though the motor virtue is neither some god nor a mind, nevertheless we must grant what the argument intends—chiefly too in the case of that motor cause which a truer philosophy brings in, namely in the case of the natural power of the bodies, because wherever and in so far as such a power is alone, it moves [a body] most regularly and in a perfect circle, and does that by the sole necessity of effort and by the everlasting simplicity of its essence. That is the case in the rotation of the solar body and in that of the Earth too in especial; for this rotation comes from one sole motor cause: whether it be a quality of the body or a sprout of the soul born with the body. For the axis with its two opposite poles stays fixed; but the body revolves around the axis most regularly and in a circle. This would be the case still, if any planetary globe were always at the same distance from the sun. For it would be carried by the sun with utmost regularity in a perfect circle, by means of the immaterial form released from the solar body set in a very regular movement of rotation. And by reason of this same very regular movement even that form from the body revolves in the amplitude of the expanse of the world, like a swift whirlpool.

But although so far we have granted the argument of the ancients, nevertheless regularity of movements in every respect does not yet follow from this. For not only do the motor virtue and the movable body come together in the movements, but also the inward rectilinear configuration of the movable body; and in proportion to its diversity of posture in relation to the sun, this configuration is affected in diverse ways in the movement: in one region it is repelled, in another it is attracted towards the inside. The axis of the magnetic movable body comes in, and so does being at rest in the parallel posture; and from that inward repose and from that revolution coming from outside there results that change of posture /575/ of the parts of the planet in relation to the sun. Finally there advenes the interval between the sun and the planet, and this interval varies with the attraction and repulsion. But when the interval has been changed and the planet comes into a denser or more rarefied virtue, then its movement too necessarily suffers intensification or remission, and the figure of its route becomes elliptical. So with reference to the concourse of so many required things, the virtue moving the planet cannot be called simple, because it moves by means of different degrees of its form.

What was the ancients' argument from place?

They considered that the region of the elements was around the centre of the world; the heavens at the surface. Therefore to the bodies made up of elements belongs rectilinear movement, which has a beginning and an end and which, disbursed according to the contrary principles of heaviness and lightness, brings any of those bodies back into its own place; and hence in proportion to different nearnesses to the natural place, or mark, there are different speeds, and finally pure rest. But the celestial bodies move everlastingly in the circular expanse of the world: and that argues that they are neither heavy nor light, and that they are not moved for the sake of rest or for the sake of occupying a place—for they are always circling in their place—but that accordingly they are moved only in order to be moved; and so their movement must be regular, and the form of their movement must be other than rectilinear, namely suited to an eternity of movement, that is, returning into itself.

What answer do you make to this third argument?

Not every irregularity of movements comes from heaviness and lightness, the properties of the elements; but some comes from the change of the distance too, as is clear in the case of the lever and the balance; and this cause produces intensification /576/ and remission of movements, as has been explained so far. We must however remark that there is nevertheless some kinship between the principles of heaviness and lightness in the elements and the natural inertia of the planetary globe with respect to movement, but no irregularity of movement is explained by this kinship.

But as regards the figure of the movement, the argument concludes nothing more than we can grant, namely that the movement bends back into itself. And not only the circular but also the elliptical are of such a kind; and so the assumptions are not denied. For in truth bodies which revolve around their axes are moved only in order that by their everlasting motion they may obey some necessity of their own globe—some bodies indeed in order that they may carry the planets around themselves in everlasting circles.

State the fourth argument of the ancients which was taken from the circular figure.

They philosophized that of all movements which return into themselves the circular is the most simple and the most perfect and that something of straightness is mixed in with all the others, such as the oval and similar figures: accordingly this circular movement is most akin to the very simple nature of the bodies, to the motors, which are divine minds—for its beauty and perfection is somehow of the mind—and finally to the heavens, which have a spherical figure.

How must this be refuted?

To this I make answer as follows: Firstly, if the celestial movements were the work of mind, as the ancients believed, then the conclusion that the routes of the planets are perfectly circular would be plausible. For then the form of movement conceived by the mind would be to the virtue a rule and mark to which the movement would be referred. But the celestial movements are not the work of mind but of nature, that is, of the natural power of the bodies, or else a work of the soul acting uniformly in accordance with those bodily powers; /577/ and that is not proved by anything more validly than by the observation of the astronomers, who, after rightfully removing the deceptions of sight, find that the elliptical figure of revolution is left in the real and very true movement of the planet; and the ellipse bears witness to the natural bodily power and to the emanation and magnitude of its form.

Then, even if we grant them their intelligences, nevertheless they do not yet obtain what they want, namely the complete perfection of the circle. For if it were a question only of the beauty of the circle, the circle would very rightly be decided upon by mind and would be suitable for any bodies whatsoever and especially the celestial, as bodies are partakers of magnitude, and the circle is the most beautiful magnitude. But because in addition to mind there was then need of natural and animal faculties also for the sake of movement; those faculties followed their own bent [*ingenium*], nor did they do everything from the dictate of mind, which they did not perceive, but they did many things from material necessity. So it is not surprising if those faculties, which are mingled together, could not attain perfection completely. The ancients themselves admit that the routes of the planets are eccentric, which seems to be a much greater deformity than the ellipse. And nevertheless they could not guard against this deformity by means of the province of those minds of theirs.

Now I have often reminded you that while I deny that the celestial movements are the work of mind; I am not at that moment speaking of the Creator's Mind, which all things indeed befit, whether circular or elliptical, whether administered and represented by minds or compelled by material necessity from the beginnings once laid down.

17.2.3   Chapter 2: On the Causes of Irregularity in Longitude

/578/ *Then what causes do you bring forward as to why, although all the routes of the primary planets are arranged around the sun, nevertheless the angles—in which as if from the centre of the sun, the different parts of the route of one planet are viewed—are not completed by the planet in proportional times?*

Two causes concur, the one optical, the other physical, and each of almost equal effect. The first cause is that the route of the planet is not described around the sun at an equal distance everywhere; but one part of it is near the sun, and the opposite part is so much the farther away from the sun. But of equal things, the near are viewed at a greater angle, and the far away, at a smaller; and of those which are viewed at an equal angle, the near are smaller, and the far away are greater.

The other cause is that the planet is really slower at its greater distance from the sun, and faster at its lesser.

Therefore if the two causes are made into one, it is quite clear that of two arcs which are equal to sight, the greater time belongs to the arc which is greater in itself, and a much greater time on account of the real slowness of the planet in that farther arc.

But could not one cause suffice, so that, because generally the orbit of the planet draws as far away from the sun on one side as it draws near on the other, we might make such a great distance that all this apparent irregularity might be explained merely by this unequal distance of the parts of the orbit?

Observations do not allow us to make the inequality of the distances as great as the inequality /579/ of the time wherein the planet makes equal angles at the sun; but they bear witness that the inequality of the distances is sufficient to explain merely half of this irregularity: therefore the remainder comes from the real acceleration and slowing up of the planet.

What are the laws and the instances of this speed and slowness?

There is a genuine instance in the lever. For there, when the arms are in equilibrium, the ratio of the weights hanging from each arm is the inverse of the ratio of the arms. For a greater weight hung from the shorter arm makes a moment equal to the moment of the lesser weight which is hung from the longer arm. And so, as the short arm is to the long, so the weight on the longer arm is to the weight on the shorter arm. And if in our mind we remove the other arm, and if instead of the weight on it we conceive at the fulcrum an equal power to lift up the remaining arm with its weight; then it is apparent that this power at the fulcrum does not have so much might over a weight which is distant as it does over the same weight when near. So too astronomy bears witness concerning the planet that the sun does not have as much power to move it and to make it revolve when the planet is farther away from the sun in a straight line, as it does when the interval is decreased. And, in brief, if on the orbit of the planet you take arcs which are equally distant, the ratio between the distances of each arc from the sun is the same as the ratio of the times which the planet spends in those arcs. Thus let the centre of the sun or world be represented by the fulcrum of the lever, and its motor power by one arm and the weight on it—and we have already given the order to dissemble the arm and the weight, and mentally to reduce them to the fulcrum; but let the planet be represented by the weight on the remaining arm, and the interval between the sun and the planet, by the arm for that weight.

/580/ Let AC be the lever, D and B the weights hanging from C and A, FE the fulcrum, and FEC and FEA right angles. As CE is to EA, so is the weight B on EA to the weight D on EC. Remove mentally EA, and let the power formed through EA by the weight B be the power of the fulcrum E; accordingly this power of the fulcrum E will keep the weight D, hung from C, in horizontal equilibrium, that is, so that FEC will be a right angle. But if this same weight, pulled away from C, approaches as near as G, then the same power of E will have more might over this weight, and will lift it up above the line EC.

Now let E be not the fulcrum but the sun, and let D be the planet; and EC and EG the different distances of the planet from the sun. Accordingly observations bear witness that as EC is to EG, so is GK, the forward movement of the planet when nearer at G, to GI or CH, the forward movement of the planet when farther away at C.

Then do you attribute weight to the planet?

It was said in the above that we must consider that instead of weight there is that natural and material resistance or inertia with respect to leaving a place once occupied; and that this inertia snatches the planet as it were out of the hands of the rotating sun, so that the planet does not yield absolutely to that force which lays hold of it.

/581/ What is the reason why the sun does not lay hold of the planet with equal strength from far away and from nearby?

The weakening of the form from the solar body is greater in a longer outflow than in a shorter; and although this weakening occurs in the ratio of the squares of the intervals, *i.e.*, both in longitude and in latitude, nevertheless it works only in the simple ratio: the reasons have been stated above.

17.3   Study Questions

QUES. 17.1 What is the true shape of the planetary orbits?

a) What, according to the ancients, is the office of the astronomer? In what sense does Kepler agree, or disagree, with the ancients?
b) What did the ancients believe regarding the shape of the orbits of the individual planets? What, on the other hand, is Kepler's view on the shape of the orbits of the planets?
c) By what four arguments did the ancients establish their opinion? How does Kepler answer each argument? Do you find Kepler's refutations convincing?

QUES. 17.2 What is the cause of the irregular motion of the individual planets?

a) What, exactly, is meant by a planet's "irregularity" in longitudinal motion? Are the irregular motions of the planets, when viewed by a spectator on earth, merely *optical effects*, or are they true irregularities?
b) What mechanism does he suggest as the cause of motion of the planets around the sun? In particular, what law (or laws) governs the motion of the planets?
c) Wherein does the power to move the planets lie? And does Kepler actually believe the planets have *weight*?

17.4   Exercises

EX. 17.1 (OPTICAL IRREGULARITY). Consider a satellite moving at a constant speed around a circular orbit with a radius of 1000 km and a period of 10 days.

Fig. 17.1 Kepler's use of the
lever, with the sun acting at
the fulcrum, to explain the
cause of a planet's irregular
motion about the sun—[*K.K.*]

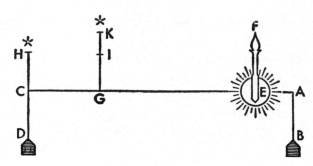

a) What is the angular speed of the planet when viewed from the center of the circle?
   (ANSWER: 1.5°/h)
b) What are the maximum (and minimum) angular speeds of the planet when viewed
   from a point displaced from the center of the circle by 500 km? (ANSWER:
   Minimum speed: 1°/h)
c) Docs Kepler believe that the observed motion of the planets can be explained by
   invoking such optical irregularities?

Ex. 17.2 (KEPLER'S FIRST LAW). Kepler mentioned that the planets follow elliptical,
rather than circular orbits. This is known as *Kepler's first law of planetary motion*.
What is the definition of an ellipse? How is an ellipse constructed? Are the major
axes of the planets' orbital ellipses all parallel? Are their eccentricities identical?
And where exactly, in a planet's orbit, is the sun located?

Ex. 17.3 (KEPLER'S SECOND LAW). Consider Fig. 17.1 as a model for the irregular
motion of Mercury about the sun. Here, the sun acts as a fulcrum of a lever with a
weight, *B*, suspended at point *A* behind it. The force this lever applies to a nearby
planet, and hence its speed, depends on its distance from the sun. When it is nearby
(as at point *G*) its speed will be greater; when it is more distant (as at point *C*), its
speed will be slower.

a) Look up the aphelial and perihelial distances of Mercury. What are they? Also
   look up the orbital speed of Mercury as it is passing through its perihelion. How
   far does it move in 1 min? (ANSWER: about 3500 km)
b) Using Kepler's model and geometrical reasoning, calculate the distance traveled
   (in kilometers) during the same time interval (1 min) by Mercury when it is passing
   through its aphelion.
c) Now look up the orbital speed of Mercury at its aphelion. Does your calculation
   agree with this value? Does Kepler's model work?
d) Calculate the area of the sector swept over by the arm of Kepler's lever during
   1 min (i) as Mercury is moving through its perihelion and (ii) as Mercury is moving
   through its aphelion. Are the areas of these sectors equal? Is this consistent with
   a modern formulation of Kepler's second law of planetary motion?

17.5 Vocabulary

1. Immutable
2. Traverse
3. Eccentric
4. Epicycle
5. Ellipse
6. Turbulence
7. Remission
8. Loadstone
9. Rectilinear
10. Advene
11. Virtue
12. Plausible
13. Emanation
14. Fulcrum
15. Inertia

Chapter 18
Mountains on the Moon

*It is most beautiful and pleasing to the eye to look upon the
lunar body, distant from us about sixty terrestrial diameters,
from so near as if it were distant by only two of these measures.*
—Galileo Galilei

18.1 Introduction

Galileo Galilei (1564–1642) was born in Pisa. As a young boy he attended a
monastery school, where he studied Latin classics and Greek. He went on to study
medicine and mathematics at the University of Pisa, but left without a degree due to
lack of funds. After this, he spent a few years doing private teaching and independent
research, then went on to serve as a lecturer at the University of Pisa before he was
appointed to chair of mathematics at the University of Padua in 1591. It was here that
he carried out the work which would be published in 1610 under the title *Sidereus
Nuncius*—the sidereal, or starry, messenger. The following text selections were translated from the latin text of Galileo's *Sidereal Messenger* by Edward Stafford Carlos
in 1880 and revised by Maurice A. Finocchiaro in 2008. Galileo's sketches of the
moon, contained herein, have been kindly provided by the History of Hydraulics
Rare Book Collection which is maintained by the IIHR at the University of Iowa.
Galileo begins this text by describing the spyglass—or telescope—with which his
ground-breaking observations were made possible.

As a result of his publication of the *Sidereal Messenger*, Galileo was appointed
Philosopher and Chief Mathematician to the grand duke of Tuscany, which provided
him with the leisure and salary conducive to carrying out extensive research and
writing. Subsequently, his strident defense of Copernican heliocentrism, coupled
with his trenchant attacks on Aristotelian doctrines, led to conflict with academic
and ecclesiastical authorities, and his eventual trial before the Roman Inquisition,
during which he was found guilty of "vehement suspicion of heresy" and sentenced
to house arrest in Florence. It was during this time that he produced his famous
work, the *Discorsi e dimostrazioni matematiche, intorno à due nuove scienze*, which
was published in 1638. In these *Dialogues Concerning Two New Sciences* Galileo
presents his views on the science of materials and projectile motion.[1]

---

[1] Selections from Galileo's *Dialogues* are included in Chaps. 1–12 of volume II.

K. Kuehn, *A Student's Guide Through the Great Physics Texts,*
Undergraduate Lecture Notes in Physics, DOI 10.1007/978-1-4939-1360-2_18,
© Springer Science+Business Media, LLC 2015

18.2   Reading: Galileo, *The Sidereal Messenger*

Finocchiaro, M. A. (Ed.), *The Essential Galileo*, Hackett Publishing Company, Indianapolis, Indiana, 2008, Chap. 1.[2]

18.2.1   To the Most Serene Cosimo II de' Medici, Fourth Grand Duke of Tuscany

[§1.1] There is certainly something very noble and humane in the intention of those who have endeavored to protect from envy the noble achievements of distinguished men, and to rescue their names, worthy of immortality, from oblivion and decay. This desire has given us the images of famous men, sculptured in marble, or fashioned in bronze, as a memorial of them to future ages; to the same feeling we owe the erection of statues, both ordinary and equestrian; hence, as the poet[3] says, has originated expenditure, mounting to the stars, upon columns and pyramids; with this desire, lastly, cities have been built, and distinguished by the names of those men, whom the gratitude of posterity thought worthy of being handed down to all ages. For the state of the human mind is such that, unless it be continually stirred by the likenesses of things obtruding themselves upon it from without, all recollection of them easily passes away from it.

Others, however, having regard for more stable and more lasting monuments, secured the eternity of the fame of great men by placing it under the protection, not of marble or bronze, but of the Muses' guardianship and the imperishable monuments of literature. But why do I mention these things? As if human wit, content with these regions, did not dare to advance further; whereas, since it well understood that all human monuments do perish at last by violence, by weather, or by age, it took a wider view and invented more imperishable signs, over which destroying Time and envious Age could claim no rights; so, betaking itself to the sky, it inscribed on the well-known orbs of the brightest stars—those everlasting orbs—the names of those who, for eminent and god-like deeds, were accounted worthy to enjoy an eternity in company with the stars. Wherefore the fame of Jupiter, Mars, Mercury, Hercules, and the rest of the heroes by whose names the stars are called, will not fade until the extinction of the splendor of the constellations themselves.

But this invention of human shrewdness, so particularly noble and admirable, /56/ has gone out of date ages ago, inasmuch as primeval heroes are in possession of those bright abodes and keep them by a sort of right. Into such company the affection of Augustus in vain attempted to introduce Julius Caesar; for when he wished that the name Julian should be given to a star that appeared in his time (one of those which the

---

[2] Cf. Galilei 1890–1909, 3: 53–96; translated by Edward Stafford Carlos (1880) from Galileo Galilei, *Sidereus nuncius* (Venice, 1610); revised by Finocchiaro.

[3] Sextus Propertius (c. 50 B.C.–c. 16 B.C.), *Elegies*, iii, 2, 17–22.

Greeks and the Latins alike name, from their hair-like tails, comets), it vanished in a short time and mocked his too-eager hope. But we are able to prophesize far truer and happier things for your highness, Most Serene Prince, for scarcely have the immortal graces of your mind begun to shine on earth, when bright stars present themselves in the heavens, like tongues to tell and celebrate your most eminent virtues to all time. Behold then, reserved for your famous name, four stars, belonging not to the ordinary and less-distinguished multitude of the fixed stars, but to the illustrious order of the planets; like genuine children of Jupiter, they accomplish their orbital revolutions around this most noble star with mutually unequal motions and with marvelous speed, and at the same time all together in common accord they also complete every 12 years great revolutions around the center of the world, certainly around the sun itself.

But the Maker of the Stars himself seemed to direct me by clear reasons to assign these new planets to the famous name of Your Highness in preference to all others. For just as these stars, like children worthy of their sire, never leave the side of Jupiter by any appreciable distance, so who does not know that clemency, kindness of heart, gentleness of manners, splendor of royal blood, nobility in public functions, wide extent of influence and power over others (all of which have fixed their common abode and seat in Your Highness),—who, I say, does not know that all these qualities, according to the providence of God, from whom all good things do come, emanate from the benign star of Jupiter? Jupiter, I maintain, at the instant of the birth of Your Highness having at length emerged from the turbulent mists of the horizon, and occupying the middle quarter of the heavens, and illuminating the eastern angle from his own royal house, from that exalted throne Jupiter looked out upon your most happy birth and poured forth into a most pure air all the brightness of his majesty, in order that your tender body and your mind (already adorned by God with still more splendid graces) might imbibe with your first breath the whole of that influence and power. But why should I use only probable arguments when I can demonstrate my conclusion with an almost necessary reason? It was the will of the Almighty God that I should be judged by your most serene parents not unworthy to be employed in teaching mathematics to Your Highness, which duty I discharged, during the four years just passed, at that time of the year when it is customary to relax from more severe studies. Wherefore, since it fell to my lot, evidently by God's will, to serve Your Highness /57/ and so to receive the rays of your incredible clemency and beneficence in a position near your person, what wonder is it if you have so warmed my heart that it thinks about scarcely anything else by day and night, but how I, who am under your dominion not only by inclination but also by my very birth and nature, may be known to be most anxious for your glory and most grateful to you? And so, inasmuch as under your auspices, Most Serene Cosimo, I have discovered these stars, which were unknown to all astronomers before me, I have, with very good right, determined to designate them with the most august name of your family. And as I was the first to investigate them, who can rightly blame me if I give them a name and call them the *Medicean Stars*, hoping that as much consideration may accrue to these stars from this title as other stars have brought to other heroes? For, not to speak of your most serene ancestors, to whose everlasting glory the monuments of all history bear witness, your virtue alone, most mighty hero, can confer on those stars an immortal name. Similarly, who can doubt that you will not only maintain and preserve the expectations, high though they be, about yourself which you have aroused by the very happy beginning of your government, but that you will also far surpass them, so that when you have conquered your peers, you may still vie with yourself and become day by day greater than yourself and your greatness?

Accept, then, Most Clement Prince, this addition to the glory of your family, reserved by the stars for you. And may you enjoy for many years those good blessings, which are sent to you not so much from the stars as from God, the Maker and Governor of the stars.

Your Highness's most devoted servant,

Galileo Galilei, Padua, 12 March 1610.

18.2.2   Astronomical Message Containing and Explaining Observations Lately Made with the Aid of a New Spyglass Regarding the Moon's Surface, the Milky Way, Nebulous Stars, an Innumerable Multitude of Fixed Stars, and Also Regarding Four Planets Never Before Seen, Which Have Been Named Medicean Stars

[§1.2] In the present small treatise I set forth some matters of great interest for all observers of natural phenomena to look at and consider. They are of great interest, I think, first, because of their intrinsic excellence; secondly because of the instrument by the aid of which they have been presented to our senses.

The number of the fixed stars which observers have been able to see without artificial powers of sight up to this day can be counted. It is therefore decidedly a great feat to add to their number, and to set distinctly before our eyes the other stars in myriads, which have never been seen before, and which surpass the old, previously known, stars in number more than ten times.

Again, it is a most beautiful and delightful sight to behold the body of the moon, which is distant from us by nearly 60 radii of the earth, as near as if it were at a distance of only two of the same measures. So the diameter of this same moon appears about 30 times larger, its surface about 900 times, and its solid mass nearly 27,000 times larger than when it is viewed only with the naked eye. And consequently anyone may know with the certainty that is due to the use of our senses that the moon certainly does not possess a smooth and polished surface, but /60/ one rough and uneven, and, just like the face of the earth itself, it is everywhere full of vast protuberances, deep chasms, and sinuosities.

Then to have got rid of disputes about the galaxy or Milky Way, and to have made its essence clear to the senses, as well as to the intellect, seems by no means a matter that ought to be considered of slight importance. In addition to this, to point out, as with one's finger, the substance of those stars which every one of the astronomers up to this time has called nebulous and to demonstrate that it is very different from what has hitherto been believed, will be pleasant and very beautiful.

But that which will excite the greatest astonishment by far, and which indeed especially moved me to call it to the attention of all astronomers and philosophers, is this: I have discovered four wandering stars, neither known nor observed by any one of the astronomers before my time; they have their orbits around a certain important star of those previously known and are sometimes in front of it, sometimes behind it, though they never depart from it beyond certain limits, like Venus and Mercury around the sun.

All these facts were discovered and observed a short time ago with the help of a spyglass[4] devised[5] by me, through God's grace first enlightening my mind. Perchance other discoveries still more excellent will be made from time to time by me and by other observers with the assistance of a similar instrument. So I will first briefly record its shape and preparation, as well as the occasion of its being devised, and then I will give an account of the observations made by me.

[§1.3] About 10 months ago, a report reached my ears that a Dutchman had constructed a spyglass, by the aid of which visible objects, although a great distance from the eye of the observer, were seen distinctly as if near; and some demonstrations of its wonderful performances were reported, which some gave credence to, but others contradicted. A few days later, I received confirmation of the report in a letter written from Paris by a noble Frenchman, Jacques Badovere. This finally determined me to give myself up first to inquire into the principle of the spyglass, and then to consider the means by which I might arrive at the invention of a similar instrument. After a little while I succeeded, through deep study of the theory of refraction. I prepared a tube, at first of lead, in the ends of which I fitted two glass lenses, both plane on one side, but on the other side one spherically convex, and the other concave. Then bringing my eye to the concave lens /61/ I saw objects satisfactorily far and near, for they appeared one-third of the distance and nine times larger than when they are seen with the natural eye alone. Shortly afterwards I constructed another more precise spyglass, which magnified objects more than 60 times. Finally, by sparing neither labor nor expense, I succeeded in constructing for myself an instrument so superior that objects seen through it appear magnified nearly 1000 times, and more than 30 times nearer than if viewed by the natural powers of sight alone.

It would be altogether a waste of time to enumerate the number and importance of the benefits which this instrument may be expected to confer when used by land or sea. But without paying attention to its use for terrestrial objects, I betook myself

---

[4] Here and in the rest of *The Sidereal Messenger* I have changed Stafford Carlos' translation of *perspicillium* as *telescope* because the latter word was not coined until 1611. For more information, see Rosen 1947; Van Helden 1989, 112; Pantin 1992, 50 no. 5; Battistini 1993, 190 no. 72.

[5] Here I retain Stafford Carlos' (1880, 9) translation of the original Latin *excogitati*. This rendition was also adopted by Drake (1983, 18). Other correct translations are *contrived* (Van Helden 1989, 36) and *conceived*, or *conçue* in French (Pantin 1992, 7). The more important point is to note that Galileo is *not* claiming to have been the *first* to *invent* the instrument, and his account in the next paragraph makes this disclaimer explicit.

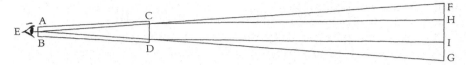

Fig. 18.1 Rays of light from an object entering the eye *via* Galileo's spyglass—[*K.K.*]

to observations of the heavenly bodies. First of all, I viewed the moon as near as if it were scarcely two radii of the earth distant. After the moon, I frequently observed other heavenly bodies, both fixed stars and planets, with incredible delight; and, when I saw their very great number, I began to consider about a method by which I might be able to measure their distances apart, and finally I found one.

Here it is fitting that all who intend to turn their attention to observations of this kind should receive certain cautions. In the first place, it is absolutely necessary for them to prepare a most perfect spyglass, one that will show very bright objects distinct and free from any mistiness and will magnify them at least 400 times and show them as if only one-twentieth of their distance. Unless the instrument be of such power, it will be in vain to attempt to view all the things that have been seen by me in the heavens, and that will be enumerated below. Then in order that one may be a little more certain about the magnifying power of the instrument, one shall fashion two circles or two square pieces of paper, one of which is 400 times greater than the other; this will happen when the diameter of the greater is twenty times the length of the diameter of the other. Then one shall view from a distance simultaneously both surfaces, fixed on the same wall, the smaller with one eye applied to the spyglass, and the larger with the other eye unassisted; for that may be done without inconvenience at one and the same instant with both eyes open. Then both figures will appear of the same size, if the instrument magnifies objects in the desired proportion.

After such an instrument has been prepared, the method of measuring distances remains for inquiry, and this shall be accomplished by the following contrivance. For the sake of being more easily understood, let $ABCD$ be the tube and E the eye of the observer (Fig. 18.1). When there are no lenses in the tube, rays from the eye to the object FG would be drawn in the straight lines ECF and EDG; /62/ but when the lenses have been inserted, the rays go in the bent lines ECH and ECI and are brought closer together, and those that originally (when unaffected by the lenses) were directed to the object FG will include only the part HI. Then, the ratio of the distance EH to the line HI being known, we shall be able to find, by means of a table of sines, the magnitude of the angle subtended at the eye by the object HI, which we shall find to contain only some minutes. Now, if we fit on the lens CD thin plates pierced some with larger and others with smaller apertures, by putting on over the lens sometimes one plate and sometimes another, as may be necessary, we shall construct at our pleasure different subtending angles of more or fewer minutes; by their help we shall be able to measure conveniently the intervals between stars separated by an angular distance of some minutes, within an error of 1 or 2 min. But let it suffice for the present to have thus slightly touched, and as it were just put

Fig. 18.2 The moon, when 4
or 5 days old. (Image
courtesy of IIHR)—[*K.K.*]

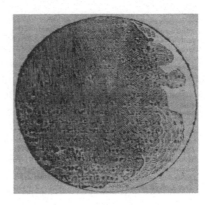

our lips to these matters, for on some other opportunity I will publish the complete
theory of this instrument.

Now let me review the observations I made during the past two months, again
calling the attention of all who are eager for true philosophy to the beginnings of
great contemplations.

[§1.4] Let me speak first of the surface of the moon that is turned toward us. For
the sake of being understood more easily, I distinguish two parts in it, which I call
respectively the brighter and the darker. The brighter part seems to surround and
pervade the whole hemisphere; but the darker part, like a sort of cloud, stains the
moon's surface and makes it appear covered with spots. Now these spots, as they are
somewhat dark and of considerable size, are plain to everyone, and every age has seen
them. Thus I shall call them *great* or *ancient* spots, to distinguish them from other
spots, smaller in size, but so thickly scattered that they sprinkle the whole surface
of the moon, especially the brighter portions of it. The latter spots have never been
observed by anyone before me. From my observation of them, often repeated, I have
been led to the opinion which I have expressed; that is, I feel sure that the surface of
the moon is not perfectly smooth, free from inequalities and exactly spherical (as a
large school of philosophers holds with regard to the moon and the other heavenly
bodies) but that on the contrary it is full of inequalities, uneven, /63/ full of hollows
and protuberances, just like the surface of the earth itself, which is varied everywhere
by lofty mountains and deep valleys. The appearances from which we may gather
this conclusion are the following.

On the 4th or 5th day after the new moon, when the moon presents itself to us
with bright horns, the boundary that divides the dark part from the bright part does
not extend smoothly in an ellipse, as would happen in the case of a purely spherical
body, but it is marked out in an irregular, uneven, and very wavy line, as represented
in the figure given (Fig. 18.2). Several bright excrescences, as they may be called,
extend beyond the boundary of light and shadow into the dark part, and on the other
hand pieces of shadow, encroach upon the bright.

Furthermore, a great quantity of small blackish spots, altogether separated from
the dark part, sprinkle everywhere almost the whole space that is at the time flooded

with the sun's light, with the exception of that part alone which is occupied by the great and ancient spots. I have noticed that the small spots just mentioned have this common characteristic always and in every case: that they have the dark part towards the sun's position, and on the side away from the sun they have brighter boundaries, as if they were crowned with shining summits. Now we have an appearance quite similar on the earth at sunrise, when we behold the valleys, not yet flooded with light, but the mountains surrounding them on the side opposite to the sun alway ablaze with the splendor of its beams; /64/ and just as the shadows in the hollows of the earth diminish in size as the sun rises higher, so also these spots on the moon lose their blackness as the illuminated part grows larger and larger.

However, not only are the boundaries of light and shadow in the moon seen to be uneven and sinuous, but—and this produces still greater astonishment—there appear very many bright points within the darkened portion of the moon, altogether divided and broken off from the illuminated area, and separated from it by no inconsiderable interval; they gradually increase in size and brightness, and after an hour or two they become joined on to the rest of the bright portion, now become somewhat larger. But in the meantime others, one here and another there, shooting up as if growing, are lighted up within the shaded portion, increase in size, and at last are linked on to the same luminous surface, now still more extended. An example of this is given in the same figure (Fig. 18.2). Now, is it not the case on the earth before sunrise that while the level plain is still in shadow, the peaks of the most lofty mountains are illuminated by the sun's rays? After a little while, does not the light spread further while the middle and larger parts of those mountains are becoming illuminated; and finally, when the sun has risen, do not the illuminated parts of the plains and hills join together? The magnitude, however, of such prominences and depressions in the moon seems to surpass the ruggedness of earth's surface, as I shall hereafter show.

And here I cannot refrain from mentioning what a remarkable spectacle I observed while the moon was rapidly approaching her first quarter, a representation of which is given in the same illustration above (Fig. 18.2). A protuberance of the shadow, of great size, indented the illuminated part in the neighborhood of the lower cusp. When I had observed this indentation a while, and had seen that it was dark throughout, finally, after about 2 h, a bright peak began to arise a little below the middle of the depression. This gradually increased, and presented a triangular shape, but was as yet quite detached and separated from the illuminated surface. Soon around it three other small points began to shine. Then when the moon was just about to set, that triangular figure, having now extended and widened, began to be connected with the rest of the illuminated part, and, still girt with the three bright peaks already mentioned, suddenly burst into the indentation of shadow like a vast promontory of light.

Moreover, at the ends of the upper /65/ and lower cusps, certain bright points, quite away from the rest of the bright part, began to rise out of the shadow, as is seen in the same illustration (Fig. 18.2). In both horns also, but especially in the lower one, there was a great quantity of dark spots, of which those that are nearer the boundary of light and shadow appear larger and darker, but those that are more remote less dark and more indistinct. In all cases, however, as I have already mentioned before, the

Fig. 18.3 The moon, before
the first quarter. (Image
courtesy of IIHR)—[*K.K.*]

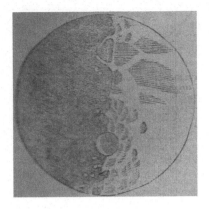

dark portion of the spot faces the direction of the sun's illumination, and a brighter
edge surrounds the darkened spot on the side away from the sun and towards the
region of the moon in shadow. This part of the surface of the moon, where it is
marked with spots like a peacock's tail with its azure eyes, looks like those glass
vases that, through being plunged while still hot from the kiln into cold water, acquire
a crackled and wavy surface, from which circumstance they are commonly called
frosted glasses.

Now, the great spots of the moon observed at the same time are not seen to be at
all similarly broken, or full of depressions and prominences, but rather to be even
and uniform; for only here and there some spaces, rather brighter than the rest, crop
up. Thus if anyone wishes to revive the old opinion of the Pythagoreans, that the
moon is another earth, so to speak, the brighter portions may very fitly represent the
surface of the land, and the darker the expanse of water; indeed, I have never doubted
that if the sphere of the earth were seen from a distance, when flooded with the sun's
rays, the part of the surface which is land would present itself to view as brighter, and
that which is water as darker in comparison. Moreover, the great spots in the moon
are seen to be more depressed than the brighter areas; for in the moon, both when
crescent and when waning, on the boundary between light and the shadow that is
seen in some places around the great spots, the adjacent regions are always brighter,
as I have indicated in drawing my illustrations; and the edges of the said spots are not
only more depressed than the brighter parts, but are more even, and are not broken
by ridges or ruggedness. But the brighter part stands out most near the spots so that
both before the first quarter and near the third quarter also, around a certain spot in
the upper part of the figure, that is, occupying the northern region of the moon, some
vast prominences on the upper and lower sides of it rise to an enormous elevation,
as the following illustrations show (see Figs. 18.3 and 18.4).

This same spot before the third quarter is seen to be walled around with boundaries
of a deeper shade, which, just like very lofty **/66/** mountain summits, appear darker
on the side away from the sun, and brighter on the side where they face the sun. But
in the case of cavities the opposite happens, for the part of them away from the sun
appears brilliant, and the part that lies nearer to the sun dark and in shadow. After a

Fig. 18.4 The moon, near
third quarter. (Image courtesy
of IIHR)—[*K.K.*]

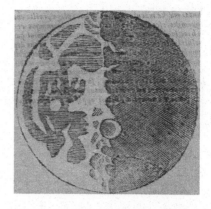

Fig. 18.5 The moon, before
the third quarter. (Image
courtesy of IIHR)—[*K.K.*]

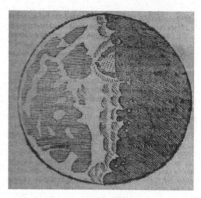

Fig. 18.6 The moon, near
third quarter. (Image courtesy
of IIHR)—[*K.K.*]

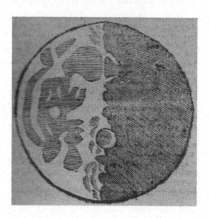

time, when the bright portion of the moon's surface has diminished in size, as soon as
the whole or nearly so of the spot already mentioned is covered with shadow, /67/ the
brighter ridges of the mountains rise high above the shade. These two appearances
are shown in the following illustrations (see Figs. 18.5 and 18.6).

There is one other point which I must on no account forget, and which I have noticed and rather wondered at. /68/ It is this. The middle of the moon, as it seems, is occupied by a certain cavity larger than all the rest, and in shape perfectly round. I have looked at this depression near both the first and third quarters, and I have represented it as well as I can in the two illustrations given above (Figs. 18.5 and 18.6). It produces the same appearance with regard to light and shade as an area like Bohemia would produce on the earth, if it were shut in on all sides by very lofty mountains arranged on the circumference of a perfect circle; for this area in the moon is walled in with peaks of such enormous height that the furthest side adjacent to the dark portion of the moon is seen bathed in sunlight before the boundary between light and shade reaches halfway across the circular space. But according to the characteristic property of the rest of the spots, the shaded portion of this too faces the sun, and the bright part is towards the dark side of the moon, which for the third time I advise to be carefully noticed as a most solid proof of the ruggedness and unevenness spread over the whole of the bright region of the moon. Of these spots, moreover, the darkest are always those that are near to the boundary line between the light and shadow, but those further off appear both smaller in size and less decidedly dark; so that finally, when the moon at opposition becomes full, the darkness of the cavities differs from the brightness of the prominences by a modest and very slight difference.

The phenomena which we have reviewed are observed in the bright areas of the moon. In the great spots, we do not see such differences of depressions and prominences as we are compelled to recognize in the brighter parts owing to the change of their shape under different degrees of illumination by the sun's rays, according to the manifold variety of the sun's position with regard to the moon. Still, in the great spots there do exist some areas rather less dark than the rest, as I have noted in the illustrations; but these areas always have the same appearance, and the depth of their shadow is neither intensified nor diminished; they do appear indeed sometimes slightly darker and sometimes slightly brighter, according as the sun's rays fall upon them more or less obliquely; and besides, they are joined to the adjacent parts of the spots with a very gradual connection, so that their boundaries mingle and melt into the surrounding region. But it is quite different with the spots that occupy the brighter parts of the moon's surface, for, just as if they were precipitous mountains with numerous rugged and jagged peaks, they have well-defined boundaries through the sharp contrast of light and shade. /69/ Moreover, inside those great spots, certain other areas are seen brighter than the surrounding region, and some of them very bright indeed; but the appearance of these, as well as of the darker areas, is always the same; there is not change of shape or brightness or depth of shadow; so it becomes a matter of certainty and beyond doubt that their appearance is due to the real dissimilarity of parts, and not to unevenness only in their configuration, changing in different ways the shadows of the same parts according to the variations of their illumination by the sun; this really happens in the case of the other smaller spots occupying the brighter portion of the moon, for day by day they change, increase, decrease, or disappear, inasmuch as they derive their origin only from the shadows of prominences.

But here I feel that some people may be troubled with grave doubt, and perhaps seized with a difficulty so serious as to compel them to feel uncertain about the conclusion just explained and supported by so many phenomena. For if that part of the moon's surface which reflects the sun's rays most brightly is full of innumerable sinuosities, protuberances, and cavities, why does the outer edge looking toward the west when the moon is waxing, and the other half-circumference looking toward the east when the moon is waning, and the whole circle at full moon appear not uneven, rugged, and irregular, but perfectly round and circular, as sharply defined as if marked out with a compass, and without the indentation of any protuberances or cavities? And most remarkably so, because the whole unbroken edge belongs to the brighter part of the moon's surface, which I have said to be full of protuberances and cavities; for not one of the great spots extends quite to the circumference, but all of them are seen to be together away from the edge. Of this phenomenon, which provides a handle for such serious doubts, I produce two causes, and so two solutions of the difficulty.

The first solution I offer is this. If the protuberances and cavities in the body of the moon existed only on the edge of the circle that bounds the hemisphere which we see, then the moon might, or rather would have to, show itself to us with the appearance of a toothed wheel, being bounded with an irregular and uneven circumference. But if instead of a single set of prominences arranged along the actual circumference only, there are many ranges of mountains with their cavities and ruggedness set one behind the other along the extreme edge of the moon (and that too not only in the hemisphere which we see but also in that which is turned away from us, but still near the boundary of the hemisphere), then the eye, viewing them from afar, will not at all be able to detect the differences of prominences and cavities; /70/ for the intervals between the mountains situated in the same circle, or in the same chain, are hidden by the jutting forward of other prominences situated in other ranges, especially if the eye of the observer is placed in the same line with the tops of the prominences mentioned. Similarly, on the earth the summits of a number of mountains close together appear situated in one plane, if the spectator is a long way off and standing at the same elevation; and when the sea is rough, the tops of the waves seem to form one plane, although between the billows there is many a gulf and chasm, so deep that not only the hulls, but even the bulwarks, masts, and sails of stately ships are hidden among them. Therefore, within the moon as well as around her circumference, there is a manifold arrangement of prominences and cavities, and the eye, viewing them from a great distance, is placed in nearly the same plane with their summits, and so no one need think it strange that they present themselves to the visual ray which just grazes them as an unbroken line quite free from unevenness.

To this explanation may be added another, namely, that there is around the body of the moon, just as around the earth, an envelope of some substance denser that the rest of the aether, which is sufficient to receive and reflect the sun's rays, although it does not possess so much opaqueness as to be able to prevent our seeing through it—especially when it is not illuminated. That envelope, when illuminated by the sun's rays, renders the body of the moon apparently larger than it really is, and would be able to stop our sight from penetrating to the solid body of the moon, if

Fig. 18.7 The varying
thickness of the atmospheric
envelope surrounding the
moon when viewed at
different angles.—[*K.K.*]

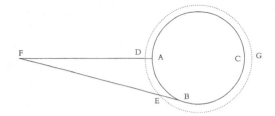

its thickness were greater. Now, it is of greater thickness around the periphery of the
moon—greater, I mean, not in actual thickness, but with reference to our sight-rays,
which cut it obliquely. And so it may stop our vision, especially when it is in a state
of brightness, and it may conceal the true circumference of the moon on the side
towards the sun. This may be understood more clearly from the following figure
(Fig. 18.7).

/71/ Here, the body of the moon, *ABC*, is surrounded by an enveloping atmosphere, *DEG*. An eye at *F* penetrates to the middle parts of the moon, as at *A*,
through a thickness, *DA*, of the atmosphere; but towards the extreme parts a mass
of atmosphere of greater depth, *EB*, shuts out its boundary from our sight. An argument in favor of this is that the illuminated portion of the moon appears of larger
circumference than the rest of the globe that is in shadow. Perhaps some will also
think that this same cause provides a very reasonable explanation why the greater
spots on the moon are not seen to reach the edge of the circumference on any side,
although it might be expected that some would be found near the edge as well as
elsewhere; it seems credible that there are spots there, but that they cannot be seen
because they are hidden by a mass of atmosphere too thick and too bright for the
sight to penetrate.

I think it has been sufficiently made clear, from the description of the phenomena
given above, that the brighter part of the moon's surface is dotted everywhere with
protuberances and cavities. It only remains for me to speak about their size, and
to show that the disparities of the earth's surface are far smaller than those of the
moon's—smaller, I mean, absolutely, so to speak, and not only smaller in proportion
to the size of the globes on which they are. And this is plainly shown thus.

I often observed in various positions of the moon with references to the sun
that some summits within the portion of the moon in shadow appeared illuminated,
although at some distance from the boundary of light. Then by comparing their
distance with the complete diameter of the moon, I learned that it sometimes exceeded
one-twentieth of the diameter. Suppose the distance to be exactly one-twentieth of
the diameter, and let the following diagram (see Fig. 18.8) represent the moon's
globe.

Here, *CAF* is a great circle, *E* its center, and *CF* a diameter, which consequently
bears to the diameter of the earth the ratio 2:7; and since the diameter of the earth,
according to the most exact observations, contains 7000 Italian miles, *CF* will be
2000, *CE* 1000, and one-twentieth of the whole *CF* will be 100 miles. Also, let *CF*
be a diameter of the great circle that divides the bright part of the moon from the

Fig. 18.8 Galileo's method
for determining the altitude of
lunar mountains.—[*K.K.*]

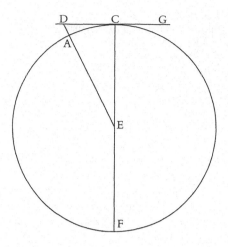

dark part, /72/ for owing to the very great distance of the sun from the moon this circle does not differ sensibly from a great one; let the distance of A from point C be one-twentieth of that diameter; let the radius EA be drawn, and let it be extended to cut the tangent line GCD (which represents a light ray) at point D. Then the arc CA, or the straight line CD, will be 100 of such units, and CE is 1000. But the sum of the squares of CD and CE is 1,010,000, and the square of ED is equal to this; thus, the whole ED will be more than 1004, and AD will be more than four of such units since CE was 1000. Therefore, the height of AD on the moon, which represents a summit reaching up to the sun's ray GCD and separated from the extremity C by the distance CD, is more than four Italian miles. But on the earth there are no mountains that reach to the perpendicular height of even 1 mile. We are therefore left to conclude that it is clear that the prominences on the moon are loftier than those on the earth.

I wish in this place to assign the cause of another lunar phenomenon well worthy of notice. This phenomenon was observed by me not lately, but many years ago, and it has been pointed out to some of my intimate friends and pupils, explained, and assigned to its true cause. Yet since the observation of it is rendered easier and more vivid by the help of a spy glass, I have considered that it would not be unsuitably introduced in this place. I wish to introduce it chiefly in order that the connection and resemblance between the moon and the earth may appear more plainly.

When the moon, both before and after conjunction, is found not far from the sun, not only does its globe show itself to our sight on the side where it is furnished with shining horns, but a slight and faint circumference is also seen to mark out the circle of the dark part (namely, the part that is turned away from the sun), and to separate it from the darker background of the sky. Now, if we examine the matter more closely, we shall see that not only is the extreme edge of the part in shadow shining with faint brightness, but that the entire face of the moon (namely, the side that does not feel the sun's glare) is whitened with a not-inconsiderable light. At first glance only a fine circumference appears shining, on account of the darker part of the sky adjacent to it; whereas, on the contrary, the rest of the surface appears dark, on account of the contiguity of the shining horns, which obscures our vision. But if one looks at the moon from such a position that by the interposition of a distant roof, chimney, or some other object the shining horns are hidden and the rest of the lunar globe is left exposed to one's view, /73/ then one will find that this part of the moon (although deprived of sunlight) also gleams with considerable light, and particularly so if the gloom of the night has already deepened from the absence of the sun; for with a darker background the same light appears brighter. Moreover, it is found that this secondary brightness of the moon, as I may call it, is greater in proportion as the moon is less distant from the sun, and that it abates more and more in proportion to the moon's distance from that body; thus, after the first quarter and before the end of the second it is found to be weak and very faint, even if it is observed in a darker sky; whereas at an angular distance of 60° or less, even during twilight, it is wonderfully bright, so bright indeed that, with the help of a good spyglass, the great spots may be distinguished in it.

This strange brightness has presented no small perplexity to philosophical minds; and some have mentioned one thing, some another, as the cause to be alleged for it. Some have said that it is the inherent and natural glow of the moon; some that it is imparted to that body by the planet Venus, or, as others maintain, by all the stars; while some have said that it comes from the sun, whose rays find a way through the solid mass of the moon. But statements of this kind are disproved without much difficulty and convincingly demonstrated to be false.

In fact, if this kind of light were the moon's own, or were contributed by the stars, the moon would retain it, particularly in eclipses, and would show it then, when she is in an unusually dark sky. But this is contrary to experience since the brightness that is seen on the moon in eclipses is far less intense, being somewhat reddish and almost copper-colored, whereas this is brighter and whiter. Moreover, the brightness seen during an eclipse is changeable and shifting, for it wanders over the face of the moon in such a way that the part near the circumference of the circle of shadow cast by the earth is bright, but the rest of the moon is always seen to be dark. From this circumstance we understand without hesitation that this brightness is due to the proximity of the sun's rays coming into contact with some denser region that surrounds the moon as an envelope; owing to this contact a sort of dawn-light is diffused over the neighboring regions of the moon, just as twilight spreads in the morning and evening on the earth. But I will treat more fully of this matter in my book *On the System of the World*.

On the other hand, to assert that this sort of light is imparted to the moon by the planet Venus is so childish as to be undeserving of an answer. For who is so ignorant as not to understand that at conjunction and within an angular distance of 60° it is quite impossible for the part of the moon turned away from the sun to be seen by the planet Venus?

However, that this light is derived from the sun penetrating with its light the solid mass of the moon, and rendering it luminous, is equally untenable. For then this light would never lessen, since a hemisphere of the moon is always illuminated by the sun, except at the moment of a lunar eclipse; but in reality it quickly decreases while the moon /74/ is drawing near to the end of her first quarter, and when she has passed her first quarter it becomes quite dull.

Now, since this kind of secondary brightness is not inherent in and characteristic of the moon, nor borrowed from the sun or any other heavenly body, and since there now remains in the whole universe no other body whatever except the earth, what, pray, must we think? What must we propose? Shall we propose that the body of the moon, or some other dark and opaque body, receives light from the earth? What is so strange about that? Look: the earth, with fair and grateful exchange, pays back to the moon an illumination like that which it receives from the moon nearly the whole time during the darkest gloom of night. Let me explain the matter more clearly.

At conjunction, when the moon occupies a position between the sun and the earth, the moon is illuminated by the sun's rays on her half towards the sun and turned away from the earth, and the other half facing the earth is covered with darkness and so does not illuminate the earth's surface in any way. When the moon has separated slightly from the sun, straightaway she is partly illuminated on the half directed towards us; she turns towards us a slender silvery crescent and illuminates the earth slightly. The sun's illumination increases upon the moon as she approaches her first quarter, and the reflection of that light increases on the earth. Next, the brightness of the moon extends beyond the semicircle, and our nights grow brighter. Then the entire face of the moon looking towards the earth is irradiated with the most intense brightness by the sun, which happens when the sun and moon are on opposite sides of the earth; then far and wide the surface of the earth shines with the flood of moonlight. After this the moon, now waning, sends out less powerful beams, and the earth is illuminated less powerfully. Finally, the moon draws near her first position of conjunction with the sun, and forthwith black night invades the earth. In such a cycle the moonlight gives us each month alternations of brighter and fainter illumination. But the benefit of her light to the earth is balanced and repaid by the benefit of the light of the earth to her; for while the moon is found near the sun about the time of conjunction, she has in front of her the entire surface of that hemisphere of the earth which is exposed to the sun and is vividly illuminated with his beams, and so she receives light reflected from the earth. Owing to such reflection, the hemisphere of the moon nearer to us, though deprived of sunlight, appears of considerable brightness. Again, when removed from the sun by a quadrant, the moon sees only one-half of the earth's illuminated hemisphere (namely the western half), for the other (the eastern) is covered with the shades of night; the moon is then less brightly illuminated by the earth, and accordingly that secondary light appears fainter to us. But if you imagine the moon to be set on the opposite side of the earth to the sun, she will see the hemisphere of the earth, now between the moon and the sun, quite dark, and steeped in the gloom of night; if, therefore, an eclipse should accompany such a position of the moon, /75/ she will receive no light at all, being deprived of the illumination of the sun and earth together. In any other position with regard to the earth and the sun, the moon receives more or less light by reflection from the earth, according as she sees a greater or smaller portion of the hemisphere of the earth illuminated by the sun; for such a law is observed between these two globes, that at whatever times the earth is most highly illuminated by the moon, at those times, on the contrary, the moon is least illuminated by the earth; and vice versa.

Let these few words on this subject suffice in this place. I will consider it more fully in my *System of the World*. There, by very many arguments and experiments, it is shown that there is a very strong reflection of the sun's light from the earth, for the benefit of those who urge that the earth must be excluded from the dance of the stars, chiefly for the reason that it has neither motion nor light. For I will prove that the earth has motion, and surpasses the moon in brightness, and is not the place where the dull refuse of the universe has settled down; and I will support these conclusions by countless arguments taken from natural phenomena.

18.3   Study Questions

QUES. 18.1 After whom does Galileo name the planetary discovery described in the *Sidereal Messenger*? For what reason does he do so?

QUES. 18.2 What type of telescope does Galileo recommend for astronomical viewing?

a) What were the shapes of the lenses? What was the magnifying power of his telescope?
b) How can the magnification of a telescope be experimentally determined? What minimum magnification does Galileo suggest for viewing astronomical objects?
c) How can one measure the angle between two stars using a telescope?

QUES. 18.3 What is most significant about Galileo's observations of the moon?

a) What are some of the specific features that he observed? And to what does he compare the surface of the moon?
b) If Galileo is right about the moon's surface, then why, when it is full, does it look so smooth?
c) How did Galileo measure the size of the lunar mountains? How tall are the tallest?
d) Is the moon a source of light? If not, then why is its outline visible when in conjunction with the sun?
e) What do Galileo's observations imply about the moon's composition? Is it comprised of aether, as believed by Aristotle? Why is this significant?

18.4   Exercises

Ex. 18.1 (ANGULAR WIDTH). Suppose that you are holding a 1986 US Silver Eagle coin, and a 1942 Mercury Dime, both at arm's length (90 cm) from your eye.

a) What is the area, angular width, and solid angle of each? (ANSWER: the angular width of the dime is 0.02 rad, or 1.14°. The solid angle of the dime is 0.0003 steradians.)

b) What magnification power would you need to employ, when looking at the dime, so that it appears identical in size to the silver dollar?

Ex. 18.2 (MOUNTAIN HEIGHT). Suppose that a bright spot appears $1/8$th of the way between the perimeter and the center of the moon when the moon appears in its first quarter. If this is the tip of a mountain, then how tall is the mountain (in moon diameters, and also in miles)?

Ex. 18.3 (FOCAL LENGTH). The objective lens of the refracting telescope at the Yerkes Observatory has a focal length of 20 m. When using an eyepiece with a focal length of 4 cm, what is the magnification of this telescope? What would be the angular size of Venus when viewed with this telescope? (ANSWER: 500 times magnification.)

Ex. 18.4 (TELESCOPE LABORATORY). In the following laboratory exercises, we will attempt to understand the design and operation of a Keplerian telescope. For these experiments, you will need a number of lenses and lens holders and an optical bench or rail.[6]

Setting Up the Telescope First, arrange two or more small, bright light sources on one side of the laboratory. These will act as "stars". On the opposite side of the laboratory, place a lens (having a focal length of approximately 30 or 40 cm) in a lens holder atop an optical table or rail. This will serve as our *objective lens*. On the side of the lens opposite the constellation, place a small rigid white screen. Move the screen forward or backwards until an image of the constellation is focused on the screen. The distance between the lens and the screen is called the *focal length* of the lens.

Carefully inspect the image on the screen. Is it larger or smaller than the actual objects being observed? By how much? You might try to make the image appear larger by moving your eye very close to the image on the screen. How close can you get your eye before you can no longer focus on the image? This is called your *near point*. Your near point is set by the ability of the ciliary muscles of your eye to squeeze the lens so as to focus light on the retina in the rear of your eye. This physiological limitation can be overcome by using a magnifying glass to inspect the image formed by the objective lens. In this way, the image formed by the objective lens acts as the object for the magnifying glass. So select another lens having a focal length between 5 and 10 cm. We'll call this magnifying glass the *eyepiece lens*. What do you see? What problems do you run into when trying to look through the eyepiece lens at the screen?

It would probably be much easier to inspect the image from behind the screen. Unfortunately, the screen is opaque, so let's replace it with a semi-transparent sheet of glass, such as a partially frosted microscope slide. Again, focus the image on the

---

[6] For example, the Basic Optics System (Model OS-8515 C), PASCO, Roseville, CA.

frosted glass using your objective lens. Incidentally, this image is called a *real image*, as opposed to a *virtual image*. A virtual image is one formed at an apparent location at which no light rays are actually converging—for instance the image formed behind a mirror.

Now mount your eyepiece lens in a holder on the side of the frosted glass opposite the objective lens. Move the eyepiece lens until you can look through it and see a focused image of the constellation on the frosted glass slide. How is the distance between the image and the eyepiece related to the focal length of the eyepiece lens? You might need to do a quick measurement to figure out the focal length of this lens.

You have now built a *Keplerian telescope*. You can remove the frosted glass and use your telescope to view distant objects. Try to use your telescope to observe distant objects, such as shapes drawn on a board on the opposite side of the laboratory. What is the magnification of your telescope? In order to determine the magnification, you might use the method described by Galileo: he observed two objects simultaneously, one through the telescope and the other with the unaided eye. Explore some of the limitations of your telescope. In particular, how do the magnification, field of view, image brightness, and image sharpness depend upon the focal lengths and diameters of the objective and eyepiece lenses? Be as quantitative as possible when exploring these relationships.

The Thin-Lens Equation Place a backlit object at one end of an optical bench, a white screen at the other end, and a long-focal-length convex lens near the middle; all three should lie along the same axis. Now without moving the lens or object, move the screen until a sharp image of the obstruction appears on the screen. Measure the distances between the object and the lens (the *object distance*, d_o) and between the lens and the screen (the *image distance*, d_i). Measure the height, h_i, of the image appearing on the screen. Compared to the object, is the image on the screen enlarged? Is it upright or inverted?

Move the object a bit closer to the lens. Now again, without moving the object, adjust the position of the screen until a sharp image is formed. Record d_o, d_i and h_i. Repeat this procedure several times. What range of vales of d_o is feasible in obtaining good images? Plot d_o vs d_i. What is the shape of the curve? Now try to plot your data in a way that yields a straight line. Hint: consider plotting the inverse of your quantities. If you know the focal length of your lens, f, how is it related to your plot? If you do not know the focal length, then how can you deduce it from your straight line plot? Write down an equation relating d_o, d_i and f. This equation is called the "thin lens equation."

Magnification The magnification of a telescope is defined as the ratio of the image size to the object size. Calculate the magnification, m, for each point on your plot. Is there a relationship between the magnification and the distances d_o and d_i? If so, can you express it mathematically? Are there any stipulations regarding the distances d_o and d_i one must use so as to obtain a magnification greater than unity?

Spherical Aberration   The *f-ratio* of an optical system (such as a telescope or camera) is the (dimensionless) ratio of the focal length to the lens diameter. First, calculate the f-ratio of your lens. Now, pick one of your data points and set up the object, lens and screen so as to obtain a focused image on the screen. Place an adjustable iris between the object and the lens. Does the image change at all as the iris is closed? If so, how? This effect is referred to as *spherical aberration*. How can this effect be minimized?

Angular Width   The perceived size of an object is determined by its angular width. What is the angular width of your fingernail, in degrees, when held at arm's length? Now move your fingernail closer to your eye; it gets bigger and bigger. Eventually, however, you will reach your "near point," closer than which you will be unable to focus on your finger nail. What is your near point? What is the angular width of your fingernail at your near point? This is the largest angular width of your fingernail achievable without using a magnifying glass. Now place a converging lens between your fingernail and your eye. Can you move your fingernail closer than the near point of your eye? Compare the angular width of your fingernail at your near point and the angular width of your fingernail when it is at the focal point of your magnifying glass. The ratio of these two quantities is the angular magnification of the magnifying glass (in the case that your eye is fully relaxed).

Refracting Telescope Theory   A refracting telescope uses two lenses, an objective and an eyepiece. Light from a distant object first passes through the objective lens. The objective lens forms a real inverted image which is somewhat smaller than the object itself. The image produced by the objective then acts as the object for the eyepiece. In a refracting telescope, the image produced by the objective lies at, or just inside, the focal point of the eyepiece. If the focal length of the eyepiece is much less than the user's near point, then he or she can use the eyepiece as a magnifying glass to inspect this image up close without straining his or her eye. What is the relationship between the focal lengths of the objective and eyepiece lenses and the total length of your telescope? What is the relationship between the focal lengths of the two lenses and the magnification of your telescope? Why is this the case? Can you explain this geometrically, using light ray diagrams?

18.5   Vocabulary

1. Oblivion
2. Equestrian
3. Posterity
4. Obtruding
5. Inscribe
6. Primeval
7. Eminent
8. Illustrious
9. Sire
10. Clemency
11. Imbibe
12. Beneficence
13. Vie
14. Hitherto
15. Myriad
16. Terrestrial
17. Contrivance
18. Inquiry
19. Pervade
20. Excrescences
21. Girt
22. Promontory
23. Azure
24. Precipitous
25. Billows
26. Bulwark
27. Manifold
28. Disparity
29. Contiguity
30. Interposition
31. Abate
32. Perplexity
33. Inherent
34. Proximity
35. Luminous
36. Untenable
37. Forthwith

Chapter 19
The Medician Stars

Although I believed them to belong to the number of fixed stars,
yet they made me wonder somewhat, because they seemed to be
arranged exactly in a straight line.

—Galileo Galilei

19.1 Introduction

In the initial sections of *The Sidereal Messenger*, Galileo focused on his telescopic
observations of the moon. Perhaps most importantly, he noticed that the moon, like
Earth, possesses mountains and valleys. This seemingly mundane observation had
great significance, as it clearly demonstrated the falsehood of Aristotle's notion that
the moon, like other heavenly bodies, must be made of a fifth—ethereal—substance,
whose natural motion was circular. Aristotle's physics, founded on the natural motion
of the four earthly elements, was becoming increasingly difficult to defend. Now, in
the following text selection, Galileo turns his attention to the stars and the planets.
What additional observations did he make in this arena? And what new problems did
his discoveries pose for the followers of Aristotle?

19.2 Reading: Galileo, *The Sidereal Messenger*

Finocchiaro, M. A. (Ed.), *The Essential Galileo*, Hackett Publishing Company,
Indianapolis, Indiana, 2008, Chap. 1.

19.2.1 The Sidereal Messenger (continued)

[§1.5] Hitherto I have spoken of the observations which I have made concerning the
moon's body. Now I will briefly announce the phenomena that have been, as yet,
seen by me with reference to the fixed stars. And first of all the following fact is
worthy of consideration.

The stars, fixed as well as wandering, when seen with a spyglass, by no means
appear to be increased in magnitude in the same proportion as other objects, and the

K. Kuehn, *A Student's Guide Through the Great Physics Texts,*
Undergraduate Lecture Notes in Physics, DOI 10.1007/978-1-4939-1360-2_19,
© Springer Science+Business Media, LLC 2015

moon herself, gain increase of size. In the case of the stars such an increase appears
much less, so that you may consider that a spyglass which (for sake of illustration)
is powerful enough to magnify other objects 100 times will scarcely render the stars
magnified four or five times. The reason for this is as follows. When stars are viewed
with our natural eyesight, they do not present themselves to us in their bare, real size,
but beaming with a certain vividness, and fringed with sparkling rays, especially
when the night is far advanced; and from this circumstance they appear much larger
than they would if they were stripped of those adventitious fringes, for the angle
which they subtend at the eye is determined not by the primary disc of the star,
but by the brightness that so widely surrounds it. Perhaps you will understand this
most clearly from the well known circumstance that when stars rise at sunset, in the
beginning of twilight they appear very small, although they may be stars of the first
magnitude; and even the planet Venus itself, on any occasion when it may present
itself to view in broad daylight, is so small to see that it scarcely seems equal to a
star of the last magnitude. It is different in the case of other objects, and even of the
moon, which, whether viewed in the light of midday or in the depth of night, always
appears the same size. We conclude, therefore, that /76/ the stars are seen at midnight
in uncurtailed glory, but their fringes are of such a nature that the daylight can cut
them off, and not only daylight, but any slight cloud that may be interposed between
a star and the eye of the observer. A dark veil or colored glass has the same effect,
for upon placing it between the eye and the stars, all the blase that surrounds them
leaves them at once. A spyglass also accomplishes the same result, for it removes
from the stars their adventitious and accidental splendors before it enlarges their true
globes (if indeed they are of that shape), and so they seem less magnified than other
objects; for example, a star of the fifth or sixth magnitude seen through a spyglass is
shown as of the first magnitude.

The difference between the appearance of the planets and of the fixed stars seems
also deserving of notice. The planets present their bodies perfectly delineated and
round, and appear as so many little moons, completely illuminated and of a globular
shape. However the fixed stars do not look to the naked eye bounded by a circular
periphery, but rather like blazes of light, shooting out beams on all sides and very
sparkling; and with a spyglass they appear of the same shape as when they are viewed
by simply looking at them, but so much larger that a star of the fifth or sixth magnitude
seems to equal the Dog Star, the largest of the fixed stars. Beyond the stars of the sixth
magnitude, you will behold through the spyglass a host of other stars that escape
the unassisted eye, so numerous as to be almost beyond belief. You may see more
than six other magnitudes, and the largest of these (which one could call stars of the
seventh magnitude, or of the first magnitude of invisible stars) appear with the aid
of the spyglass larger and brighter than stars of the second magnitude seen with the
unassisted sight. In order that you may see one or two proofs of the inconceivable
manner in which they are crowded together, I have wanted to give you drawings of
two star clusters, so that from them as specimen you may decide about the rest.

As my first example (see Fig. 19.1), I had determined to depict the entire constellation of Orion, but I was overwhelmed by the vast quantity of stars and by want
of time, and so I have deferred attempting this to another occasion; for there are

Fig. 19.1 Constellation of
Orion's Belt and Sword

Fig. 19.2 Constellation of the
Pleiades

adjacent to, or scattered among, the old stars more than 500 new stars within the
limits of one or two degrees. For this reason I have selected the three stars in Orion's
Belt and the six in his Sword, which have long been well known groups, and I have
added eighty other stars recently discovered in their vicinity, and I have preserved
as exactly as possible the intervals between them. The well-known or old stars, for
the sake of distinction, I have depicted of larger size, and I have outlined them with
a double line; the others, invisible to the naked eye, I have marked smaller and with
one line only. I have also preserved the differences in magnitude as much as I could.

/78/ As a second example (see Fig. 19.2), I have depicted the six stars of the
constellation Taurus, called the Pleiades (I say *six* intentionally, since the seventh is
scarcely ever visible). This is a group of stars that appear in the heavens within very
narrow limits. Near these there lie more than forty others invisible to the naked eye,
no one of which is much more than half a degree off any of the aforesaid six. In my
diagram, I have marked only 36 stars; I have preserved their intervals, magnitudes,
and the distinction between the old and the new stars, just as in the case of the
constellation Orion.

The third thing which I have observed is the essence or substance of the Milky
Way. By the aid of a spyglass anyone may behold this in a manner which so distinctly
appeals to the senses that all the disputes which have tormented philosophers through

Fig. 19.3 Nebula of Orion

Fig. 19.4 Nebula of Praesepe

so many ages are exploded at once by the indubitable evidence of our eyes, and we are freed from wordy disputes upon this subject. In fact, the galaxy is nothing but a mass of innumerable stars planted together in clusters. For upon whatever part of it you direct the spyglass, straightaway a vast crowd of stars presents itself to view; many of them are tolerably large and extremely bright, but the number of small ones is quite beyond determination.

That milky brightness, like the brightness of a white cloud, is seen not only in the Milky Way, but also in several spots of a similar color that shine here and there in the heavens. If you turn the spyglass upon any of them, you will find a cluster of stars packed close together. Furthermore—and you will be surprised at this—the stars that have been called by every one of the astronomers up to this day *nebulous* are groups of small stars set thick together in a wonderful way. Although each star escapes our sight on account of its smallness, or of its immense distance from us, from the commingling of their rays there arises that brightness which has hitherto been believed to be the denser part of the heavens, able to reflect the rays of the stars or the sun **[79]**.

I have observed some of these, and I wish to reproduce the star clusters of two of these nebulas. First you have a diagram of the nebula called Orion's Head, in which I have counted 21 stars (Fig. 19.3). The second cluster contains the nebula called Praesepe, which is not a single star but a mass of more than forty small stars; besides

East   *   *   O   *   West

Fig. 19.5 Configuration of Jupiter and the Medicean stars on Jan. 7, 1610.—[*K.K.*]

East   O   *   *   *   West

Fig. 19.6 Configuration of Jupiter and the Medicean stars on Jan. 8, 1610.—[*K.K.*]

the two Aselli, I have marked thirty-six stars, arranged as in the following diagram (Fig. 19.4).

[§1.6] I have not finished my brief account of the observations which I have thus far made with regard to the moon, the fixed stars, and the galaxy. There remains the matter that seems to me to deserve to be considered the most important in this work. That is, I should disclose and publish to the world the occasion of discovering and observing four planets never seen from the beginning of the world up to our own times, their positions, and the observations made during /80/ the last two months about their movements and their changes of magnitude. And I summon all astronomers to apply themselves to examine and determine their periodic times, which it has not been permitted me to achieve up to this day owing to the restriction of my time. However, I give them warning again that they will need a very accurate spyglass, such as I have described at the beginning of this account, so that they may not approach such an inquiry to no purpose.

On the seventh day of January of the present year, 1610, at the first hour of the following night, when I was viewing the constellations of the heavens through a spyglass, the planet Jupiter presented itself to my view. As I had prepared for myself a very excellent instrument, I noticed a circumstance which I had never been able to notice before, owing to want of power in my other spyglass. That is, three little stars, small but very bright, were near the planet. Although I believed them to belong to the number of fixed stars, yet they made me wonder somewhat, because they seemed to be arranged exactly in a straight line parallel to the ecliptic, and to be brighter than the rest of the stars equal to them in magnitude. Their position with reference to one another and to Jupiter follows (Fig. 19.5).

On the east side there were two stars, and a single one towards the west. The star that was furthest towards the east, and the western star, appeared rather larger than the third. I scarcely troubled at all about the distance between them and Jupiter, for, as I have already said, at first I believed them to be fixed stars.

However, when on 8 January, led by some fatality, I turned again to look at the same part of the heavens, I found a very different state of things, There were three little stars all west of Jupiter, and nearer together than on the previous night; and they were separated from one another by equal intervals as the following illustration shows (Fig. 19.6).

At this point, although I gave no thought at all to the fact that the stars appeared closer to one another, yet I began to wonder how Jupiter could one day be found to the east of all the aforesaid fixed stars when the day before it had been west of two of them. And forthwith I wondered whether the planet might have been moving with

East   *   *   O   West

Fig. 19.7 Configuration of Jupiter and the Medicean stars on Jan. 10, 1610.—[*K.K.*]

East   *   *   O   West

Fig. 19.8 Configuration of Jupiter and the Medicean stars on Jan. 11, 1610.—[*K.K.*]

East   *   * O   *   West

Fig. 19.9 Configuration of Jupiter and the Medicean stars on Jan. 12, 1610.—[*K.K.*]

direct motion, contrary to the calculation of astronomers, and so might have passed those stars by its own proper motion. I therefore waited for the next night with the most intense longing, but I was disappointed in my hope, for the sky was covered with clouds in every direction.

/81/ But on 10 January, the stars appeared in the following position with regard to Jupiter (Fig. 19.7).

There were only two, and both on the east side of Jupiter, the third, as I thought, being hidden by the planet. They were situated, just as before, exactly in the same straight line with Jupiter, and along the zodiac. After seeing this, I understood that the corresponding changes of position could not by any means belong to Jupiter. Moreover, I knew that the stars I saw had always been the same, for there were no others either in front or behind, within a great distance along the zodiac. Finally, changing from perplexity to amazement, I became certain that the observed interchange of position was due not to Jupiter but to the said stars. Thus I thought that henceforth they ought to be observed with more attention and precision.

Accordingly, on 11 January, I saw an arrangement of the following kind (Fig. 19.8).

That is, there were only two stars to the east of Jupiter, the nearer of which was three times as far from it as from the star further to the east; and the star furthest to the east was nearly twice as large as the other one. But on the previous night they had appeared nearly of equal magnitude. I therefore concluded, and decided unhesitatingly, that there were three stars in the heavens moving around Jupiter, like Venus and Mercury around the sun. This was finally established as clear as daylight by numerous other subsequent observations. These observations also established that there are not only three, but four, wandering sidereal bodies performing their revolutions around Jupiter. The following account will report on the observations of these changes of position made with more exactness on succeeding nights. I have also measured the intervals between them with the spyglass, in the manner already explained. Besides this, I have given the times of observation, especially when several were made in the same night, for the revolutions of these planets are so swift that an observer may generally get differences of position every hour.

On 12 January, at the first hour of the next night, I saw these heavenly bodies arranged in this manner (Fig. 19.9).

East   *   O   *   *   West

Fig. 19.10 Configuration of Jupiter and the Medicean stars on Jan. 13, 1610.—[*K.K.*]

The star furthest to the east was greater than the one furthest to the west, but both were very conspicuous and bright. The distance of each one from Jupiter was two minutes. /82/ A third star, certainly not in view before, began to appear at the third hour; it nearly touched Jupiter on the east side and was exceedingly small. They were all arranged in the same straight line, along the ecliptic.

On 13 January, for the first time, four stars were in view in the following position with regard to Jupiter (Fig. 19.10).

There were three to the west, and one to the east. They made almost a straight line, but the middle star of those to the west deviated a little from the straight line towards the north. The star furthest to the east was at a distance of two minutes from Jupiter. There were intervals of only one minute between Jupiter and the nearest star, and between the stars themselves, west of Jupiter. All the stars appeared of the same size, and though small they were very brilliant and far outshone the fixed stars of the same magnitude. . .[1]

. . .I have wanted to report these comparisons of the position of Jupiter and its adjacent planets to a fixed star so that anyone may be able to understand from them that the movements of these planets both in longitude and in latitude agree exactly with the motions derived from the tables.

These are my observations of the four Medicean Planets, recently discovered for the first time by me. Although I am not yet able to deduce by calculation from these observations the orbits of these bodies, I may be allowed to make some statements based upon them, well worthy of attention. In the first place, since they are sometimes behind and sometimes before Jupiter at like distances and deviate from this planet towards the east and towards the west only within very narrow limits of divergence, and since they accompany this planet when its motion is retrograde as well as when it is direct, no one can doubt that they perform their revolutions around this planet while at the same time they all together accomplish orbits of 12 years' duration around the center of the world. Moreover, they revolve in unequal circles, which is evidently the conclusion /95/ from the fact that I never saw two planets in conjunction when their distance from Jupiter was great, whereas near Jupiter two, three, and sometimes all four have been found closely packed together. Furthermore, it may be deduced that the revolutions of the planets that describe smaller circles around Jupiter are more rapid, for the satellites nearer to Jupiter are often seen in the east when the day before they have appeared in the west, and vice versa; also the satellite moving in the greatest orbit seems to me, after carefully weighing the timing of its returning to positions previously noticed, to have a periodic time of half a month.

Additionally, we have a notable and splendid argument to remove the scruple of those who can tolerate the revolution of the planets around the sun in the Copernican

---

[1] Several diagrams have been omitted for the sake of brevity.—[*K.K.*].

system, but are so disturbed by the motion of one moon around the earth (while both accomplish an orbit of a year's length around the sun) that they think this constitution of the universe must be rejected as impossible. For now we have not just one planet revolving around another while both traverse a vast orbit around the sun, but four planets which our sense of sight presents to us circling around Jupiter (like the moon around the earth) while the whole system travels over a mighty orbit around the sun in the period of 12 years.

Lastly, I must not pass over the consideration of the reason why it happens that the Medicean Stars, in performing very small revolutions around Jupiter, seem sometimes more than twice as large as at other times. We can by no means look for an explanation in the mists of the earth's atmosphere, for they appear increased or diminished while the discs of Jupiter and the neighboring fixed stars are seen quite unaltered. It seems altogether untenable that they approach and recede from the earth at the points of their revolutions nearest to and furthest from the earth to such an extent as to account for such great changes, for a strict circular motion can by no means produce those phenomena; and an elliptical motion (which in this case would be almost rectilinear) seems to be both unthinkable and by no means in harmony with the observed phenomena. But I gladly offer the explanation that has occurred to me upon this subject, and I submit it to the judgment and criticism of all true philosophers. It is known that when atmospheric mists intervene, the sun and moon appear larger, but the fixed stars and planets smaller; hence the former luminaries, when near the horizon, are larger than at other times, but stars appear smaller and are frequently scarcely visible; and they are still more diminished if those mists are bathed in light; so stars appear very small by day and in the twilight, but the moon does not appear so, as I have previously remarked. Moreover, it is certain that not only the earth, but also the moon, has its own vaporous sphere enveloping it, /96/ for reasons which I have previously mentioned, and especially for those that shall be stated more fully in my *System*; and we may accordingly decide that the same is true with regard to the rest of the planets; so it seems to be by no means an untenable opinion to place also around Jupiter an atmosphere denser than the rest of the aether, around which, like the moon around the sphere of the elements, the Medicean Planets revolve; then by the interposition of this atmosphere, they appear smaller when they are at apogee; but when in perigee, through the absence or attenuation of that atmosphere, they appear larger. Lack of time prevents me form going further into these matters; my readers may expect further remarks upon these subjects in a short time.

19.3   Study Questions

QUES. 19.1 What is most significant about Galileo's observations of the Milky Way?

a) Why doesn't Galileo's telescope magnify the stars as much as it does the moon or the planets? And how do the appearances of stars and planets differ?

b) What did he see when looking through his spyglass at Orion, the Pleiades, and at regions of the Milky Way? What are the implications of these observations?

QUES. 19.2 What is most significant about Galileo's observations of Jupiter?

a) What object(s) did he discover, and what type of motion did they execute?
b) What did Galileo's observations imply about their nature? Are they stars? How do you know?
c) What are the implications of Galileo's discovery, especially for the geocentric and heliocentric world-views?

19.4 Exercises

Ex. 19.1 (JUPITER'S MOONS). Look up the names and orbital periods of Jupiter's four largest moons. Can you identify each of these moons in Figs. 19.5–19.10 in the text? What makes this procedure difficult?

Ex. 19.2 (JUPITER OBSERVATIONS). In this field exercise you will use a telescope to measure the angular width of Jupiter and to understand the configuration and motion of its four largest moons.

Preparation You will need a moderately high-powered telescope with a motorized drive.[2] Before taking the telescope into the field, you should learn to set up and tear down the telescope in a bright and comfortable environment. You should also familiarize yourself with the controls for changing the declination and right ascension of the telescope.

Calibration At your outdoor observing site, begin by setting up and polar aligning the telescope. Then point your telescope at Jupiter and turn on the motorized drive so as to track the motion of the planet in the westward direction. You may wish to alternately turn on and off the drive so as to ascertain the westward direction in your field of view when looking through the eyepiece of the telescope. You can determine the angular field of view of your eyepiece by turning off the drive and timing how long an object takes to drift across the entire field of view.

Measurements After calibrating the field of view of your eyepiece, determine the angular width of Jupiter by measuring how long it takes for the crosshairs in the eyepiece to drift across the diameter of Jupiter. Then determine the angular separation of each of Jupiter's moons from Jupiter itself. From observations on several subsequent evenings, you can attempt to determine the period of revolution of Jupiter's moons.

---

[2] I have used the Celestar Deluxe 8, a Schmidt-Cassegrain telescope with an 8 inch aperture and an 80 inch focal length manufactured by Celestron. The Schmidt-Cassegrain optical system uses a combination of mirrors and lenses so as to allow for a very powerful telescope in a portable package. A motorized drive allows the unit to track celestial objects and also allows the user to change the declination and right ascension with a hand-control unit. A special adapter is also available to attach a digital camera to the eyepiece for short exposure prime focus digital astrophotography.

19.5   Vocabulary

1. Adventitious	13. Commingling
2. Fringe	14. Praesepe
3. Subtend	15. Aselli
4. Twilight	16. Ecliptic
5. Magnitude	17. Zodiac
6. Curtail	18. Sidereal
7. Delineated	19. Conspicuous
8. Globular	20. Divergence
9. Periphery	21. Retrograde
10. Dog star	22. Conjunction
11. Pleiades	23. Scruple
12. Indubitable	24. Untenable

Chapter 20
The Luminosity of Variable Stars

> *A remarkable relation between the brightness of these variables*
> *and the length of their periods will be noticed.*
> —Henrietta Leavitt

20.1   Introduction

Before proceeding into the next reading let us pause to make a few historical remarks.
The reader will perhaps notice that the previous reading selection, Galileo's *Sidereal
Messenger*, and the present one, Henrietta Leavitt's record of variable stars, are separated in time by three centuries. During this period a great deal was accomplished;
most notably, Newton published his *Mathematical Principles of Natural Philosophy*
in 1687. The *Principia*, as it is often called, contained (among other things) his three
laws of motion and his universal theory of gravitation. The reader is encouraged to
study Chaps. 19–28 of volume II, where these topics are treated in detail. Suffice
it to say here that Newton's *Principia* extended the ancient concept of gravity so
that it applied to all bodies, not just ones here on Earth. This generalization allowed
him to explain the observed elliptical orbits of the planets as the combined effect
of (i) the gravitational force pulling the planets inward towards the sun, and (ii) the
inertia of the planets carrying them around so as to prevented them from simply
falling into the sun. This hypothesis applied just as well to the orbit of the moon
around Earth. In fact, Newton's careful examination of the moon's motion is what
led him most directly to his *universal law of gravitation*, which may be expressed
succinctly as

$$F = G\frac{m_1 m_2}{r^2}. \tag{20.1}$$

Here, F is the gravitational force of attraction acting between any two bodies, m_1
and m_2 are their masses, r is the distance between their respective centers of mass,
and G is a fundamental constant of nature which was not measured until a century
after the time of Newton.[1]

---

[1] See the discussion of Henry Cavendish's famous torsion balance experiment in Ex. 27.4 of
volume II.

K. Kuehn, *A Student's Guide Through the Great Physics Texts*,
Undergraduate Lecture Notes in Physics, DOI 10.1007/978-1-4939-1360-2_20,
© Springer Science+Business Media, LLC 2015

Newton speculated, in his *Principia*, that the fixed stars were not all equidistant from the Sun and confined to the surface of a vast celestial sphere, as had been supposed by most earlier thinkers. Unfortunately, the astronomical tools at his disposal lacked the precision to measure the distance to even the closest stars. Such a feat would not be possible until over a century later, when telescopic instrumentation had advanced enough for Friedrich Bessel to accurately measure the distance to a nearby star, using the method of stellar parallax, from his observatory in Königsberg in 1838. According to Bessel's measurements, 61 Cygni has an annual parallax of 0.314 arc-seconds, which corresponds to a distance of about 10 light years from Earth. This means that the apparent angular position of 61 Cygni varied by a total of 0.314 s of arc during the time earth moved halfway through its orbit around the sun. With Bessel's observations, a new period in cosmology began. For now the vast realms of space, and the distribution of stars and nebulae within it, was opened to exploration and quantitative measurement.

So now let us jump ahead another half a century to a discovery that would eventually yield a much more versatile method for measuring astronomical distances. Henrietta Leavitt (1888–1921) was born in Lancaster, Massachusetts. After attending Oberlin College and graduating from Radcliff College, she took a position at the Harvard College Observatory. There, she worked under Edward Pickering analyzing the brightness of stars appearing on the observatory's photographic plates. In so doing, she discovered the relationship between the luminosity and the period of oscillation of Cepheid variables located in the Magellanic Clouds. The reading that follows, which describes Leavitt's observations, was published in the 1912 *Harvard College Observatory Circular*.

20.2   Reading: Leavitt, *Periods of 25 variable stars in the small magellanic cloud*

Leavitt, H., Periods of 25 variable stars in the small magellanic cloud, *Harvard College Observatory Circular*, *173*, 1912.

20.2.1   *Periods of 25 Variable Stars in the Small Magellanic Cloud*

The following statement regarding the periods of 25 variable stars in the Small Magellanic Cloud has been prepared by Miss Leavitt.

A Catalogue of 1777 variable stars in the two Magellanic Clouds is given in H.A. 60, No. 4. The measurement and discussion of these objects present problems of unusual difficulty, on account of the large area covered by the two regions, the extremely crowded distribution of the stars contained in them, the faintness of the variables, and the shortness of their periods. As many of them never become brighter than the 15th magnitude, while very few exceed the 13th magnitude at maximum,

Table 20.1 Periods of variable stars in the small magellanic cloud.—[*K.K.*]

H.	Max.	Min.	Epoch	Period	Res. M.	Res. m.	H.	Max.	Min.	Epoch	Period	Res. M.	Res. m.
1505	14.8	16.1	0.02	1.25336	−0.6	−0.5	1400	14.1	14.8	4.0	6.650	+0.2	−0.3
1436	14.8	16.4	0.02	1.6637	−0.3	+0.1	*1355*	14.0	14.8	4.8	7.483	+0.2	−0.2
1446	14.8	16.4	1.38	1.7620	−0.3	+0.1	1374	13.9	15.2	6.0	8.397	+0.2	−0.3
1506	15.1	16.3	1.08	1.87502	+0.1	+0.1	818	13.6	14.7	4.0	10.336	0.0	0.0
1413	14.7	15.6	0.35	2.17352	−0.2	−0.5	*1610*	13.4	14.6	11.0	11.645	0.0	0.0
1460	14.4	15.7	0.00	2.913	−0.3	−0.1	*1365*	13.8	14.8	9.6	12.417	+0.4	+0.2
1422	14.7	15.9	0.6	3.501	+0.2	+0.2	*1351*	13.4	14.4	4.0	13.08	+0.1	−0.1
842	14.6	16.1	2.61	4.2897	+0.3	+0.6	827	13.4	14.3	11.6	13.47	+0.1	−0.2
1425	14.3	15.3	2.8	4.547	0.0	−0.1	822	13.0	14.6	13.0	16.75	−0.1	+0.3
1742	14.3	15.5	0.95	4.9866	+0.1	+0.2	823	12.2	14.1	2.9	31.94	−0.3	+0.4
1646	14.4	15.4	4.30	5.311	+0.3	+0.1	824	11.4	12.8	4.0	65.8	−0.4	−0.2
1649	14.3	15.2	5.05	5.323	+0.2	−0.1	821	11.2	12.1	97.0	127.0	−0.1	−0.4
1492	13.8	14.8	0.6	6.2926	−0.2	−0.4							

long exposures are necessary, and the number of available photographs is small. The determination of absolute magnitudes for widely separated sequences of comparison stars of this degree of faintness may not be satisfactorily completed for some time to come. With the adoption of an absolute scale of magnitudes for stars in the North Polar Sequence, however, the way is open for such a determination.

Fifty-nine of the variables in the Small Magellanic Cloud were measured in 1904, using a provisional scale of magnitudes, and the periods of 17 of them were published in H.A. 60, No. 4, Table VI. They resemble the variables found in globular clusters, diminishing slowly in brightness, remaining near minimum for the greater part of the time, and increasing very rapidly to a brief maximum. Table 20.1 gives all the periods which have been determined thus far, 25 in number, arranged in the order of their length. The first five columns contain the Harvard Number, the brightness at maximum and at minimum as read from the light curve, the epoch expressed in days following J.D. 2,410,000, and the length of the period expressed in days. The Harvard Numbers in the first column are placed in italics, when the period has not been published hitherto. A remarkable relation between the brightness of these variables and the length of their periods will be noticed. In H.A. 60, No. 4, attention was called to the fact that the brighter variables have the longer periods, but at that time it was felt that the number was too small to warrant the drawing of general conclusions. The periods of eight additional variables which have been determined since that time, however, conform to the same law.

The relation is shown graphically in Fig. 20.1, in which the abscissas are equal to the periods, expressed in days, and the ordinates are equal to the corresponding magnitudes at maxima and at minima. The two resulting curves, one for maxima and one for minima, are surprisingly smooth, and of remarkable form. In Fig. 20.2, the abscissas are equal to the logarithms of the periods, and the ordinates to the corresponding magnitudes, as in Fig. 20.1. A straight line can readily be drawn among each of the two series of points corresponding to maxima and minima, thus showing that there is a simple relation between the brightness of the variables and

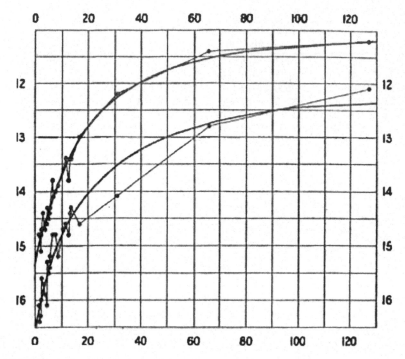

Fig. 20.1 Magnitudes of several cepheid variables as a function of period.—[*K.K.*]

their periods. The logarithm of the period increases by about 0.48 for each increase of one magnitude in brightness. The residuals of the maximum and minimum of each star from the lines in Fig. 20.2 are given in the sixth and seventh columns of Table 20.1. It is possible that the deviations from a straight line may become smaller when an absolute scale of magnitudes is used, and they may even indicate the corrections that need to be applied to the provisional scale. It should be noticed that the average range, for bright and faint variables alike, is about 1.2 magnitudes. Since the variables are probably at nearly the same distance from the Earth, their periods are apparently associated with their actual emission of light, as determined by their mass, density, and surface brightness.

The faintness of the variables in the Magellanic Clouds seems to preclude the study of their spectra, with our present facilities. A number of brighter variables have similar light curves, as UY Cygni, and should repay careful study. The class of spectrum ought to be determined for as many such objects as possible. It is to be hoped, also, that the parallaxes of some variables of this type may be measured. Two fundamental questions upon which light may be thrown by such inquiries are whether there are definite limits to the mass of variable stars of the cluster type, and if the spectra of such variables having long periods differ from those of variables whose periods are short.

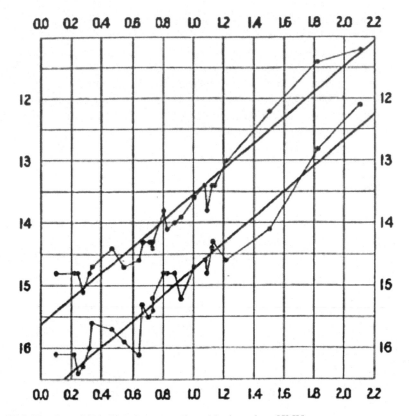

Fig. 20.2 The data of Fig. 20.1 shown on a logarithmic scale.—[*K.K.*]

The facts known with regard to these 25 variables suggest many other questions with regard to distribution, relations to star clusters and nebulae, differences in the forms of the light curves, and the extreme range of the length of the periods. It is hoped that a systematic study of the light changes of all the variables, nearly two thousand in number, in the two Magellanic Clouds may soon be undertaken at this Observatory.

Edward C. Pickering, March 3, 1912

20.3 Study Questions

Ques. 20.1 What did Henrietta Leavitt find when carefully studying the Magellanic clouds?

a) On what astronomical objects were Leavitt's studies focused? What rendered such studies difficult?

b) Explain the columns in Table 20.1. In particular, what is meant by terms such as "brightness", "epoch", "period" and "residuals"? To what does "J.D." refer?[2]

c) What is plotted in Figs. 20.1 and 20.2? What significant feature of the data do these plots highlight?

d) What did Leavitt assume about the distance to these objects? What conclusion does she draw, based upon this assumption? Was this a good assumption?

e) According to Leavitt, upon what properties does the actual emission of light depend? What further observations and analysis does this suggest?

20.4   Exercises

Ex. 20.1 (VARIABILITY-LUMINOSITY). Consider two Cepheid variables, A and B, both located within the small Magellanic cloud. If A is 100 times as bright as B (on average), then which has the longer period of variability? By how much?

20.5   Vocabulary

1. Magellanic   6. Ordinate
2. Globular cluster   7. Residual
3. Period   8. Provisional
4. Epoch   9. Spectra
5. Abscissa

---

[2] See, for instance, the detailed discussion of astronomical and civil calendar conventions in Chap. 18 of Herschel, J. F. W., *Outlines of Astronomy*, Longmans, Green, and Co., 1893.

Chapter 21
Galactic Spectra

These objects have since been found to be possessed of
extraordinary motions.

—Vesto Slipher

21.1   Introduction

Vesto Slipher (1875–1969) was born in Mulberry, Indiana. He received his undergraduate degree in mechanics and astronomy at Indiana University before taking a position at the Lowell Observatory in Flagstaff, Arizona in 1901, where he would spend the rest of his career. In the mean time, he received his doctorate from Indiana University in 1909. While at the Lowell Observatory, Slipher's work focused on employing and improving spectrographic techniques to study planetary atmospheres and spiral nebulae.[1] Slipher's discovery of the doppler shift of galactic spectra, and the enormous galactic velocities which these measurements implied, was instrumental in revealing the expansion of the universe and in the formulation of Hubble's Law. The reading that follows was published in 1917 in the *Proceedings of the American Philosophical Society*.

21.2   Reading: Slipher, *Nebulae*

Slipher, V., Nebulae, *Proceedings of the American Philosophical Society*, 56(5), 403–409, 1917.

21.2.1   Nebulae

In addition to the planets and comets of our solar system and the countless stars of our stellar system there appear on the sky many cloud-like masses—the nebulae. These for a long time have been generally regarded as presenting an early stage in

---

[1] For more on the technique of spectroscopy, see Ex. 21.2 at the end of the present chapter.

K. Kuehn, *A Student's Guide Through the Great Physics Texts*,
Undergraduate Lecture Notes in Physics, DOI 10.1007/978-1-4939-1360-2_21,
© Springer Science+Business Media, LLC 2015

the evolution of stars and of our solar system, and they have been carefully studied and something like 10,000 of them catalogued.

Keeler's classical investigation of the nebulae with the Crossley reflector by photographic means revealed unknown nebulae in great numbers. He estimated that such plates as his if they were made to cover the whole sky would contain at least 120,000 nebulae, an estimate which later observations show to be considerably too small. He made also the surprising discovery that more than half of all nebulae are spiral in form; and he expressed the opinion that the spiral nebulae might prove to be of particular interest in questions concerning cosmogony.

I wish to give at this time a brief account of a spectrographic investigation of the spiral nebulae which I have been conducting at the Lowell Observatory since 1912. Observations had been previously made, notably by Fath at the Lick and Mount Wilson Observatories, which yielded valuable information on the character of the spectra of the spiral nebulae. These objects have since been found to be possessed of extraordinary motions and it is the observation of these that will be discussed here.

In their general features nebular spectra may for convenience be placed under two types characterized as (I) bright-line and (II) dark-line. The gaseous nebulae, which included the planetary and some of the irregular nebulae, are of the first type; while the much more numerous family of spiral nebulae are, in the main, of the second type. But the two are not mutually exclusive and in the spirals are sometimes found both types of spectra. This is true of the nebulae numbered 598, 1068 and 5236 of the "New General Catalogue" of the nebulae.

Some of the gaseous nebulae are relatively bright and their spectra are especially so since their light is all concentrated in a few bright spectral lines. These have been successfully observed for a long time. Keeler in his well-known determination of the velocities of 13 gaseous nebulae was able to employ visually more than 20 times the dispersion usable on the spiral nebulae.

Spiral nebulae are intrinsically very faint. The amount of their light admitted by the narrow slit of the spectrograph is only a small fraction of the whole and when it is dispersed by the prism it forms a continuous spectrum of extreme weakness. The faintness of these spectra has discouraged their investigation until recent years. It will only be emphasizing the fact that their faintness still imposes a very serious obstacle to their spectrographic study when it is pointed out, for example, that an excellent spectrogram of the Virgo spiral N.G.C. 4594 secured with the great Mount Wilson reflector by Pease was exposed 80 h.

A large telescope has some advantages in this work, but unfortunately no choice of telescope either of aperture or focal-length will increase the brightness of the nebular surface. It is chiefly influenced by the spectrograph whose camera alone practically determines the efficiency of the whole equipment. The camera of the Lowell spectrograph has a lens working at a speed ratio of about 1:2.5. The dispersion piece of the spectrograph has generally been a 64° prism of dense glass, but for two of the nebulae a dispersion of two 64° prisms was used. The spectrograph was attached to the 24-in refractor.

With this equipment I have secured between 40 and 50 spectrograms of 25 spiral nebulae. The exposures are long—generally from 20 to 40 h. It is usual to continue

Table 21.1 Radial velocities
of 25 spiral nebulae

Nebula		Vel. (km)	Nebula		Vel. (km)
N.G.C.	221	−300	N.G.C.	4526	+580
	224	−300		4565	+1100
	598	−260		4565	+580
	1023	+300		4594	+1100
	1068	+1100		4736	+290
	2683	+400		4826	+150
	3031	−30		5005	+900
	3115	+600		5055	+450
	3379	+780		5194	+270
	3521	+730		5236	+500
	3623	+800		5866	+650
	3627	+650		7331	+500
	4258	+500			

the exposure through several nights but occasionally it may run into weeks owing to unfavorable weather or the telescope's use in other work. Besides the exposures cannot be continued in the presence of bright moonlight and this seriously retards the accumulation of observations.

The iron-vanadium spark comparison spectrum is exposed a number of times during the nebular exposure in order to insure that the comparison lines are subjected to the same influences as the nebular lines. The spectrograph is electrically maintained at a constant temperature which avoids the ill effects of the usual fall of the night temperature.

The equivalent slit-width is usually about 0.06 mm.

The linear dispersion of the spectra is about 140 tenth-meters per millimeter in the violet of the spectrum which is sufficient to detect and measure the velocities of the spiral nebulae. As the objects yet to be observed are fainter than those already observed the prospects of increasing the accuracy by employing greater dispersion are not now promising.

The plates are measured under the Hartmann spectrocomparator in which one optically superposes the nebular plate of unknown velocity upon one of a like dark-line spectrum of known velocity, used as a standard. A micrometer screw, which shifts one plate relatively to the other, is read when the dark lines of the nebula and the standard spectrum coincide; and again when the comparison lines of the two plates coincide. The difference of the two screw readings with the known dispersion of the spectrum gives the velocity of the nebula. By this method weak lines and groups of lines can be utilized that otherwise would not be available because of faintness or uncertainty of wavelength.

In Table 21.1 are given the velocities for the 25 spiral nebulae thus far observed. In the first column is the New General Catalogue number of the nebula and in the second the velocity. The plus sign denotes the nebula is receding, the minus sign that it is approaching.

Generally the value of the velocity depends upon a single plate which, in many instances, was underexposed and some of the values for these reasons may be in error by as much as 100 km. This however is not so discreditable as at first it might seem to be. The arithmetic mean of the velocities is 570, and 100 km is hence scarcely 20 %

Table 21.2 Velocities of nebulae by different observers

Nebula	Vel. (km)	Observers
N.G.C. 224	−300	Slipher, mean from several plates
Great Andromeda Nebula	−304	Wright, Lick Observatory, one plate
	−329	Pease, Mt. Wilson Observatory, one plate
	−300 to 400	Wolf, Heidelberg, one plate, approx
N.G.C. 598	−278	Pease, Mt. Wilson, from bright lines
Great Spiral of Triangulum	−263	Slipher, from bright lines
N.G.C. 1068	+1100	Slipher, from dark and bright lines
	+765	Pease, from two bright lines
	+910	Moore, Lick Observatory, from three bright lines
N.G.C. 4584	+1100	Slipher
	+1180	Pease, Mt. Wilson Observatory

Table 21.3 Velocities of spiral nebulae grouped

Face view spirals		Inclined spirals		Edge view spirals	
N.G.C.	Vel. (km)	N.G.C.	Vel. (km)	N.G.C.	Vel. (km)
598	−260	224	−300	2683	+400
4736	+290	3623	+800	3115	600
5194	+270	3627	+650	4565	+1100
5236	+500	4826	+300	4594	+1100
		5005	+920	5866	+600
		5055	+450		
		7331	+500		
Mean ...	330	...	560	...	760

of the quantity measured. Thus owing to the very high magnitude of the velocity of the spiral nebulae the percentage error in its observation is comparable with that of star velocity measurements.

Since the earlier publication of my preliminary velocities for a part of this list of spiral nebulae, observations have been made elsewhere of four objects with results in fair agreement with mine, as shown in Table 21.2.

Referring to the table of velocities again: the average velocity 570 km. is about 30 times the average velocity of the stars. And it is so much greater than that known of any other class of celestial bodies as to set the spiral nebulae aside in a class to themselves. Their distribution over the sky likewise shows them to be unique—they shun the Milky Way and cluster about its poles.

The mean of the velocities with regard to sign is positive, implying the nebulae are receding with a velocity of nearly 500 km. This might suggest that the spiral nebulae are scattering but their distribution on the sky is not in accord with this since they are inclined to cluster. A little later a tentative explanation of the preponderance of positive velocities will be suggested.

Grouping the nebulae as in Table 21.3, there appears to be some evidence that spiral nebulae move edge forward.

The form of the spiral nebulae strongly suggests rotational motion. In the spring of 1913, I obtained spectrograms of the spiral nebulae N.G.C. 4594 the lines of which were inclined after the manner of those in the spectrum of Jupiter, and, later, spectrograms which showed rotation or internal motion in the Great Andromeda Nebula and in the two in Leo N.G.C. 3623 and 3627 and in nebulae N.G.C. 5005 and 2683—less well in the last three. The motion in the Andromeda nebula and in 3623 is possibly more like that in the system of Saturn. It is greatest in nebula N.G.C. 4594. The rotation in this nebula has been verified at the Mt. Wilson Observatory.

Because of its bearing on the evolution of spiral nebulae it is desirable to know the direction of rotation relative to the arms of the spirals. But this requires us to know which edge of the nebula is the nearer us, and we have not as yet by direct means succeeded in determining even the distance of the spiral nebulae. However, indirect means, I believe, may here help us. It is well known that spiral nebulae presenting their edge to us are commonly crossed by a dark band. This coincides with the equatorial plane and must belong to the nebula itself. It doubtless has its origin in dark or deficiently illuminated matter on our edge of the nebula, which absorbs (or occults) the light of the more brightly illumined inner part of the nebula. If now we imagine we view such a nebula from a point somewhat outside its plane the dark band would shift to the side and render the nebula unsymmetrical—the deficient edge being of course the one nearer us. This appears to be borne out by the nebulae themselves for the inclined ones commonly show this typical dissymmetry. Thus we may infer their deficient side to be the one toward us.

When the result of this reasoning was applied to the above cases of rotation it turned out that the direction of rotation relative to the spiral arms was the same for all. (The nebula N.G.C. 4594 is unfortunately not useful in this as it is not inclined enough to show clearly the arms.) The central part—which is all of the nebulae the spectrograms record—turns into the spiral arms as a spring turns in winding up. This agreement in direction of rotation furnishes a favorable check on the conclusion as to the nearer edge of the nebulae, for of course we should expect that dynamically all spiral nebulae rotate in the same direction with reference to the spiral arms. The character and rapidity of the rotation of the Virgo nebula N.G.C. 4594 suggests the possibility that it is expanding instead of contracting under the influence of gravitation, as we have been wont to think.

As noted before the majority of the nebulae here discussed have positive velocities, and they are located in the region of sky near right ascension 12 h which is rich in spiral nebulae. In the opposite point of the sky some of the spiral nebulae have negative velocities, *i.e.*, are approaching us; and it is to be expected that when more are observed there, still others will be found to have approaching motion. It is unfortunate that the 25 observed objects are not more uniformly distributed over the sky as then the case could be better dealt with. It calls to mind the radial velocities of the stars which, in the sky about Orion, are receding and in the opposite part of the sky are approaching. This arrangement of the star velocities is due to the motion of the solar system relative to the stars. Prof. Campbell at the Lick Observatory has accumulated a vast store of star velocities and has determined the motion of our sun with reference to those stars.

We may in like manner determine our motion relative to the spiral nebulae, when sufficient material becomes available. A preliminary solution of the material at present available indicates that we are moving in the direction of right-ascension

22 h and declination −22° with a velocity of about 700 km. While the number of nebulae is small and their distribution poor this result may still be considered as indicating that we have some such drift through space. For us to have such motion and the stars not show it means that our whole stellar system moves and carries us with it. It has for a long time been suggested that the spiral nebulae are stellar systems seen at great distances. This is the so-called "island universe" theory, which regards our stellar system and the Milky Way as a great spiral nebula which we see from within. This theory, it seems to me, gains favor in the present observations.

It is beyond the scope of this paper to discuss the different theories of the spiral nebulae in the face of these and other observed facts. However, it seems that, if our solar system evolved from a nebula as we have long believed, that nebula was probably not one of the class of spirals here dealt with.

Our lamented Dr. Lowell was deeply interested in this investigation as he was in all matters touching upon the evolution of our solar system and I am indebted to him for his constant encouragement.

Lowell Observatory, April, 1917.

21.3   Study Questions

QUES. 21.1 What is the nature and composition of nebulae? For example, what do they look like? How many of them are there? And can they be classified according to their spectra?

QUES. 21.2 What notable feature(s) did the spectra of the spiral nebulae contain?

a) Was Slipher the first to observe the spectrum of a nebula? What makes nebular spectra particularly difficult to study? How did Slipher carry out his observations?
b) How did Slipher measure the velocity of the nebulae using their spectra? What did his measurements imply?
c) Was Slipher able to observe the internal motion of spiral nebulae? What makes this determination challenging? And what does Slipher conclude form his studies?

QUES. 21.3 What do Slipher's observations imply about the nature and structure of the universe?

a) Is our stellar system in motion? If so, with respect to what? How do you know?
b) What is the theory of "island universes"? Do Slipher's observations support this theory?
c) What conclusions does Slipher draw regarding the history of our solar system? Are these conclusions reasonable? Are they correct?

Fig. 21.1 Project Star
portable low-cost
spectrometer and activity
booklet

21.4 Exercises

Ex. 21.1 (GALAXY ROTATION). The Whirlpool Galaxy, NGC 5194 (M51), resides 9.6 MegaParsecs distant from our sun. Suppose that the H1 emission line (21 cm wavelength) of M51 displays the following features: (i) the H1-line from light from one edge of the galaxy is redshifted by about 0.22 μm; (ii) the H1-line from light from the other edge is redshifted by 0.43 μm; (iii) on average, the red-shift of the H1 line from M51 is 0.33 μm. Now, what is the speed of M51 with respect to the earth? You will need to look up the doppler shift formula which relates the velocity of an object to the wavelength shift of its emitted light. And what exactly does the variation in the red-shift when looking at different parts of M51 imply about is motion? (ANSWER: 460 m/s)

Ex. 22.2 (SPECTROSCOPY LABORATORY). In this laboratory exercise, we will learn what the color of light emitted (or absorbed) by objects can tell us about the object itself. In particular, we will observe various light sources using a hand-held spectrometer, such as the one shown in Fig. 21.1. A spectrometer allows you to quantitatively measure the wavelength of the colors of light emitted by these sources.[2] It consists of a slit at one end which admits a thin vertical ribbon of light into the spectrometer tube, and a diffraction grating (or prism) at the other end which separates the admitted light into vertical fringes of color called a spectrum. The precise angular position of each of these fringes of color is determined by its wavelength. If the spectrometer has a calibrated scale, the wavelength (or energy) of the light may be read directly from the scale. As an example, the emission spectrum of a mercury vapor light source, as viewed with a hand-held spectrometer, is shown in Fig. 21.2. Begin by familiarizing yourself with your spectrometer. Can you read the wavelength (and energy) scales? Then, observe various light sources in the laboratory or around your neighborhood

---

[2] A low-cost, portable spectrometer, the Project Star Spectrometer, is manufactured by Science First, Inc., Yulee, FL. A user manual, written by William Luzader, comes with this spectrometer. It contains details on the construction and the theory of operation of the spectrometer, as well as several exercises.

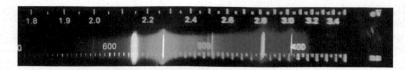

Fig. 21.2 The emission spectrum of a mercury vapor lamp

using your spectrometer. Try incandescent lights, fluorescent lights, light emitting diodes, neon lights, Xenon automobile headlamps, and chemical glow-sticks. You might also observe the light from an incandescent bulb after it passes through transparent colored objects such as dyes or filters. What does this tell you? What about the light scattered from the sky or the clouds during daytime, or from the moon at night? For each of these activities, sketch the observed spectrum and identify emission lines. If you have access to a drill, some screws, and an appropriately sized sheet of wood or sturdy plexiglass, you can construct a mounting base for both the spectrometer and a digital camera. This will allow you to "digi-scope" your spectrometer so as to acquire digital images of your observed spectra, as shown in Fig. 21.2.

21.5   Vocabulary

1. Nebulae
2. Cosmogony
3. Spectrograph
4. Velocity
5. Dispersion
6. Intrinsically
7. Prism
8. Focal-length

9. Micrometer
10. Wavelength
11. Arithmetic mean
12. Celestial
13. Cluster
14. Preponderance
15. Andromeda
16. Dissymmetry

Chapter 22
Measuring Astronomical Distances

The procedure is very simple, once the period-luminosity
relation is set up and accurately calibrated.

—Harlow Shapley

22.1 Introduction

Harlow Shapley (1885–1972) was born in Nashville, Missouri. He studied astronomy
at the University of Missouri and did graduate work at Princeton University where he
performed photometric measurements of eclipsing binary stars. Thereafter, he took
a position as a regular observer at the Mount Wilson Observatory. His work on the
distances of globular clusters, the Small Magellanic Cloud and the Milky Way led to
a drastic revision of astronomers' understanding of the size and structure of the Milky
Way galaxy. The reading selection that follows was taken from a chapter of Shapley's
book entitled *Galaxies*. Herein, he describes a method for measuring astronomical
distances using the period-luminosity relationship discovered by Henrietta Leavitt.

22.2 Reading: Shapley, *Galaxies*

Reprinted by permission of the publisher from "The Astronomical Toolhouse" in
Galaxies, Third Edition by Harlow Shapley, revised by Paul W. Hodge, pp. 57–67,
Cambridge, MA: Harvard University Press, Copyright © 1943, 1961, 1972 by the
President and Fellows of Harvard College. Copyright © renewed 1971 by Harlow
Shapely. Copyright © renewed 1989 by Willis H. Shapley.

22.2.1 The Astronomical Toolhouse

The two Clouds of Magellan, as remarked in the preceding chapter, are satisfactorily
located in space for the effective study of many properties of galaxies, even though
they are inconveniently far south for easy exploitation by the majority of astronomers.
Their distance of about 160,000 light-years gives easy access to all of their giant and
supergiant stars. Their considerable angular separations from the star clouds of the

Milky Way keep them clear not only of most of the light-scattering dust near the galactic plane, but also of the confusingly rich foreground of stars and nebulosity near the Milky Way. They are nicely isolated.

During the past half century high profit has accrued from our studies of these nearby galaxies, for they have turned out to be veritable treasure chests of sidereal knowledge, and astronomical tool houses of great merit. We shall see that the hypotheses, deductions, and techniques that arise from studies of the stars and nebulae of the Magellanic galaxies can be used to explore our own surrounding system, and also the more distant galaxies.

The usefulness of the Magellanic Clouds in the larger problems of cosmography can be illustrated by presenting, without stopping now to explain the meaning of the items or their significance, a partial list of the contributions to our knowledge of stars and galaxies that have already come from studies of the Clouds, or are on the way:

1. The period-luminosity relation
2. The general luminosity curve, that is, the relative number of stars in successive intervals of intrinsic brightness
3. Measures of the internal motions of irregular galaxies
4. A comparison of the sizes, luminosities, and types of open star clusters
5. The frequency of cepheid variation, shown by the number of cepheid variables compared with the numbers of other types of giant stars of approximately the same mass and brightness
6. The spread of the lengths of period of cepheid variables
7. The dependence of various characteristics of the light curves of cepheid variables on the length of period
8. The dependence of a cepheid's period on location in a galaxy
9. The total absolute magnitudes of globular star clusters, and the maximum luminosity of numerous special types of stars
10. The demonstration of the "star haze" and "hydrogen haze" surrounding some if not all galaxies

It seems inevitable that additional discoveries will reward the future investigators of these two external systems that can be studied objectively and in detail because of their nearness and externality.

Nearly all of the subjects listed can be investigated more successfully in the Magellanic Clouds than elsewhere. And many can be read about in the technical reports better than here. Some involve the problems of stellar evolution; others, of galactic dimensions and structure. Several of the items will have their use chiefly in the future, rather than in the past; and although all are important in astronomy, only a few can be considered fully in this chapter.

22.2.2   The Abundance of Cepheid Variables

The outstanding phenomenon associated with the Magellanic Clouds is undoubtedly the relatively great number of giant variable stars, of which a majority are of the

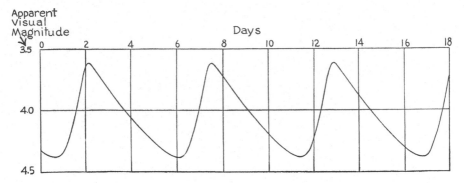

Fig. 22.1 An 18-day section of the light curve of the typical cepheid variable Delta Cephei, which for indefinite centuries will faithfully and monotonously repeat the 5.37-day oscillation.

cepheid class (Fig. 22.1). They are easily available for detailed investigation since they stand out conspicuously among the brighter stars.

In each of the Clouds there are more classical cepheid variables than are as yet known in our own much larger Galaxy. The survey in the Clouds approaches completeness; the survey in the galactic system is fragmentary and seriously hindered by the interstellar dust along the Milky Way where classical cepheids are concentrated. Probably fewer than half of the cepheid variables of our Milky Way system have been detected.

Of the variable stars in the Magellanic Clouds that have been worked up, about 80 % are classical cepheids. In the neighborhood of the sun there are only a few of these pulsating stars; among them are Polaris and Delta Cephei; the latter being the star that gives a name to the class. In the solar neighborhood, as elsewhere in the galactic system, variables of other types are considerably more numerous than cepheids; for instance, here there are hundreds of eclipsing binaries, whereas only a few score are known in the Magellanic Clouds. Also we have found in our Galaxy more than 3000 "cluster" (RR Lyrae) variables, which are variables with periods less than a day, but only a few have been identified with certainty in the Magellanic Clouds. In the galactic system there are a great many long-period variables—the kind of stars that are carefully watched by the organized variable-star observers—but such stars are not yet abundant in the records of the Clouds.

Does this richness in the Magellanic Clouds of classical cepheid variables, with periods between 1 and 50 days, indicate that the population differs fundamentally in such irregular galaxies from that in the Milky Way spiral? Not necessarily so. The relative scarcity of cluster-type cepheids, long-period variables, and eclipsing stars in the present records of the Clouds is best accounted for by the relatively low candlepower of variable stars of those types. Even at maximum, such variables are not quite bright enough to get numerously into our 18th-magnitude picture of the Magellanic Clouds. Until recently we have photographed almost exclusively the giants that are 200 times or more brighter than the sun. The larger reflectors are beginning to explore among the fainter stars and possibly will soon reveal many cluster variables at the nineteenth magnitude, and eventually get down to stars of the sun's brightness.

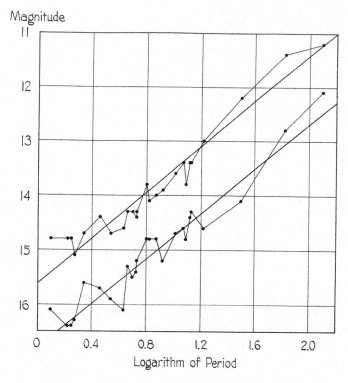

Fig. 22.2 Miss Leavitt's original diagram showing, separately for the maxima and the minima of 25 variable stars in the Small Cloud, the relation between photographic magnitude (vertical coordinate) and the logarithm of the period (horizontal coordinate)

22.2.3   The Period-Luminosity Relation and the Light Curves of Cepheids

Some years after Miss Leavitt had discovered and published 1777 variable stars in the two Clouds, she presented the results of a study of the periods of some of the variables. For the investigation she had selected the brightest of the variables as well as a few fainter ones. At once there appeared the interesting fact that, when the average brightness of a given variable is high, the period, which is the time interval separating successive maxima of brightness, is long compared with the intervals for fainter stars. The fainter the variable, the shorter the period.

The graph of her results for 25 variables is reproduced in Fig. 22.2.

It is of historic significance. Miss Leavitt and Prof. Pickering recognized at once that if the periods of variation depend on the brightness they must also be associated with other physical characteristics of the stars, such as mass and density and size. But apparently they did not foresee that this relation between brightness and period for cepheids in the Small Cloud would be the preliminary blueprint of one of astronomy's most potent tools for measuring the universe; nor did they, in fact, identify these variables of the Magellanic Cloud with the already well-known cepheid variables of the solar neighborhood. They simply had found a curiosity among the variables of the Small Magellanic Cloud.

Soon after Miss Leavitt's announcement of the period–magnitude relation for this small fraction of the variables that she had discovered in the Small Magellanic Cloud, Ejnar Hertzsprung and others pointed out that the nearby cepheid variable stars of the Milky Way are giants—a fact that was readily deduced from their small cross motions and from spectral peculiarities. Therefore, if the galactic cepheids and the Magellanic variables are closely comparable in luminosities, these 15th- and 16th-magnitude objects in the Clouds must also be giants; and in order to appear so faint, they must be very remote, and so also must be the Clouds.

Shapley and others pursued the inquiry and supplemented Miss Leavitt's work by studies of the variable stars that Bailey and others had detected in the globular star clusters. The many variables of the globular clusters are mostly cepheids of the cluster type with periods less than a day. But also in clusters are a few longer-period cepheids, and it was eventually possible to bring together all the data necessary for a practical but tentative period-luminosity curve. The new investigation appeared to connect the typical or "classical" cepheids with the cluster variables. Shapley then derived a zero point from trigonometric measures of the distances of the nearby cepheids and thereby changed the Leavitt relation from period and apparent magnitude to period and absolute luminosity, thus making distances determinable from light measures only, as will be shown below.

Using the apparently similar cepheid variables found in globular star clusters as guides to the zero point, Shapley derived the first period-luminosity relation for cepheid variables in 1917 (Fig. 22.3). He was able to establish a true and absolute luminosity for cepheids by taking advantage of the fact that what appeared to be normal cepheids existed in globular star clusters together with the cluster type RR Lyrae variables. The latter are fairly common in our Galaxy and there are enough near the sun that the distances to the RR Lyrae variables could be established by statistical considerations based on their motions as seen over the sky. Therefore, it was well known that the RR Lyrae variables all have absolute magnitudes of approximately zero and thus Shapley was able to establish magnitudes for the cepheid variables in the globular clusters and, by comparison, also in Magellanic Clouds.

More recent studies of the true luminosity of RR Lyrae stars have confirmed rather well these early results utilized by Shapley more than 50 years ago. The accurate absolute luminosities obtained for globular star clusters for which we can measure distances by comparison of their main-sequence stars with main-sequence stars in nearby clusters, such as the Hyades, have led to the conclusion that in the mean, RR Lyrae variables have an absolute magnitude of approximately +0.5, with a spread from cluster to cluster of 0.2 or 0.3 magnitude. However, it is an entirely different story with regard to the longer-period cepheids found in the globular clusters. In the 1950s it was established that there are two kinds of cepheid variables, one belonging to the Population I, young, spiral-arm component of the Galaxy and the other belonging to the Population II component, which includes the halo and the globular star clusters. The Population II cepheids were found on the average to be 1.5 magnitudes fainter for a given period than the Population I cepheids and therefore

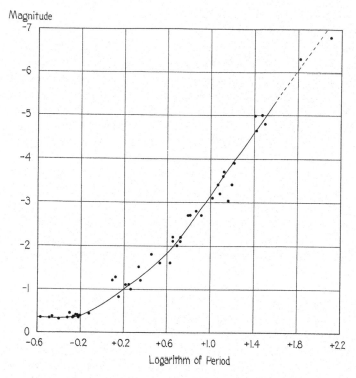

Fig. 22.3 The early Harvard period-luminosity relation, based on the 25 Small Cloud variables and cepheids from globular clusters in the local galaxy. The ordinate gives visual magnitudes on the absolute scale

the period-luminosity relation for the two have a very different zero point. This meant that the cepheid variables in the Magellanic Clouds, which were clearly Population I cepheids, were very much more luminous than had originally been thought and therefore the distance to the Clouds must be twice as great as had been computed in 1917.

It may be well at this point to show how one uses the period-luminosity curve of Fig. 22.4 to measure the distances of the classical cepheids in our Milky Way, or the distance to some remote external galaxy, like the Andromeda Nebula. The procedure is very simple, once the period-luminosity relation is set up and accurately calibrated. First must come the discovery of a periodic variable star, and then, through the making of a hundred or so observations of the brightness at scattered times, comes the verification, from the shape of the mean light curve, that the variable belongs to the cepheid class. On a correct magnitude scale we next determine the amplitude (range) of variation, and the value of the magnitude half-way between maximum and minimum. The *median apparent magnitude*, $\dot{m}$, which is now almost always determined photographically or photoelectrically, constitutes one half of the needed observational material. The other half, namely the period P, is also determined from the observations of magnitudes.

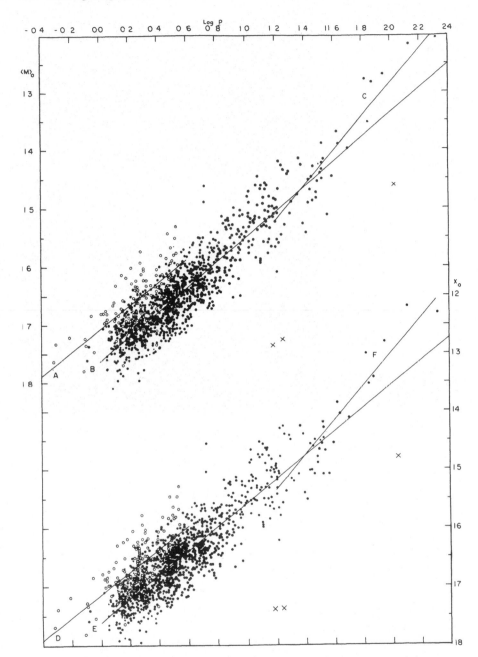

Fig. 22.4 The period-luminosity relation for cepheid variables in the Small Magellanic Cloud. The points are plotted twice, in the *upper curve* according to luminosity at mean magnitude ($< M >_0$) and in the *lower curve* according to a mean magnitude (X_0), figured differently. *Lines* labeled *A*, *B*, *C*, *D*, *E*, *F* are various solutions for the equation for the period-luminosity relation. (From C. H. Payne-Gaposchkin and Sirgay Gaposhkin, *Smithsonian Contributions to Astrophysics*)

Fig. 22.5 The globular star cluster Messier 3, one of the most conspicuous in the northern sky, renowned for its nearly 200 RR Lyrae variables. (Palomar photograph, 200-in telescope)

With the period and its logarithm known, the relative absolute luminosity, M, is then derived directly. For example, the simple formula

$$\dot{M} = -1.78 - 1.74 \log P \qquad (22.1)$$

is satisfactory for getting the relative absolute magnitudes of all cepheids with periods between 1.2 and 40 days.

When we have thus derived the relative absolute magnitude from the period, we compute the distance d from the equally simple relation

$$\log d = 0.2(\dot{m} - \dot{M} - \delta m) + 1 \qquad (22.2)$$

where the distance is expressed in parsecs (1 parsec $= 3.26$ light-years, or about 19 trillion miles), and δm is the correction one must make to the observed median magnitude because of the scattering and absorption of starlight by the dust and gas of interstellar space. (The derivation of this standard formula is given by Bart J. Bok and Priscilla F. Bok in *The Milky Way* and in various general textbooks.)

If space is essentially transparent, as in directions toward the poles of the Galaxy, δm can be set equal to zero. Such is the case for Messier 3 (Fig. 22.5). In directions where scattering is appreciable, we are frequently in trouble because δm is not zero and is difficult to determine. When we ignore the correction we have an upper limit for the distance. Thus, for cepheids in the Milky Way star clouds, where there is much dimming from dust, we can from this simple procedure, when scattering is ignored, determine only that the cepheids are not more remote than the computed distance; if δm is 1.5 magnitudes, they are actually only half as distant.

For a cepheid in a galaxy well away from the dust-filled Milky Way star clouds, we can safely assume that δm is less than 0.3 and rather accurately compute the distance of the cepheid from the formulas above. We then have not only the distance of the cepheid, whose $\dot{M}$ we get from P and whose $\dot{m}$ and P we get from the measures of magnitude, but also, without further measurement, we have the distance of the whole galaxy of a thousand million stars or more.

In summary, this simple but powerful photometric method based on cepheid variables involves only the observational determination of the periods and apparent magnitudes, followed by a direct calculation of absolute magnitudes and distances; we assume that our cepheid is of the classical type and not of the globular-cluster type.

Since cepheids with median photographic magnitudes as faint as the 21st can be discovered and studied with existing telescopes, and since such cepheids may have periods of 40 days, we can with the period-luminosity relation readily measure enormous distances. For example, a period of 40 days gives

$$\dot{M} = -1.78 - 1.74 \times 1.60 = -4.56 \tag{22.3}$$

according to the first formula above. Then, away from the Milky Way absorption, the second formula gives

$$\log d = 0.2(21.0 + 4.56) + 1 = 6.112 \tag{22.4}$$

The distance that is measurable with this supergiant cepheid in high latitudes is therefore $d = 1,300,000$ parsecs, or approximately 4,200,000 light-years. The uncertainty in the result, on a percentage basis, is distinctly less than that in a measurement locally of 500 light-years by the older trigonometric method.

22.3 Study Questions

QUES. 22.1 What is Shapley's method for determining astronomical distances?

a) How can one distinguish whether a star is nearby and dim or far away and bright?
b) What is the relationship between the period and absolute luminosity of a variable star?
c) How is the luminosity of a variable star measured, since it is variable?
d) What is the relationship between absolute and apparent luminosity of a star?
e) How, exactly, does this method enable one to measurement astronomical distances?

Fig. 22.6 Stellar parallax

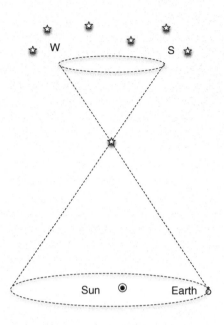

22.4   Exercises

Ex. 22.1 (ASTRONOMICAL DISTANCE: STELLAR PARALLAX). Suppose that during the course of one Earth year a particular star, α-Centauri, appears to shift position with respect to its surrounding stars. This phenomenon, known as stellar parallax, is illustrated in Fig. 22.6. During the winter α-Centauri appears among the very distant stars near W; during the summer it appears among the very distant stars near S. If the parallax angle is measured to be 0.742 arcseconds then how distant is α-Centauri from the sun (i) in astronomical units and (ii) in light years?[1] How would the observed trajectory of α-Centauri differ if it lied in Earth's orbital plane rather than above it as shown in Fig. 22.6? And what do you think finally limits the effectiveness of this parallax method of astronomical distance measurement?

Ex. 22.2 (ASTRONOMICAL DISTANCE: APPARENT MAGNITUDE). Consider a cepheid variable with an apparent magnitude of 11 and a period of 10 days. What is its (average) absolute magnitude? How many parsecs away from Earth is it?[2] Is it nearer or farther than the Magellanic cloud? What do you think finally limits the effectiveness of Shapley's method of astronomical distance measurement?

---

[1] One astronomical unit (AU) is defined as the mean distance between the centers of Earth and the sun.

[2] One parsec is defined as the distance to an object which exhibits 1 arcsecond of parallax during Earth's orbit around the sun.

22.5   Vocabulary

1. Sidereal
2. Galaxy
3. Cosmogony
4. Cepheid variable
5. Conspicuous
6. Eclipsing binary
7. Cluster variable
8. Candlepower
9. Medium
10. Apparent magnitude
11. Parsec
12. Median

Chapter 23
A New Theory of Gravity

In general, rays of light are propagated curvilinearly in gravitational fields.

—Albert Einstein

23.1 Introduction[1]

Albert Einstein (1879–1955) was born at Ulm, in Württemberg, Germany. He received a diploma in 1901 from the Swiss Federal Polytechnic School in Zurich, where he studied physics and mathematics. After seeking in vain to acquire a teaching position, he worked as a technical assistant in the Swiss Patent Office in Bern, Switzerland. He received his doctorate in 1905, was appointed lecturer at Bern University in 1908, associate professor of theoretical physics at Zurich University in 1909, Prof. of Theoretical Physics at the German University of Prague in 1911, and Prof. of theoretical physics at the Federal Institute of Technology of Zurich in 1912. After returning to Zurich, he went on to become the director of the Kaiser Wilhelm Physical Institute and Prof. at the University of Berlin. In 1933, he moved to the United States to become Prof. of Theoretical Physics at Princeton.[2] He retired in 1945 and became a leading advocate for the creation of a single world-wide government with global jurisdiction.[3] After the death of Chaim Weizmann, he was offered the presidency of the state of Israel by Prime Minister Ben Gurion in 1952, which he politely declined.

During his brief time at the Swiss patent office, Einstein published a number of ground-breaking theoretical articles on the subjects of Brownian motion, electromagnetic radiation, the photoelectric effect, the principle of relativity, and the equivalence of mass and energy. Although Einstein was later awarded a Nobel Prize in physics for his viewpoint concerning the photoelectric effect, his name is most

---

[1] These notes are also included in the introduction to Chap. 29 of Vol. II.

[2] A detailed account of Einstein's life and work is provided in Rosenkranz, Z., *The Einstein Scrapbook*, The Johns Hopkins University Press, 2002.

[3] See Einstein's essay entitled "Towards a World Government" in Einstein, A., *The Einstein Reader*, Citadel Press, 2006.

K. Kuehn, *A Student's Guide Through the Great Physics Texts*,
Undergraduate Lecture Notes in Physics, DOI 10.1007/978-1-4939-1360-2_23,
© Springer Science+Business Media, LLC 2015

often associated with his theory of relativity. Einstein describes this theory as a "logically rigid" and "astoundingly simple combination and generalization" of earlier work by men such as Newton, Maxwell, Mach, Lorentz, and Poincaré. In order to reconcile Maxwell's theory of light with Newton's theory of motion, Einstein introduced a connection between the space and time coordinates of events observed from different, but non-accelerating, frames of reference. This *special theory of relativity* was based essentially on the assumption that the speed of light is independent of the speed of its source relative to the observer. With the 1916 publication of his *general theory of relativity*, Einstein expanded his previous work to account for observations made from accelerating frames of reference. These considerations led to a complete reconceptualization of the relationship between mass, gravity, space and time.

Accounts of both the special and the general theories are included in Einstein's popular book entitled *Relativity*, which was translated from German into English by Robert W. Lawson and published by the Henry Holt Company in 1920. Herein, Einstein provides "an exact insight into the theory of Relativity to those readers who, from a general scientific and philosophical point of view, are interested in the theory, but who are not conversant with the mathematical apparatus of theoretical physics."[4] Part II of the book, from which the next set of readings are extracted, focuses on Einstein's general theory of relativity. He begins this part by reiterating the special principle of relativity, which he first introduced in Part I.[5] What does the special principle of relativity state? How does he generalize it so as to explain the motion of falling masses? And what does this generalization imply regarding the effect of gravity upon light, which itself (presumably) lacks mass?

23.2   Reading: Einstein, *Relativity*

Einstein, A., *Relativity*, Great Minds, Prometheus Books, Amherst, NY, 1995. Part II, The General Theory of Relativity.

23.2.1   *XVIII: Special and General Principle of Relativity*

The basal principle, which was the pivot of all our previous considerations, was the *special* principle of relativity, *i.e.* the principle of the physical relativity of all *uniform* motion. Let us once more analyse its meaning carefully.

It was at all times clear that, from the point of view of the idea it conveys to us, every motion must be considered only as a relative motion. Returning to the illustration we have frequently used of the embankment and the railway carriage, we

---

[4] This quotation is from Einstein's preface to *Relativity*.

[5] See Chaps. 29–32 of Vol. II.

can express the fact of the motion here taking place in the following two forms, both of which are equally justifiable:

(a) The carriage is in motion relative to the embankment.
(b) The embankment is in motion relative to the carriage.

In (a) the embankment, in (b) the carriage, serves as the body of reference in our statement of the motion taking place. If it is simply a question of detecting or of describing the motion involved, it is in principle immaterial to what reference-body we refer the motion. As already mentioned, this is self-evident, but it must not be confused with the much more comprehensive statement called "the principle of relativity," which we have taken as the basis of our investigations.

The principle we have made use of not only maintains that we may equally well choose the carriage or the embankment as our reference-body for the description of any event (for this, too, is self-evident). Our principle rather asserts what follows: If we formulate the general laws of nature as they are obtained from experience, by making use of

(a) the embankment as reference-body,
(b) the railway carriage as reference-body,

then these general laws of nature (*e.g.* the laws of mechanics or the law of the propagation of light in *vacuo*) have exactly the same form in both cases. This can also be expressed as follows: For the *physical* description of natural processes, neither of the reference bodies K, K' is unique (*lit.* "specially marked out") as compared with the other. Unlike the first, this latter statement need not of necessity hold *a priori*; it is not contained in the conceptions of "motion" and "reference-body" and derivable from them; only *experience* can decide as to its correctness or incorrectness.

Up to the present, however, we have by no means maintained the equivalence of *all* bodies of reference K in connection with the formulation of natural laws. Our course was more on the following lines. In the first place, we started out from the assumption that there exists a reference-body K, whose condition of motion is such that the Galileian law holds with respect to it: A particle left to itself and sufficiently far removed from all other particles moves uniformly in a straight line. With reference to K (Galileian reference-body) the laws of nature were to be as simple as possible. But in addition to K, all bodies of reference K' should be given preference in this sense, and they should be exactly equivalent to K for the formulation of natural laws, provided that they are in a state of *uniform rectilinear and non-rotary motion* with respect to K; all these bodies of reference are to be regarded as Galileian reference-bodies. The validity of the principle of relativity was assumed only for these reference-bodies, but not for others (*e.g.* those possessing motion of a different kind). In this sense we speak of the special principle of relativity, or special theory of relativity. In contrast to this we wish to understand by the "general principle of relativity" the following statement: All bodies of reference K, K', etc., are equivalent for the description of natural phenomena (formulation of the general laws of nature), whatever may be their state of motion. But before proceeding farther, it ought to be pointed out that this formulation must be replaced later by a more abstract one, for reasons which will become evident at a later stage.

Since the introduction of the special principle of relativity has been justified, every intellect which strives after generalisation must feel the temptation to venture the step towards the general principle of relativity. But a simple and apparently quite reliable consideration seems to suggest that, for the present at any rate, there is little hope of success in such an attempt. Let us imagine ourselves transferred to our old friend the railway carriage, which is travelling at a uniform rate. As long as it is moving uniformly, the occupant of the carriage is not sensible of its motion, and it is for this reason that he can without reluctance interpret the facts of the case as indicating that the carriage is at rest, but the embankment in motion. Moreover, according to the special principle of relativity, this interpretation is quite justified also from a physical point of view.

If the motion of the carriage is now changed into a non-uniform motion, as for instance by a powerful application of the brakes, then the occupant of the carriage experiences a correspondingly powerful jerk forwards. The retarded motion is manifested in the mechanical behaviour of bodies relative to the person in the railway carriage. The mechanical behaviour is different from that of the case previously considered, and for this reason it would appear to be impossible that the same mechanical laws hold relatively to the nonuniformly moving carriage, as hold with reference to the carriage when at rest or in uniform motion. At all events it is clear that the Galileian law does not hold with respect to the non-uniformly moving carriage. Because of this, we feel compelled at the present juncture to grant a kind of absolute physical reality to nonuniform motion, in opposition to the general principle of relativity. But in what follows we shall soon see that this conclusion cannot be maintained.

23.2.2   XIX: The Gravitational Field

"If we pick up a stone and then let it go, why does it fall to the ground?" The usual answer to this question is: "Because it is attracted by the earth." Modern physics formulates the answer rather differently for the following reason. As a result of the more careful study of electromagnetic phenomena, we have come to regard action at a distance as a process impossible without the intervention of some intermediary medium. If, for instance, a magnet attracts a piece of iron, we cannot be content to regard this as meaning that the magnet acts directly on the iron through the intermediate empty space, but we are constrained to imagine—after the manner of Faraday—that the magnet always calls into being something physically real in the space around it, that something being what we call a "magnetic field." In its turn this magnetic field operates on the piece of iron, so that the latter strives to move towards the magnet. We shall not discuss here the justification for this incidental conception, which is indeed a somewhat arbitrary one. We shall only mention that with its aid electromagnetic phenomena can be theoretically represented much more satisfactorily than without it, and this applies particularly to the transmission of electromagnetic waves.

The effects of gravitation also are regarded in an analogous manner. The action of the earth on the stone takes place indirectly. The earth produces in its surroundings a gravitational field, which acts on the stone and produces its motion of fall. As we know from experience, the intensity of the action on a body diminishes according to a quite definite law, as we proceed farther and farther away from the earth. From our point of view this means: The law governing the properties of the gravitational field in space must be a perfectly definite one, in order correctly to represent the diminution of gravitational action with the distance from operative bodies. It is something like this: The body (*e.g.* the earth) produces a field in its immediate neighbourhood directly: the intensity and direction of the field at points farther removed from the body are thence determined by the law which governs the properties in space of the gravitational fields themselves.

In contrast to electric and magnetic fields, the gravitational field exhibits a most remarkable property, which is of fundamental importance for what follows. Bodies which are moving under the sole influence of a gravitational field receive an acceleration, *which does not in the least depend either on the material or on the physical state of the body.* For instance, a piece of lead and a piece of wood fall in exactly the same manner in a gravitational field (*in vacuo*), when they start off from rest or with the same initial velocity. This law, which holds most accurately, can be expressed in a different form in the light of the following consideration. According to Newton's law of motion, we have

$$\text{(Force)} = \text{(inertial mass)} \times \text{(acceleration)}, \tag{23.1}$$

where the "inertial mass" is a characteristic constant of the accelerated body. If now gravitation is the cause of the acceleration, we then have

$$\text{(Force)} = \text{(gravitational mass)} \times \text{(intensity of the gravitational field)}, \tag{23.2}$$

where the "gravitational mass" is likewise a characteristic constant for the body. From these two relations follows:

$$\text{(acceleration)} = \frac{\text{(gravitational mass)}}{\text{(interial mass)}} \times \text{(intensity of the gravitational field)}. \tag{23.3}$$

If now, as we find from experience, the acceleration is to be independent of the nature and the condition of the body and always the same for a given gravitational field, then the ratio of the gravitational to the inertial mass must likewise be the same for all bodies. By a suitable choice of units we can thus make this ratio equal to unity. We then have the following law: The *gravitational* mass of a body is equal to its *inertial* mass.

It is true that this important law had hitherto been recorded in mechanics, but it had not been *interpreted.* A satisfactory interpretation can be obtained only if we recognise the following fact: *The same* quality of a body manifests itself according to circumstances as "inertia" or as "weight" (*lit.* "heaviness"). In the following section, we shall show to what extent this is actually the case, and how this question is connected with the general postulate of relativity.

23.2.3   XX: The Equality of Inertial and Gravitational Mass as an Argument for the General Postulate of Relativity

We imagine a large portion of empty space, so far removed from stars and other appreciable masses, that we have before us approximately the conditions required by the fundamental law of Galilei. It is then possible to choose a Galileian reference-body for this part of space (world), relative to which points at rest remain at rest and points in motion continue permanently in uniform rectilinear motion. As reference body let us imagine a spacious chest resembling a room with an observer inside who is equipped with apparatus. Gravitation naturally does not exist for this observer. He must fasten himself with strings to the floor, otherwise the slightest impact against the floor will cause him to rise slowly towards the ceiling of the room.

To the middle of the lid of the chest is fixed externally a hook with rope attached, and now a "being" (what kind of a being is immaterial to us) begins pulling at this with a constant force. The chest together with the observer then begin to move "upwards" with a uniformly accelerated motion. In course of time their velocity will reach unheard-of values—provided that we are viewing all this from another reference-body which is not being pulled with a rope.

But how does the man in the chest regard the process? The acceleration of the chest will be transmitted to him by the reaction of the floor of the chest. He must therefore take up this pressure by means of his legs if he does not wish to be laid out full length on the floor. He is then standing in the chest in exactly the same way as anyone stands in a room of a house on our earth. If he release a body which he previously had in his hand, the acceleration of the chest will no longer be transmitted to this body, and for this reason the body will approach the floor of the chest with an accelerated relative motion. The observer will further convince himself *that the acceleration of the body towards the floor of the chest is always of the same magnitude, whatever kind of body he may happen to use for the experiment.*

Relying on his knowledge of the gravitational field (as it was discussed in the preceding section), the man in the chest will thus come to the conclusion that he and the chest are in a gravitational field which is constant with regard to time. Of course he will be puzzled for a moment as to why the chest does not fall in this gravitational field. Just then, however, he discovers the hook in the middle of the lid of the chest and the rope which is attached to it, and he consequently comes to the conclusion that the chest is suspended at rest in the gravitational field.

Ought we to smile at the man and say that he errs in his conclusion? I do not believe we ought to if we wish to remain consistent; we must rather admit that his mode of grasping the situation violates neither reason nor known mechanical laws. Even though it is being accelerated with respect to the "Galileian space" first considered, we can nevertheless regard the chest as being at rest. We have thus good grounds for extending the principle of relativity to include bodies of reference which are accelerated with respect to each other, and as a result we have gained a powerful argument for a generalised postulate of relativity.

We must note carefully that the possibility of this mode of interpretation rests on the fundamental property of the gravitational field of giving all bodies the same acceleration, or, what comes to the same thing, on the law of the equality of inertial and gravitational mass. If this natural law did not exist, the man in the accelerated chest would not be able to interpret the behaviour of the bodies around him on the supposition of a gravitational field, and he would not be justified on the grounds of experience in supposing his reference body to be "at rest."

Suppose that the man in the chest fixes a rope to the inner side of the lid, and that he attaches a body to the free end of the rope. The result of this will be to stretch the rope so that it will hang "vertically" downwards. If we ask for an opinion of the cause of tension in the rope, the man in the chest will say: "The suspended body experiences a downward force in the gravitational field, and this is neutralised by the tension of the rope; what determines the magnitude of the tension of the rope is the *gravitational mass* of the suspended body." On the other hand, an observer who is poised freely in space will interpret the condition of things thus: "The rope must perforce take part in the accelerated motion of the chest, and it transmits this motion to the body attached to it. The tension of the rope is just large enough to effect the acceleration of the body. That which determines the magnitude of the tension of the rope is the *inertial mass* of the body." Guided by this example, we see that our extension of the principle of relativity implies the *necessity* of the law of the equality of inertial and gravitational mass. Thus we have obtained a physical interpretation of this law.

From our consideration of the accelerated chest we see that a general theory of relativity must yield important results on the laws of gravitation. In point of fact, the systematic pursuit of the general idea of relativity has supplied the laws satisfied by the gravitational field. Before proceeding farther, however, I must warn the reader against a misconception suggested by these considerations. A gravitational field exists for the man in the chest, despite the fact that there was no such field for the co-ordinate system first chosen. Now we might easily suppose that the existence of a gravitational field is always only an *apparent* one. We might also think that, regardless of the kind of gravitational field which may be present, we could always choose another reference-body such that *no* gravitational field exists with reference to it. This is by no means true for all gravitational fields, but only for those of quite special form. It is, for instance, impossible to choose a body of reference such that, as judged from it, the gravitational field of the earth (in its entirety) vanishes.

We can now appreciate why that argument is not convincing, which we brought forward against the general principle of relativity at the end of Section XVIII. It is certainly true that the observer in the railway carriage experiences a jerk forwards as a result of the application of the brake, and that he recognises in this the non-uniformity of motion (retardation) of the carriage. But he is compelled by nobody to refer this jerk to a "real" acceleration (retardation) of the carriage. He might also interpret his experience thus: "My body of reference (the carriage) remains permanently at rest. With reference to it, however, there exists (during the period of application of the brakes) a gravitational field which is directed forwards and which is variable with respect to time. Under the influence of this field, the embankment together with the earth moves non-uniformly in such a manner that their original velocity in the backwards direction is continuously reduced."

23.2.4 *XXI: In what Respects are the Foundations of Classical Mechanics and of the Special Theory of Relativity Unsatisfactory?*

We have already stated several times that classical mechanics starts out from the following law: Material particles sufficiently far removed from other material particles continue to move uniformly in a straight line or continue in a state of rest. We have also repeatedly emphasised that this fundamental law can only be valid for bodies of reference K which possess certain unique states of motion, and which are in uniform translational motion relative to each other. Relative to other reference-bodies K the law is not valid. Both in classical mechanics and in the special theory of relativity we therefore differentiate between reference-bodies K relative to which the recognised "laws of nature" can be said to hold, and reference-bodies K relative to which these laws do not hold.

But no person whose mode of thought is logical can rest satisfied with this condition of things. He asks: "How does it come that certain reference-bodies (or their states of motion) are given priority over other reference-bodies (or their states of motion)? *What is the reason for this preference?* In order to show clearly what I mean by this question, I shall make use of a comparison.

I am standing in front of a gas range. Standing alongside of each other on the range are two pans so much alike that one may be mistaken for the other. Both are half full of water. I notice that steam is being emitted continuously from the one pan, but not from the other. I am surprised at this, even if I have never seen either a gas range or a pan before. But if I now notice a luminous something of bluish colour under the first pan but not under the other, I cease to be astonished, even if I have never before seen a gas flame. For I can only say that this bluish something will cause the emission of the steam, or at least *possibly* it may do so. If, however, I notice the bluish something in neither case, and if I observe that the one continuously emits steam whilst the other does not, then I shall remain astonished and dissatisfied until I have discovered some circumstance to which I can attribute the different behaviour of the two pans.

Analogously, I seek in vain for a real something in classical mechanics (or in the special theory of relativity) to which I can attribute the different behaviour of bodies considered with respect to the reference systems K and K'.[6] Newton saw this objection and attempted to invalidate it, but without success. But E. Mach recognised

---

[6] The objection is of importance more especially when the state of motion of the reference-body is of such a nature that it does not require any external agency for its maintenance, *e.g.* in the case when the reference-body is rotating uniformly.

it most clearly of all, and because of this objection he claimed that mechanics must be placed on a new basis. It can only be got rid of by means of a physics which is conformable to the general principle of relativity, since the equations of such a theory hold for every body of reference, whatever may be its state of motion.

23.2.5   XXII: A Few Inferences from the General Principle of Relativity

The considerations of Section XX show that the general principle of relativity puts us in a position to derive properties of the gravitational field in a purely theoretical manner. Let us suppose, for instance, that we know the space-time "course" for any natural process whatsoever, as regards the manner in which it takes place in the Galileian domain relative to a Galileian body of reference K. By means of purely theoretical operations (*i.e.* simply by calculation) we are then able to find how this known natural process appears, as seen from a reference-body K' which is accelerated relatively to K. But since a gravitational field exists with respect to this new body of reference K', our consideration also teaches us how the gravitational field influences the process studied.

For example, we learn that a body which is in a state of uniform rectilinear motion with respect to K (in accordance with the law of Galilei) is executing an accelerated and in general curvilinear motion with respect to the accelerated reference-body K' (chest). This acceleration or curvature corresponds to the influence on the moving body of the gravitational field prevailing relatively to K'. It is known that a gravitational field influences the movement of bodies in this way, so that our consideration supplies us with nothing essentially new.

However, we obtain a new result of fundamental importance when we carry out the analogous consideration for a ray of light. With respect to the Galileian reference-body K, such a ray of light is transmitted rectilinearly with the velocity c. It can easily be shown that the path of the same ray of light is no longer a straight line when we consider it with reference to the accelerated chest (reference-body K'). From this we conclude, *that, in general, rays of light are propagated curvilinearly in gravitational fields*. In two respects this result is of great importance.

In the first place, it can be compared with the reality. Although a detailed examination of the question shows that the curvature of light rays required by the general theory of relativity is only exceedingly small for the gravitational fields at our disposal in practice, its estimated magnitude for light rays passing the sun at grazing incidence is nevertheless 1.7 s of arc. This ought to manifest itself in the following way. As seen from the earth, certain fixed stars appear to be in the neighbourhood of the sun, and are thus capable of observation during a total eclipse of the sun. At such times, these stars ought to appear to be displaced outwards from the sun by an amount indicated above, as compared with their apparent position in the sky when the sun is situated at another part of the heavens. The examination of the correctness or otherwise of this deduction is a problem of the greatest importance, the early solution of which is to be expected of astronomers.[7]

In the second place our result shows that, according to the general theory of relativity, the law of the constancy of the velocity of light *in vacuo*, which constitutes one of the two fundamental assumptions in the special theory of relativity and to which we have already frequently referred, cannot claim any unlimited validity. A curvature of rays of light can only take place when the velocity of propagation of light varies with position. Now we might think that as a consequence of this, the special theory of relativity and with it the whole theory of relativity would be laid in the dust. But in reality this is not the case. We can only conclude that the special theory of relativity cannot claim an unlimited domain of validity; its results hold only so long as we are able to disregard the influences of gravitational fields on the phenomena (*e.g.* of light).

Since it has often been contended by opponents of the theory of relativity that the special theory of relativity is overthrown by the general theory of relativity, it is perhaps advisable to make the facts of the case clearer by means of an appropriate comparison. Before the development of electrodynamics the laws of electrostatics were looked upon as the laws of electricity. At the present time we know that electric fields can be derived correctly from electrostatic considerations only for the case, which is never strictly realised, in which the electrical masses are quite at rest relatively to each other, and to the co-ordinate system. Should we be justified in saying that for this reason electrostatics is overthrown by the field-equations of Maxwell in electrodynamics? Not in the least. Electrostatics is contained in electrodynamics as a limiting case; the laws of the latter lead directly to those of the former for the case in which the fields are invariable with regard to time. No fairer destiny could be allotted to any physical theory, than that it should of itself point out the way to the introduction of a more comprehensive theory, in which it lives on as a limiting case.

In the example of the transmission of light just dealt with we have seen that the general theory of relativity enables us to derive theoretically the influence of a gravitational field on the course of natural processes, the laws of which are already known when a gravitational field is absent. But the most attractive problem, to the solution of which the general theory of relativity supplies the key, concerns the investigation of the laws satisfied by the gravitational field itself. Let us consider this for a moment.

We are acquainted with space-time domains which behave (approximately) in a "Galileian" fashion under suitable choice of reference-body, *i.e.* domains in which gravitational fields are absent. If we now refer such a domain to a reference-body K' possessing any kind of motion, then relative to K' there exists a gravitational field

---

[7] By means of the star photographs of two expeditions equipped by a Joint Committee of the Royal and Royal Astronomical Societies, the existence of the deflection of light demanded by theory was first confirmed during the solar eclipse of 29th May, 1919. (Cf. Appendix III of Einstein, A., *Relativity*, Great Minds, Prometheus Books, Amherst, NY, 1995.)

which is variable with respect to space and time.[8] The character of this field will of course depend on the motion chosen for K'. According to the general theory of relativity, the general law of the gravitational field must be satisfied for all gravitational fields obtainable in this way. Even though by no means all gravitational fields can be produced in this way, yet we may entertain the hope that the general law of gravitation will be derivable from such gravitational fields of a special kind. This hope has been realised in the most beautiful manner. But between the clear vision of this goal and its actual realisation it was necessary to surmount a serious difficulty, and as this lies deep at the root of things, I dare not withhold it from the reader. We require to extend our ideas of the space-time continuum still farther.

23.2.6   XXIII: Behaviour of Clocks and Measuring-Rods on a Rotating Body of Reference

Hitherto I have purposely refrained from speaking about the physical interpretation of space- and time-data in the case of the general theory of relativity. As a consequence, I am guilty of a certain slovenliness of treatment, which, as we know from the special theory of relativity, is far from being unimportant and pardonable. It is now high time that we remedy this defect; but I would mention at the outset, that this matter lays no small claims on the patience and on the power of abstraction of the reader. We start off again from quite special cases, which we have frequently used before. Let us consider a spacetime domain in which no gravitational field exists relative to a reference-body K whose state of motion has been suitably chosen. K is then a Galileian reference-body as regards the domain considered, and the results of the special theory of relativity hold relative to K. Let us suppose the same domain referred to a second body of reference K', which is rotating uniformly with respect to K. In order to fix our ideas, we shall imagine K' to be in the form of a plane circular disc, which rotates uniformly in its own plane about its centre. An observer who is sitting eccentrically on the disc K' is sensible of a force which acts outwards in a radial direction, and which would be interpreted as an effect of inertia (centrifugal force) by an observer who was at rest with respect to the original reference-body K. But the observer on the disc may regard his disc as a reference-body which is "at rest"; on the basis of the general principle of relativity he is justified in doing this. The force acting on himself, and in fact on all other bodies which are at rest relative to the disc, he regards as the effect of a gravitational field. Nevertheless, the space-distribution of this gravitational field is of a kind that would not be possible on Newton's theory of gravitation.[9] But since the observer believes in the general theory of relativity, this does not disturb him; he is quite in the right when he believes

---

[8] This follows from a generalisation of the discussion in Section XX.

[9] The field disappears at the centre of the disc and increases proportionally to the distance from the centre as we proceed outwards.

that a general law of gravitation can be formulated—a law which not only explains the motion of the stars correctly, but also the field of force experienced by himself.

The observer performs experiments on his circular disc with clocks and measuring-rods. In doing so, it is his intention to arrive at exact definitions for the signification of time- and space-data with reference to the circular disc K', these definitions being based on his observations. What will be his experience in this enterprise?

To start with, he places one of two identically constructed clocks at the centre of the circular disc, and the other on the edge of the disc, so that they are at rest relative to it. We now ask ourselves whether both clocks go at the same rate from the standpoint of the non-rotating Galileian reference-body K. As judged from this body, the clock at the centre of the disc has no velocity, whereas the clock at the edge of the disc is in motion relative to K in consequence of the rotation. According to a result obtained in Section XII, it follows that the latter clock goes at a rate permanently slower than that of the clock at the centre of the circular disc, *i.e.* as observed from K. It is obvious that the same effect would be noted by an observer whom we will imagine sitting alongside his clock at the centre of the circular disc. Thus on our circular disc, or, to make the case more general, in every gravitational field, a clock will go more quickly or less quickly, according to the position in which the clock is situated (at rest). For this reason it is not possible to obtain a reasonable definition of time with the aid of clocks which are arranged at rest with respect to the body of reference. A similar difficulty presents itself when we attempt to apply our earlier definition of simultaneity in such a case, but I do not wish to go any farther into this question.

Moreover, at this stage the definition of the space co-ordinates also presents insurmountable difficulties. If the observer applies his standard measuring-rod (a rod which is short as compared with the radius of the disc) tangentially to the edge of the disc, then, as judged from the Galileian system, the length of this rod will be less than 1, since, according to Section XII, moving bodies suffer a shortening in the direction of the motion. On the other hand, the measuring-rod will not experience a shortening in length, as judged from K, if it is applied to the disc in the direction of the radius. If, then, the observer first measures the circumference of the disc with his measuring-rod and then the diameter of the disc, on dividing the one by the other, he will not obtain as quotient the familiar number $\pi = 3.14\ldots$, but a larger number,[10] whereas of course, for a disc which is at rest with respect to K, this operation would yield π exactly. This proves that the propositions of Euclidean geometry cannot hold exactly on the rotating disc, nor in general in a gravitational field, at least if we attribute the length 1 to the rod in all positions and in every orientation. Hence the idea of a straight line also loses its meaning. We are therefore not in a position to define exactly the co-ordinates x, y, z relative to the disc by means of the method

---

[10] Throughout this consideration we have to use the Galileian (non-rotating) system K as reference-body, since we may only assume the validity of the results of the special theory of relativity relative to K (relative to K' a gravitational field prevails).

used in discussing the special theory, and as long as the co-ordinates and times of events have not been defined, we cannot assign an exact meaning to the natural laws in which these occur.

Thus all our previous conclusions based on general relativity would appear to be called in question. In reality we must make a subtle detour in order to be able to apply the postulate of general relativity exactly. I shall prepare the reader for this in the following paragraphs.

23.3 Study Questions

QUES. 23.1 What is the principle of relativity?

a) What is meant by the term "laws of nature"? What are some examples of such laws?

b) Is the principle of relativity correct *a priori*? That is: can it be accepted as true before carrying out any observations of nature? If not, is *any* principle at all correct a priori?

c) What is uniform motion? And what does it mean for two systems of reference to be equivalent? Is a state of uniform motion discernible from a state of rest? What about nonuniform motion? What does this suggest?

d) With what type of reference bodies does the special theory of relativity deal? How does this differ from the general theory?

QUES. 23.2 Is gravity a mediated action, or does it act "at-a-distance"?

a) How is the question "why do dropped rocks fall to the ground" typically answered?

b) What is action-at-a-distance? How does modern physics avoid this concept in explaining magnetism?

c) How is the action produced by gravitational fields fundamentally different than that produced by electric and magnetic fields?

QUES. 23.3 Why do all masses fall with the same acceleration in a vacuum?

a) Is there a conceptual difference between inertial mass and gravitational mass? In what context does each arise?

b) How does the general principle of relativity make sensible, or perhaps even necessary, the seeming equivalence of inertial and gravitational mass?

c) Can a state of acceleration ever be distinguished from the presence of a gravitational field?

d) Is it always possible to choose a reference body with respect to which a gravitational field entirely vanishes?

e) Is it plausible to derive the properties of gravitational fields in a purely theoretical manner—that is—without recourse to experimental observation?

QUES. 23.4 Does light have weight?

a) Does light always travel in a straight line in empty space? In particular, does the sun deflect passing light rays? If so, what does this imply about the speed of light in a vacuum?

b) If the speed of light were not constant, would this necessarily imply that the special theory of relativity is false?

QUES. 23.5 What is the ratio of the circumference to the diameter of a stationary plane circular disc? A spinning disc? Does your answer depend on whether you accept Einstein's general principle of relativity?

23.4   Exercises

Ex. 23.1 (FALLING ROCKS AND LIGHT). Suppose that you awake to find yourself inside a room-sized chest which is attached to a rocket ship by a hook on the roof of the chest. The rocket ship appears to be accelerating in a straight line so that its speed increases by 2 m/s every second.

a) If you were to release a rock from your hand at a height of 1 m above the floor, how long would it take the rock to hit the floor? (ANSWER:1 s.)

b) If you were to aim a flashlight horizontally 1 m above the floor and turn it on briefly, what would happen? Would the emitted light strike the opposite wall (which is 2 m away) at the same height as the flashlight? If not, how far would it fall? (ANSWER: It would strike the wall about 4.4×10^{-17} m below the flashlight.)

c) Which has mass: the rock? the light? both? neither? Explain your reasoning.

d) From your measurements, can you be sure that the chest is, in fact, accelerating? Is there an alternative possibility which would explain your observations?

e) If you were to briefly shine the flashlight straight upwards through a skylight in the roof of the chest what would happen? Would the light ever fall back into the chest? What problem does this raise?

Ex. 23.2 (SPINNING CAROUSEL AND GENERAL RELATIVITY). A stationary circular carousel has a diameter of 8 m. What is the carousel's circumference? What (standard) value of π did you use to calculate this circumference? Now consider the same carousel when it is spinning about its axis at an astounding rate of 10 million times per second. Try to answer the following questions.

a) First, what is the speed of a point on the circumference as measured by an observer, K, who is watching the carousel spin from a vantage point directly above the axis of the carousel?

b) According to Einstein's special theory of relativity, the length of a moving body is contracted (in its direction of motion) by a factor of

$$\gamma = \frac{1}{\sqrt{1 - \frac{v^2}{c^2}}} \qquad (23.4)$$

Here, v is the speed of the body with respect to the observer and c is the speed of light in a vacuum.[11] According to observer K what is the value of π? In answering this question, note that the length of the circumference (but not the diameter) is contracted by a factor of γ. (ANSWER: $\pi = 1.72$)

c) Observer G is tenaciously clinging to the edge of the carousel. Can he reasonably conclude that he is stationary and that the rest of the world is in fact spinning around him? To what principle must he appeal in order to do so?

d) Suppose that observer G measures the circumference and diameter of the carousel and calculates the same value of π as K did. If G maintains that he is at rest, then can he legitimately attribute this value to relativistic length contraction (Eq. 23.4), as does K? If not, then to what might G attribute its strange value?

e) According to G, is the force he feels drawing him outward a fictitious (centrifugal) force or a true (gravitational) force? Does K agree with G on this interpretation? If not, which observer is correct regarding the true nature of the force felt by G?

23.5 Vocabulary

1. Basal
2. Immaterial
3. Electromagnetic
4. Action-at-a-distance
5. Diminution
6. Perforce
7. Electrodynamics
8. Electrostatics
9. Surmount
10. Continuum
11. Slovenliness
12. Insurmountable
13. Tangential
14. Postulate

---

[11] For a detailed treatment of relativistic length contraction and time dilation, see the discussion of Einstein's theory of special relativity included in Chaps. 30 and 31 of the present volume.

Chapter 24
Euclid, Gauss and Mercury's Orbit

> *The method of Cartesian co-ordinates must then be discarded,*
> *and replaced by another which does not assume the validity of*
> *Euclidean geometry for rigid bodies.*
>
> —Einstein

24.1 Introduction

In the previous reading selection from Part II of his book *Relativity*, Einstein introduced his general principle of relativity as a reasonable extension of special relativity. Whereas the special principle of relativity claims that all inertial (*i.e.* uniformly moving) observers are equivalent—meaning that none of them have a privileged perspective—the general principle claims that non-inertial observers are all equivalent too. For Einstein recognized that an observer who is in a state of uniform acceleration will witness precisely the same motion of nearby objects as if he or she were at rest in the presence of a gravitational field. For example, consider an observant parent standing in the park watching his child clinging tenaciously to the rim of a spinning carousel. The parent can legitimately claim that the force which the child is experiencing is not a real force, but rather a fictitious "centrifugal force" which arises merely because the child is riding in a rotating (*i.e.* accelerating) frame of reference. The child, on the other hand, can legitimately claim that he is at rest and that his parent—along with the park and the rest of the world—are in fact spinning around him. The force that he feels which is drawing him outwards must be caused by a gravitational field arising from this spinning world. And when this world stops spinning around him, the gravitational field dies away. Note that the child and the parent agree that the child experiences a force; they simply disagree on the cause of the force. Most importantly, by the general principle of relativity, *neither the parent's nor child's perspective is to be preferred*—they are equivalent. Now, in the next reading selection, Einstein goes on to explain how this general principle of relativity may be expressed mathematically using concepts from non-Euclidean geometry. He begins simply: by explaining how a Cartesian coordinate system may be constructed on a "flat" surface, namely, one in which Euclid's rules of geometry are valid.

K. Kuehn, *A Student's Guide Through the Great Physics Texts,*
Undergraduate Lecture Notes in Physics, DOI 10.1007/978-1-4939-1360-2_24,
© Springer Science+Business Media, LLC 2015

24.2   Reading: Einstein, *Relativity*

Einstein, A., *Relativity*, Great Minds, Prometheus Books, Amherst, NY, 1995. Part II, The General Theory of Relativity.

24.2.1   XXIV: Euclidean and non-Euclidean Continuum

The surface of a marble table is spread out in front of me. I can get from any one point on this table to any other point by passing continuously from one point to a "neighbouring" one, and repeating this process a (large) number of times, or, in other words, by going from point to point without executing "jumps." I am sure the reader will appreciate with sufficient clearness what I mean here by "neighbouring" and by "jumps" (if he is not too pedantic). We express this property of the surface by describing the latter as a continuum.

Let us now imagine that a large number of little rods of equal length have been made, their lengths being small compared with the dimensions of the marble slab. When I say they are of equal length, I mean that one can be laid on any other without the ends overlapping. We next lay four of these little rods on the marble slab so that they constitute a quadrilateral figure (a square), the diagonals of which are equally long. To ensure the equality of the diagonals, we make use of a little testing-rod. To this square we add similar ones, each of which has one rod in common with the first. We proceed in like manner with each of these squares until finally the whole marble slab is laid out with squares. The arrangement is such, that each side of a square belongs to two squares and each corner to four squares.

It is a veritable wonder that we can carry out this business without getting into the greatest difficulties. We only need to think of the following. If at any moment three squares meet at a corner, then two sides of the fourth square are already laid, and, as a consequence, the arrangement of the remaining two sides of the square is already completely determined. But I am now no longer able to adjust the quadrilateral so that its diagonals may be equal. If they are equal of their own accord, then this is an especial favour of the marble slab and of the little rods, about which I can only be thankfully surprised. We must needs experience many such surprises if the construction is to be successful.

If everything has really gone smoothly, then I say that the points of the marble slab constitute a Euclidean continuum with respect to the little rod, which has been used as a "distance" (line-interval). By choosing one corner of a square as "origin," I can characterise every other corner of a square with reference to this origin by means of two numbers. I only need state how many rods I must pass over when, starting from the origin, I proceed towards the "right" and then "upwards," in order to arrive at the corner of the square under consideration. These two numbers are then the "Cartesian co-ordinates" of this corner with reference to the "Cartesian co-ordinate system" which is determined by the arrangement of little rods.

By making use of the following modification of this abstract experiment, we recognise that there must also be cases in which the experiment would be unsuccessful. We shall suppose that the rods "expand" by an amount proportional to the increase of temperature. We heat the central part of the marble slab, but not the periphery, in which case two of our little rods can still be brought into coincidence at every position on the table. But our construction of squares must necessarily come into disorder during the heating, because the little rods on the central region of the table expand, whereas those on the outer part do not.

With reference to our little rods—defined as unit lengths—the marble slab is no longer a Euclidean continuum, and we are also no longer in the position of defining Cartesian co-ordinates directly with their aid, since the above construction can no longer be carried out. But since there are other things which are not influenced in a similar manner to the little rods (or perhaps not at all) by the temperature of the table, it is possible quite naturally to maintain the point of view that the marble slab is a "Euclidean continuum." This can be done in a satisfactory manner by making a more subtle stipulation about the measurement or the comparison of lengths.

But if rods of every kind (*i.e.*, of every material) were to behave *in the same way* as regards the influence of temperature when they are on the variably heated marble slab, and if we had no other means of detecting the effect of temperature than the geometrical behaviour of our rods in experiments analogous to the one described above, then our best plan would be to assign the distance *one* to two points on the slab, provided that the ends of one of our rods could be made to coincide with these two points: for how else should we define the distance without our proceeding being in the highest measure grossly arbitrary? The method of Cartesian co-ordinates must then be discarded, and replaced by another which does not assume the validity of Euclidean geometry for rigid bodies.[1] The reader will notice that the situation depicted here corresponds to the one brought about by the general postulate of relativity (Sect. XXIII).

24.2.2 XXV: Gaussian Co-ordinates

According to Gauss, this combined analytical and geometrical mode of handling the problem can be arrived at in the following way. We imagine a system of arbitrary curves (see Fig. 24.1) drawn on the surface of the table. These we designate as

---

[1] Mathematicians have been confronted with our problem in the following form. If we are given a surface (an ellipsoid) in Euclidean three-dimensional space, then there exists for this surface a two-dimensional geometry just as much as for a plane surface. Gauss undertook the task of treating this two-dimensional geometry from first principles without making use of the fact that the surface belongs to a Euclidean continuum of three dimensions. If we imagine constructions to be made with rigid rods *in the surface* (similar to that above with the marble slab) we should find that different laws hold for these from those resulting on the basis of Euclidean plane geometry. The surface is not a Euclidean continuum with respect to the rods, and we cannot define Cartesian co-ordinates *in the surface*. Gauss indicated the principles according to which we can treat the geometrical relationships in the surface, and thus pointed out the way to the method of Riemann of treating multi-dimensional, non-Euclidean *continua*. Thus it is that mathematicians long ago solved the formal problems to which we are led by the general postulate of relativity.

Fig. 24.1 A system of
arbitrary curves for treating a
non-Euclidean
surface.—[K.K.]

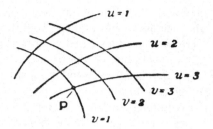

u-curves, and we indicate each of them by means of a number. The curves $u = 1$,
$u = 2$ and $u = 3$ are drawn in the diagram. Between the curves $u = 1$ and $u = 2$
we must imagine an infinitely large number to be drawn, all of which correspond to
real numbers lying between 1 and 2. We have then a system of u-curves, and this
"infinitely dense" system covers the whole surface of the table. These u-curves must
not intersect each other, and through each point of the surface one and only one curve
must pass. Thus a perfectly definite value of u belongs to every point on the surface
of the marble slab. In like manner we imagine a system of v-curves drawn on the
surface. These satisfy the same conditions as the u-curves, they are provided with
numbers in a corresponding manner, and they may likewise be of arbitrary shape.
It follows that a value of u and a value of v belong to every point on the surface
of the table. We call these two numbers the co-ordinates of the surface of the table
(Gaussian co-ordinates). For example, the point P in the diagram has the Gaussian
co-ordinates $u = 3$, $v = 1$. Two neighbouring points P and P' on the surface then
correspond to the co-ordinates

$$P: \qquad u, v$$

$$P': \qquad u + \mathrm{d}u, v + \mathrm{d}v$$

where $\mathrm{d}u$ and $\mathrm{d}v$ signify very small numbers. In a similar manner we may indicate
the distance (line interval) between P and P', as measured with a little rod, by means
of the very small number $\mathrm{d}s$. Then according to Gauss we have

$$\mathrm{d}s^2 = g_{11}\,\mathrm{d}u^2 + 2g_{12}\,\mathrm{d}u\,\mathrm{d}v + g_{22}\,\mathrm{d}v^2, \tag{24.1}$$

where g_{11}, g_{12}, g_{22}, are magnitudes which depend in a perfectly definite way on u
and v. The magnitudes g_{11}, g_{12} and g_{22} determine the behaviour of the rods relative
to the u-curves and v-curves, and thus also relative to the surface of the table. For
the case in which the points of the surface considered form a Euclidean continuum
with reference to the measuring-rods, but only in this case, it is possible to draw
the u-curves and v-curves and to attach numbers to them, in such a manner, that we
simply have:

$$\mathrm{d}s^2 = \mathrm{d}u^2 + \mathrm{d}v^2, \tag{24.2}$$

Under these conditions, the u-curves and v-curves are straight lines in the sense of
Euclidean geometry, and they are perpendicular to each other. Here the Gaussian co-
ordinates are simply Cartesian ones. It is clear that Gauss co-ordinates are nothing

more than an association of two sets of numbers with the points of the surface considered, of such a nature that numerical values differing very slightly from each other are associated with neighbouring points "in space."

So far, these considerations hold for a continuum of two dimensions. But the Gaussian method can be applied also to a continuum of three, four or more dimensions. If, for instance, a continuum of four dimensions be supposed available, we may represent it in the following way. With every point of the continuum we associate arbitrarily four numbers, x_1, x_2, x_3, x_4, which are known as "co-ordinates." Adjacent points correspond to adjacent values of the co-ordinates. If a distance ds is associated with the adjacent points P and P', this distance being measurable and well defined from a physical point of view, then the following formula holds:

$$ds^2 = g_{11}\, dx_1^2 + 2g_{12}\, dx_1\, dx_2 \cdots + g_{44}\, dx_4^2, \qquad (24.3)$$

where the magnitudes g_{11}, *etc.*, have values which vary with the position in the continuum. Only when the continuum is a Euclidean one is it possible to associate the co-ordinates $x_1 \cdots x_4$ with the points of the continuum so that we have simply

$$ds^2 = dx_1^2 + dx_2^2 + dx_3^2 + dx_4^2. \qquad (24.4)$$

In this case relations hold in the four-dimensional continuum which are analogous to those holding in our three-dimensional measurements.

However, the Gauss treatment for ds^2 which we have given above is not always possible. It is only possible when sufficiently small regions of the continuum under consideration may be regarded as Euclidean continua. For example, this obviously holds in the case of the marble slab of the table and local variation of temperature. The temperature is practically constant for a small part of the slab, and thus the geometrical behaviour of the rods is *almost* as it ought to be according to the rules of Euclidean geometry. Hence the imperfections of the construction of squares in the previous section do not show themselves clearly until this construction is extended over a considerable portion of the surface of the table.

We can sum this up as follows: Gauss invented a method for the mathematical treatment of continua in general, in which "size-relations" ("distances" between neighbouring points) are defined. To every point of a continuum are assigned as many numbers (Gaussian coordinates) as the continuum has dimensions. This is done in such a way, that only one meaning can be attached to the assignment, and that numbers (Gaussian coordinates) which differ by an indefinitely small amount are assigned to adjacent points. The Gaussian coordinate system is a logical generalisation of the Cartesian co-ordinate system. It is also applicable to non-Euclidean continua, but only when, with respect to the defined "size" or "distance," small parts of the continuum under consideration behave more nearly like a Euclidean system, the smaller the part of the continuum under our notice.

24.2.3   XXVI: The Space–time Continuum of the Special Theory of Relativity Considered as a Euclidean Continuum

We are now in a position to formulate more exactly the idea of Minkowski, which was only vaguely indicated in Sect. XVII. In accordance with the special theory of relativity, certain co-ordinate systems are given preference for the description of the four-dimensional, space–time continuum. We called these "Galileian co-ordinate systems." For these systems, the four co-ordinates x, y, z, t, which determine an event or—in other words—a point of the four-dimensional continuum, are defined physically in a simple manner, as set forth in detail in the first part of this book. For the transition from one Galileian system to another, which is moving uniformly with reference to the first, the equations of the Lorentz transformation are valid. These last form the basis for the derivation of deductions from the special theory of relativity, and in themselves they are nothing more than the expression of the universal validity of the law of transmission of light for all Galileian systems of reference.

Minkowski found that the Lorentz transformations satisfy the following simple conditions. Let us consider two neighbouring events, the relative position of which in the four-dimensional continuum is given with respect to a Galileian reference-body K by the space co-ordinate differences dx, dy, dz and the time-difference dt. With reference to a second Galileian system we shall suppose that the corresponding differences for these two events are dx', dy', dz', dt'. Then these magnitudes always fulfil the condition[2]

$$dx^2 + dy^2 + dz^2 - c^2\, dt^2 = dx'^2 + dy'^2 + dz'^2 - c^2\, dt'^2. \qquad (24.5)$$

The validity of the Lorentz transformation follows from this condition. We can express this as follows: the magnitude

$$ds^2 = dx^2 + dy^2 + dz^2 - c^2\, dt^2, \qquad (24.6)$$

which belongs to two adjacent points of the four-dimensional space–time continuum, has the same value for all selected (Galileian) reference-bodies. If we replace $x, y, z, \sqrt{-1}\,ct$, by x_1, x_2, x_3, x_4, we also obtain the result that

$$ds^2 = dx_1^2 + dx_2^2 + dx_3^2 + dx_4^2 \qquad (24.7)$$

is independent of the choice of the body of reference. We call the magnitude ds the "distance" apart of the two events or four-dimensional points.

Thus, if we choose as time-variable the imaginary variable $\sqrt{-1}\,ct$ instead of the real quantity t, we can regard the space–time continuum—in accordance with the special theory of relativity—as a "Euclidean" four-dimensional continuum, a result which follows from the considerations of the preceding section.

---

[2] Cf. Appendices I and II [included in Einstein, A., *Relativity*, Great Minds, Prometheus Books, Amherst, NY, 1995.] The relations which are derived there for the co-ordinates themselves are valid also for co-ordinate differences; and thus also for co-ordinate differentials (indefinitely small differences).

24.2.4   XXVII: The Space–time Continuum of the General Theory of Relativity is not a Euclidean Continuum

In the first part of this book we were able to make use of space–time co-ordinates which allowed of a simple and direct physical interpretation, and which, according to Sect. XXVI, can be regarded as four-dimensional Cartesian co-ordinates. This was possible on the basis of the law of the constancy of the velocity of light. But according to Sect. XXI, the general theory of relativity cannot retain this law. On the contrary, we arrived at the result that according to this latter theory the velocity of light must always depend on the co-ordinates when a gravitational field is present. In connection with a specific illustration in Sect. XXIII, we found that the presence of a gravitational field invalidates the definition of the co-ordinates and the time, which led us to our objective in the special theory of relativity.

In view of the results of these considerations we are led to the conviction that, according to the general principle of relativity, the space–time continuum cannot be regarded as a Euclidean one, but that here we have the general case, corresponding to the marble slab with local variations of temperature, and with which we made acquaintance as an example of a two-dimensional continuum. Just as it was there impossible to construct a Cartesian co-ordinate system from equal rods, so here it is impossible to build up a system (reference body) from rigid bodies and clocks, which shall be of such a nature that measuring-rods and clocks, arranged rigidly with respect to one another, shall indicate position and time directly. Such was the essence of the difficulty with which we were confronted in Sect. XXIII.

But the considerations of Sects. XXV and XXVI show us the way to surmount this difficulty. We refer the four-dimensional space–time continuum in an arbitrary manner to Gauss co-ordinates. We assign to every point of the continuum (event) four numbers, x_1, x_2, x_3, x_4 (co-ordinates), which have not the least direct physical significance, but only serve the purpose of numbering the points of the continuum in a definite but arbitrary manner. This arrangement does not even need to be of such a kind that we must regard x_1, x_2, x_3 as "space" co-ordinates and x_4 as a "time" co-ordinate.

The reader may think that such a description of the world would be quite inadequate. What does it mean to assign to an event the particular co-ordinates x_1, x_2, x_3, x_4, if in themselves these co-ordinates have no significance? More careful consideration shows, however, that this anxiety is unfounded. Let us consider, for instance, a material point with any kind of motion. If this point had only a momentary existence without duration, then it would be described in space–time by a single system of values x_1, x_2, x_3, x_4. Thus its permanent existence must be characterised by an infinitely large number of such systems of values, the co-ordinate values of which are so close together as to give continuity; corresponding to the material point, we thus have a (uni-dimensional) line in the four-dimensional continuum. In the same way, any such lines in our continuum correspond to many points in motion. The only statements having regard to these points which can claim a physical existence are in reality the statements about their encounters. In our mathematical treatment, such an encounter is expressed in the fact that the two lines which represent the motions of the points in question have a particular system of co-ordinate values, x_1, x_2, x_3, x_4, in common. After mature consideration the reader will doubtless admit that in reality such encounters constitute the only actual evidence of a time–space nature with which we meet in physical statements.

When we were describing the motion of a material point relative to a body of reference, we stated nothing more than the encounters of this point with particular points of the reference-body. We can also determine the corresponding values of the time by the observation of encounters of the body with clocks, in conjunction with the observation of the encounter of the hands of clocks with particular points on the dials. It is just the same in the case of space-measurements by means of measuring-rods, as a little consideration will show.

The following statements hold generally: every physical description resolves itself into a number of statements, each of which refers to the space–time coincidence of two events A and B. In terms of Gaussian co-ordinates, every such statement is expressed by the agreement of their four co-ordinates x_1, x_2, x_3, x_4. Thus in reality, the description of the time–space continuum by means of Gauss co-ordinates completely replaces the description with the aid of a body of reference, without suffering from the defects of the latter mode of description; it is not tied down to the Euclidean character of the continuum which has to be represented.

24.2.5   *XXVIII: Exact Formulation of the General Principle of Relativity*

We are now in a position to replace the provisional formulation of the general principle of relativity given in Sect. XVIII by an exact formulation. The form there used, "All bodies of reference K, K', etc., are equivalent for the description of natural phenomena (formulation of the general laws of nature), whatever may be their state of motion," cannot be maintained, because the use of rigid reference-bodies, in the sense of the method followed in the special theory of relativity, is in general not possible in space–time description. The Gauss co-ordinate system has to take the place of the body of reference. The following statement corresponds to the fundamental idea of the general principle of relativity: *"All Gaussian co-ordinate systems are essentially equivalent for the formulation of the general laws of nature."*

We can state this general principle of relativity in still another form, which renders it yet more clearly intelligible than it is when in the form of the natural extension of the special principle of relativity. According to the special theory of relativity, the equations which express the general laws of nature pass over into equations of the same form when, by making use of the Lorentz transformation, we replace the space–time variables x, y, z, t, of a (Galileian) reference-body K by the space–time variables x', y', z', t', of a new reference-body K'. According to the general theory of relativity, on the other hand, by application of *arbitrary substitutions* of the Gauss variables x_1, x_2, x_3, x_4 the equations must pass over into equations of the same form; for every transformation (not only the Lorentz transformation) corresponds to the transition of one Gauss co-ordinate system into another.

If we desire to adhere to our "old-time" three-dimensional view of things, then we can characterise the development which is being undergone by the fundamental idea of the general theory of relativity as follows: the special theory of relativity has reference to Galileian domains, *i.e.* to those in which no gravitational field exists. In this connection a Galileian reference-body serves as body of reference, *i.e.* a rigid body the state of motion of which is so chosen that the Galileian law of the uniform rectilinear motion of "isolated" material points holds relatively to it.

Certain considerations suggest that we should refer the same Galileian domains to *non-Galileian* reference-bodies also. A gravitational field of a special kind is then present with respect to these bodies (*cf.* Sects. XX and XXIII).

In gravitational fields there are no such things as rigid bodies with Euclidean properties; thus the fictitious rigid body of reference is of no avail in the general theory of relativity. The motion of clocks is also influenced by gravitational fields, and in such a way that a physical definition of time which is made directly with the aid of clocks has by no means the same degree of plausibility as in the special theory of relativity.

For this reason non-rigid reference-bodies are used, which are as a whole not only moving in any way whatsoever, but which also suffer alterations in form *ad lib.* during their motion. Clocks, for which the law of motion is of any kind, however irregular, serve for the definition of time. We have to imagine each of these clocks fixed at a point on the non-rigid reference-body. These clocks satisfy only the one condition, that the "readings" which are observed simultaneously on adjacent clocks (in space) differ from each other by an indefinitely small amount. This non-rigid reference-body, which might appropriately be termed a "reference-mollusc" is in the main equivalent to a Gaussian four-dimensional co-ordinate system chosen arbitrarily.

That which gives the "mollusc" a certain comprehensibility as compared with the Gauss co-ordinate system is the (really unjustified) formal retention of the separate existence of the space co-ordinates as opposed to the time co-ordinate. Every point on the mollusc is treated as a space-point, and every material point which is at rest relatively to it as at rest, so long as the mollusc is considered as reference-body. The general principle of relativity requires that all these molluscs can be used as reference-bodies with equal right and equal success in the formulation of the general laws of nature; the laws themselves must be quite independent of the choice of mollusc.

The great power possessed by the general principle of relativity lies in the comprehensive limitation which is imposed on the laws of nature in consequence of what we have seen above.

24.2.6   *XXIX: The Solution of the Problem of Gravitation on the Basis of the General Principle of Relativity*

If the reader has followed all our previous considerations, he will have no further difficulty in understanding the methods leading to the solution of the problem of gravitation.

We start off from a consideration of a Galileian domain, a domain in which there is no gravitational field relative to the Galileian reference-body K. The behaviour of measuring-rods and clocks with reference to K is known from the special theory of relativity, likewise the behaviour of "isolated" material points; the latter move uniformly and in straight lines.

Now let us refer this domain to a random Gauss coordinate system or to a "mollusc" as reference-body K'. Then with respect to K' there is a gravitational field G (of a particular kind). We learn the behaviour of measuring-rods and clocks and also of freely moving material points with reference to K' simply by mathematical transformation. We interpret this behaviour as the behaviour of measuring-rods, clocks and material points under the influence of the gravitational field G. Hereupon we introduce a hypothesis: that the influence of the gravitational field on measuring-rods, clocks and freely moving material points continues to take place according to the same laws, even in the case where the prevailing gravitational field is *not* derivable from the Galileian special case, simply by means of a transformation of co-ordinates.[3]

The next step is to investigate the space–time behaviour of the gravitational field G, which was derived from the Galileian special case simply by transformation of the co-ordinates. This behaviour is formulated in a law, which is always valid, no matter how the reference-body (mollusc) used in the description may be chosen.

This law is not yet the *general* law of the gravitational field, since the gravitational field under consideration is of a special kind. In order to find out the general law-of-field of gravitation we still require to obtain a generalisation of the law as found above. This can be obtained without caprice, however, by taking into consideration the following demands:

(a) The required generalisation must likewise satisfy the general postulate of relativity.
(b) If there is any matter in the domain under consideration, only its inertial mass, and thus according to Section XV only its energy is of importance for its effect in exciting a field.
(c) Gravitational field and matter together must satisfy the law of the conservation of energy (and of impulse).

Finally, the general principle of relativity permits us to determine the influence of the gravitational field on the course of all those processes which take place according to known laws when a gravitational field is absent, which have already been fitted into the frame of the special theory of relativity. In this connection we proceed in principle according to the method which has already been explained for measuring-rods, clocks and freely moving material points.

The theory of gravitation derived in this way from the general postulate of relativity excels not only in its beauty; nor in removing the defect attaching to classical mechanics which was brought to light in Sect. XXI; nor in interpreting the empirical

---

[3] The procedure described in this paragraph is the one we used when doing Ex. 23.2 in the previous chapter of this volume.—[*K.K.*]

law of the equality of inertial and gravitational mass; but it has also already explained a result of observation in astronomy, against which classical mechanics is powerless.

If we confine the application of the theory to the case where the gravitational fields can be regarded as being weak, and in which all masses move with respect to the co-ordinate system with velocities which are small compared with the velocity of light, we then obtain as a first approximation the Newtonian theory. Thus the latter theory is obtained here without any particular assumption, whereas Newton had to introduce the hypothesis that the force of attraction between mutually attracting material points is inversely proportional to the square of the distance between them. If we increase the accuracy of the calculation, deviations from the theory of Newton make their appearance, practically all of which must nevertheless escape the test of observation owing to their smallness.

We must draw attention here to one of these deviations. According to Newton's theory, a planet moves round the sun in an ellipse, which would permanently maintain its position with respect to the fixed stars, if we could disregard the motion of the fixed stars themselves and the action of the other planets under consideration. Thus, if we correct the observed motion of the planets for these two influences, and if Newton's theory be strictly correct, we ought to obtain for the orbit of the planet an ellipse, which is fixed with reference to the fixed stars. This deduction, which can be tested with great accuracy, has been confirmed for all the planets save one, with the precision that is capable of being obtained by the delicacy of observation attainable at the present time. The sole exception is Mercury, the planet which lies nearest the sun. Since the time of Leverrier, it has been known that the ellipse corresponding to the orbit of Mercury, after it has been corrected for the influences mentioned above, is not stationary with respect to the fixed stars, but that it rotates exceedingly slowly in the plane of the orbit and in the sense of the orbital motion. The value obtained for this rotary movement of the orbital ellipse was 43 arcsec per century, an amount ensured to be correct to within a few seconds of arc. This effect can be explained by means of classical mechanics only on the assumption of hypotheses which have little probability, and which were devised solely for this purpose.

On the basis of the general theory of relativity, it is found that the ellipse of every planet round the sun must necessarily rotate in the manner indicated above; that for all the planets, with the exception of Mercury, this rotation is too small to be detected with the delicacy of observation possible at the present time; but that in the case of Mercury it must amount to 43 arcsec per century, a result which is strictly in agreement with observation.

Apart from this one, it has hitherto been possible to make only two deductions from the theory which admit of being tested by observation, to wit, the curvature of light rays by the gravitational field of the sun,[4] and a displacement of the spectral lines of light reaching us from large stars, as compared with the corresponding lines for light produced in an analogous manner terrestrially (by the same kind of atom).[5] These two deductions from the theory have both been confirmed.

---

[4] First observed by Eddington and others in 1919.

[5] For a discussion of light falling in a gravitational field see Ex. 23.1.—[*K.K.*].

24.3   Study Questions

QUES. 24.1 What is the difference between a Euclidean and a non-Euclidean continuum?

a) What is meant by a continuum, or a continuous space? Are all spaces continuous? How, in practice, can one lay out a Cartesian coordinate system in such a continuous space?
b) In what cases, if any, is this method not feasible? What specific feature of space renders the establishment of a Cartesian coordinate system impossible?

QUES. 24.2 How are non-Euclidean spaces described mathematically?

a) What are gaussian coordinates? In what way are they more flexible than cartesian coordinates?
b) How many gaussian coordinates are required to label a particular point in space? Does it depend on the number of dimensions of space?
c) What is the distance between two nearby points in a two-dimensional continuous Euclidean space? How is this expressed mathematically?
d) What is the distance between two nearby points in a continuous Euclidean space having three, four, or even five-dimensions?
e) How can the distances between nearby points be expressed for non-Euclidean spaces?

QUES. 24.3 Is the space–time of special relativity continuous? Is it Euclidean?

a) What is a Lorentz transformation? What does it transform between? And what measurable quantity remains constant during the course of this transformation?
b) Are space coordinates and time coordinates treated identically in the theory of special relativity?
c) What does it mean to say that the space–time of special relativity is a Euclidean four-dimensional continuum? How is this expressed mathematically?

QUES. 24.4 Is our own space–time continuum a Euclidean one?

a) What effect do gravitational fields have on the speed of light? What does this effect imply regarding the feasibility of constructing a Cartesian coordinate system?
b) How is this difficulty surmounted? Is the description of events in our space–time continuum using gaussian coordinates adequate?

QUES. 24.5 What is the general principle of relativity? More specifically, under what types of transformations are the equations which express the general laws of nature invariant, or unchanged?

QUES. 24.6 Whose theory of gravity is better: Newton's or Einstein's?

a) In what way(s) does Einstein claim that his theory of gravity excels? Are these good reasons to adopt his theory?
b) What hypothesis is assumed in Newton's theory of gravity? Does Einstein's theory of gravity make the same assumption?

c) Under what conditions is Einstein's theory of gravity equivalent to Newton's? Are the predictions of the two theories always consistent?

d) What specific planetary phenomenon does Einstein's theory of gravity predict, which Newton's theory does not? Has this phenomenon ever been observed?

e) What other testable deductions are derived from Einstein's theory? Have either of these phenomena been observed?

24.4 Exercises

Ex. 24.1 (Geometry on a flat surface). Consider the surface of a flat sheet of paper.

a) What is the distance, ds, between two nearby points which are separated by dx in the x-direction and dy in the y-direction?

b) If you consider dx and dy to be gaussian coordinates on the space formed by the flat sheet of paper, what are the values of g_{xx}, g_{xy} and g_{yy} which describe this space?

c) Can you draw two parallel lines on the flat paper which never intersect? Is the ratio if the circumference to the diameter of a circle drawn on the flat paper equal to π? More generally, is the surface of the flat paper a Euclidean space?

Ex. 24.2 (Geometry on a sphere). Consider the surface of a sphere of radius R. The distance between two nearby points on the surface of the sphere is given by

$$ds^2 = R^2(d\theta^2 + \sin^2\theta\, d\phi^2). \tag{24.8}$$

Here, θ is the polar angle (measured down from the axis of the sphere), and ϕ is the azimuthal angle (measured around the equator of the sphere).

a) If you consider dθ and dϕ to be gaussian coordinates on the spherical surface, then what are the values of $g_{\theta\theta}$, $g_{\theta\phi}$ and $g_{\phi\phi}$?

b) Can you draw two parallel lines on the surface of the sphere which never intersect? Is the ratio of the circumference to the diameter of a circle drawn on the surface of the sphere equal to π? Is the surface of the sphere a Euclidean space?

24.5 Vocabulary

1. Continuum	7. Deduction
2. Pedantic	8. Conjunction
3. Quadrilateral	9. Provisional
4. Veritable	10. Plausible
5. Stipulation	11. Caprice
6. Analogous	12. Ellipse

Chapter 25
A Finite Universe with No Boundary

The geometrical properties of space are not independent, but they are determined by matter.

—Einstein

25.1 Introduction

In Parts I and II of his book, *Relativity*, Einstein explained his special and his general principles of relativity. In so doing, he derived a novel theory of gravity which was able to explain the slow precession of Mercury's elliptical orbit around the sun and also why light falls in a gravitational field. Newton's theory of gravity was powerless to explain these experimental observations. Now, in Part III, Einstein applies his theory of gravity to the problem of the overall size and structure of the universe. He begins by laying before the reader a fundamental difficulty which is encountered if one takes seriously Newton's theory of gravitation. How does he address this problem? To what conclusion is he led? And do you find his conclusions sensible?

25.2 Reading: Einstein, *Relativity*

Einstein, A., *Relativity*, Great Minds, Prometheus Books, Amherst, NY, 1995. Part III, Considerations on the Universe as a Whole.

25.2.1 *XXX: Cosmological Difficulties of Newton's Theory*

Apart from the difficulty discussed in Section XXI, there is a second fundamental difficulty attending classical celestial mechanics, which, to the best of my knowledge, was first discussed in detail by the astronomer Seeliger. If we ponder over the question as to how the universe, considered as a whole, is to be regarded, the first answer that suggests itself to us is surely this: As regards space (and time) the universe is infinite. There are stars everywhere, so that the density of matter, although very variable in detail, is nevertheless on the average everywhere the same. In other words: However

K. Kuehn, *A Student's Guide Through the Great Physics Texts*,
Undergraduate Lecture Notes in Physics, DOI 10.1007/978-1-4939-1360-2_25,
© Springer Science+Business Media, LLC 2015

far we might travel through space, we should find everywhere an attenuated swarm of fixed stars of approximately the same kind and density.

This view is not in harmony with the theory of Newton. The latter theory rather requires that the universe should have a kind of centre in which the density of the stars is a maximum, and that as we proceed outwards from this centre the group-density of the stars should diminish, until finally, at great distances, it is succeeded by an infinite region of emptiness. The stellar universe ought to be a finite island in the infinite ocean of space.[1]

This conception is in itself not very satisfactory. It is still less satisfactory because it leads to the result that the light emitted by the stars and also individual stars of the stellar system are perpetually passing out into infinite space, never to return, and without ever again coming into interaction with other objects of nature. Such a finite material universe would be destined to become gradually but systematically impoverished. In order to escape this dilemma, Seeliger suggested a modification of Newton's law, in which he assumes that for great distances the force of attraction between two masses diminishes more rapidly than would result from the inverse square law. In this way it is possible for the mean density of matter to be constant everywhere, even to infinity, without infinitely large gravitational fields being produced. We thus free ourselves from the distasteful conception that the material universe ought to possess something of the nature of a centre. Of course we purchase our emancipation from the fundamental difficulties mentioned, at the cost of a modification and complication of Newton's law which has neither empirical nor theoretical foundation. We can imagine innumerable laws which would serve the same purpose, without our being able to state a reason why one of them is to be preferred to the others; for any one of these laws would be founded just as little on more general theoretical principles as is the law of Newton.

25.2.2 XXXI: The Possibility of a "finite" and yet "unbounded" Universe

But speculations on the structure of the universe also move in quite another direction. The development of non-Euclidean geometry led to the recognition of the fact, that we can cast doubt on the *infiniteness* of our space without coming into conflict with the laws of thought or with experience (Riemann, Helmholtz). These questions have

---

[1] *Proof*—According to the theory of Newton, the number of "lines of force" which come from infinity and terminate in a mass m is proportional to the mass m. If, on the average, the mass-density ρ_o is constant throughout the universe, then a sphere of volume V will enclose the average mass $\rho_o V$. Thus the number of lines of force passing through the surface F of the sphere into its interior is proportional to $\rho_o V$. For unit area of the surface of the sphere the number of lines of force which enters the sphere is thus proportional to $\rho_o \frac{V}{F}$ or to $\rho_o R$. Hence the intensity of the field at the surface would ultimately become infinite with increasing radius R of the sphere, which is impossible.

already been treated in detail and with unsurpassable lucidity by Helmholtz and Poincaré, whereas I can only touch on them briefly here.

In the first place, we imagine an existence in two-dimensional space. Flat beings with flat implements, and in particular flat rigid measuring-rods, are free to move in a *plane*. For them nothing exists outside of this plane: that which they observe to happen to themselves and to their flat "things" is the all-inclusive reality of their plane. In particular, the constructions of plane Euclidean geometry can be carried out by means of the rods, the lattice construction, considered in Section XXIV. In contrast to ours, the universe of these beings is two-dimensional; but, like ours, it extends to infinity. In their universe there is room for an infinite number of identical squares made up of rods, its volume (surface) is infinite. If these beings say their universe is "plane," there is sense in the statement, because they mean that they can perform the constructions of plane Euclidean geometry with their rods. In this connection the individual rods always represent the same distance, independently of their position.

Let us consider now a second two-dimensional existence, but this time on a spherical surface instead of on a plane. The flat beings with their measuring-rods and other objects fit exactly on this surface and they are unable to leave it. Their whole universe of observation extends exclusively over the surface of the sphere. Are these beings able to regard the geometry of their universe as being plane geometry and their rods withal as the realisation of "distance"? They cannot do this. For if they attempt to realise a straight line, they will obtain a curve, which we "three-dimensional beings" designate as a great circle, a self-contained line of definite finite length, which can be measured up by means of a measuring-rod. Similarly, this universe has a finite area that can be compared with the area of a square constructed with rods. The great charm resulting from this consideration lies in the recognition of the fact that *the universe of these beings is finite and yet has no limits*.

But the spherical-surface beings do not need to go on a world-tour in order to perceive that they are not living in a Euclidean universe. They can convince themselves of this on every part of their "world," provided they do not use too small a piece of it. Starting from a point, they draw "straight lines" (arcs of circles as judged in three-dimensional space) of equal length in all directions. They will call the line joining the free ends of these lines a "circle." For a plane surface, the ratio of the circumference of a circle to its diameter, both lengths being measured with the same rod, is, according to Euclidean geometry of the plane, equal to a constant value π, which is independent of the diameter of the circle. On their spherical surface our flat beings would find for this ratio the value

$$\pi \, \frac{\sin\left(\frac{r}{R}\right)}{\left(\frac{r}{R}\right)},$$

a smaller value than π, the difference being the more considerable, the greater is the radius of the circle in comparison with the radius R of the "world-sphere." By means of this relation the spherical beings can determine the radius of their universe ("world"), even when only a relatively small part of their world-sphere is available for their measurements. But if this part is very small indeed, they will no longer be able to demonstrate that they are on a spherical "world" and not on a Euclidean plane, for a small part of a spherical surface differs only slightly from a piece of a plane of the same size.

Thus if the spherical-surface beings are living on a planet of which the solar system occupies only a negligibly small part of the spherical universe, they have no means of determining whether they are living in a finite or in an infinite universe, because the "piece of universe" to which they have access is in both cases practically plane, or Euclidean. It follows directly from this discussion, that for our sphere-beings the circumference of a circle first increases with the radius until the "circumference of the universe" is reached, and that it thenceforward gradually decreases to zero for still further increasing values of the radius. During this process the area of the circle continues to increase more and more, until finally it becomes equal to the total area of the whole "world-sphere."

Perhaps the reader will wonder why we have placed our "beings" on a sphere rather than on another closed surface. But this choice has its justification in the fact that, of all closed surfaces, the sphere is unique in possessing the property that all points on it are equivalent. I admit that the ratio of the circumference c of a circle to its radius r depends on r, but for a given value of r it is the same for all points of the "world-sphere"; in other words, the "world-sphere" is a "surface of constant curvature."

To this two-dimensional sphere-universe there is a three-dimensional analogy, namely, the three-dimensional spherical space which was discovered by Riemann. Its points are likewise all equivalent. It possesses a finite volume, which is determined by its "radius" $(2\pi^2 R^3)$. Is it possible to imagine a spherical space? To imagine a space means nothing else than that we imagine an epitome of our "space" experience, of experience that we can have in the movement of "rigid" bodies. In this sense we *can* imagine a spherical space.

Suppose we draw lines or stretch strings in all directions from a point, and mark off from each of these the distance r with a measuring-rod. All the free endpoints of these lengths lie on a spherical surface. We can specially measure up the area (F) of this surface by means of a square made up of measuring-rods. If the universe is Euclidean, then $F = 4\pi r^2$; if it is spherical, then F is always less than $4\pi r^2$. With increasing values of r, F increases from zero up to a maximum value which is determined by the "world-radius," but for still further increasing values of r, the area gradually diminishes to zero. At first, the straight lines which radiate from the starting point diverge farther and farther from one another, but later they approach each other, and finally they run together again at a "counter-point" to the starting point. Under such conditions they have traversed the whole spherical space. It is easily seen that the three-dimensional spherical space is quite analogous to the two-dimensional spherical surface. It is finite (of finite volume), and has no bounds.

It may be mentioned that there is yet another kind of curved space: "elliptical space." It can be regarded as a curved space in which the two "counter-points" are identical (indistinguishable from each other). An elliptical universe can thus be considered to some extent as a curved universe possessing central symmetry.

It follows from what has been said, that closed spaces without limits are conceivable. From amongst these, the spherical space (and the elliptical) excels in its simplicity, since all points on it are equivalent. As a result of this discussion, a most interesting question arises for astronomers and physicists, and that is whether the universe in which we live is infinite, or whether it is finite in the manner of the spherical universe. Our experience is far from being sufficient to enable us to answer this question. But the general theory of relativity permits of our answering it with a moderate degree of certainty, and in this connection the difficulty mentioned in Section XXX finds its solution.

25.2.3   XXXII: The Structure of Space According to the General Theory of Relativity

According to the general theory of relativity, the geometrical properties of space are not independent, but they are determined by matter. Thus we can draw conclusions about the geometrical structure of the universe only if we base our considerations on the state of the matter as being something that is known. We know from experience that, for a suitably chosen co-ordinate system, the velocities of the stars are small as compared with the velocity of transmission of light. We can thus as a rough approximation arrive at a conclusion as to the nature of the universe as a whole, if we treat the matter as being at rest.

We already know from our previous discussion that the behaviour of measuring-rods and clocks is influenced by gravitational fields, *i.e.* by the distribution of matter. This in itself is sufficient to exclude the possibility of the exact validity of Euclidean geometry in our universe. But it is conceivable that our universe differs only slightly from a Euclidean one, and this notion seems all the more probable, since calculations show that the metrics of surrounding space is influenced only to an exceedingly small extent by masses even of the magnitude of our sun. We might imagine that, as regards geometry, our universe behaves analogously to a surface which is irregularly curved in its individual parts, but which nowhere departs appreciably from a plane: something like the rippled surface of a lake. Such a universe might fittingly be called a quasi-Euclidean universe. As regards its space it would be infinite. But calculation shows that in a quasi-Euclidean universe the average density of matter would necessarily be *nil*. Thus such a universe could not be inhabited by matter everywhere; it would present to us that unsatisfactory picture which we portrayed in Section XXX.

If we are to have in the universe an average density of matter which differs from zero, however small may be that difference, then the universe cannot be quasi-Euclidean. On the contrary, the results of calculation indicate that if matter be distributed uniformly, the universe would necessarily be spherical (or elliptical). Since in reality the detailed distribution of matter is not uniform, the real universe will deviate in individual parts from the spherical, the universe will be quasi-spherical.

But it will be necessarily finite. In fact, the theory supplies us with a simple connection[2] between the space-expanse of the universe and the average density of matter in it.

25.3 Study Questions

QUES. 25.1 What is the cosmological difficulty of Newton's theory of gravity?

a) What is the expected distribution of stars throughout an infinite universe? Is this expectation consistent with what one might predict based on Newton's theory of gravity?
b) How did Seeliger attempt to remedy this apparent problem? Is this a feasible solution? On what grounds does Einstein criticize Seeliger's solution?

QUES. 25.2 What is the size and the structure of our universe?

a) What is meant by non-Euclidean geometry? What are its distinguishing features?
b) Is it possible for two-dimensional beings living on a surface to determine the overall shape of the surface upon which they dwell? If so, how?
c) What does it mean that a sphere is a surface of "constant curvature"? Are there any other shapes of constant curvature?
d) How does Einstein generalize these considerations to three-dimensional beings? Is a three-dimensional space which is limitless yet closed conceivable?
e) What effect does matter have on the geometrical properties of space? And why does Einstein claim that our universe is like the "rippled surface of a lake?"
f) What are Einstein's conclusions regarding the size and structure of our universe? Do you believe him? Why or why not?

25.4 Vocabulary

1. Emancipation	6. Lucidity
2. Innumerable	7. Conceivable
3. Epitome	8. *Nil*
4. Speculation	9. Quasi-Euclidean
5. Non-Euclidean	

---

[2] For the "radius" R of the universe we obtain the equation

$$R^2 = \frac{2}{\kappa\rho}.$$

The use of the C.G.S. system in this equation gives $\frac{2}{\kappa} = 1.08 \times 10^{27}$; ρ is the average density of the matter.

Chapter 26
The Structure of the Universe

On this interpretation the nebulae are rushing away from us,
and the farther away they are, the faster they are traveling.
—Edwin Hubble

26.1 Introduction

Edwin Hubble (1889–1953) was born in Marshfield, Missouri. He received his Bachelor of Science degree from the University of Chicago in 1910, where he concentrated on astronomy, mathematics and philosophy. As a Rhodes scholar at The Queens College, Oxford, he studied jurisprudence, literature and Spanish. He returned to his interest in astronomy after serving as a high-school teacher for a few years, earning his doctorate from the University of Chicago in 1917. There, he carried out photographic studies of faint nebulae at the Yerkes observatory in Williams Bay, Wisconsin—home to the largest refracting telescope in the world (see Fig. 26.1). After serving in the United States Army in World War I, he went to work at the Mount Wilson Observatory, where he studied Cepheid variables in spiral nebulae. Building upon the work of Henrietta Leavitt, Ejnar Hertzsprung, Vesto Slipher and Harlow Shapley, he was able to establish a relationship between the recessional velocities of galaxies and their distances from the earth—what is now known as Hubble's Law. The Hubble Space Telescope, named after the famous cosmologist, was launched into low-earth orbit in 1990 and has subsequently produced thousands of popular and beautiful images of galaxies and other astronomical objects.[1] In the following reading selection, which appeared in *The Scientific Monthly* in 1934, Hubble begins by describing how nebulae appear when viewed through powerful telescopes, and also how their distribution throughout the observable regions of the universe is calculated. After this he proceeds to explain his famous velocity-distance relationship. As you study this text, try to identify any assumptions made by Hubble. Are his reasonable assumptions? Is there a better approach?

26.2 Reading: Hubble, *The Realm of the Nebulae*

Hubble, E., The realm of the nebulae, *The Scientific Monthly*, *39*(3), 193–202, 1934.

---

[1] See, for instance, www.hubblesite.org or www.nasa.gov/mission_pages/hubble.

K. Kuehn, *A Student's Guide Through the Great Physics Texts*,
Undergraduate Lecture Notes in Physics, DOI 10.1007/978-1-4939-1360-2_26,
© Springer Science+Business Media, LLC 2015

Fig. 26.1 The 40-inch refracting telescope at the Yerkes Observatory in Williams Bay, Wisconsin

26.2.1   The Exploration of Space

I propose to discuss some of the recent explorations in the realm of the nebulae which bear directly on the structure of the universe. The earth we inhabit is a member of the solar system. The sun with its family of planets seems isolated in space, but the sun is merely a star—one of the millions which populate our particular region of the universe. The stars are scattered about at enormous intervals, but on a still greater scale they are found to form a definite system, again isolated in space. On the grand scale we may picture the stellar system drifting through the universe as a swarm of bees drifting through the air.

From our position somewhere within the system, we look out through the swarm of stars, past the borders, into the universe beyond. It is empty for the most part—vast stretches of empty space. But here and there, at immense intervals, we find other stellar systems, comparable with our own. They are so distant that in general we do not see the individual stars. They appear as faint patches of light and hence are called nebulae, *i.e.*, clouds.

The nebulae are great beacons, scattered through the depth of space. We see a few that appear large and bright. These are the nearer nebulae. Then we find them smaller and fainter in constantly increasing numbers, and we know we are reaching out into space farther and ever farther until, with the faintest nebulae that can be detected

Fig. 26.2 The "Whirlpool Nebula" (M. 51 in Canes Venatic). This, the first nebula in which the spiral structure was discerned, is about 1 million light years from the earth. (Taken at Mt. Wilson Observatory of Carnegie Institution)

with the greatest telescope, we have reached the frontiers of the known universe. This last horizon defines the Observable Region—the region of space which can be explored with existing telescopes. It is a vast sphere, some 600 million light years in diameter, throughout which are scattered 100 million nebulae (Fig. 26.2).

The question immediately arises as to whether the nebulae form an isolated super-system, analogous to the system of the stars but on a still grander scale. Actually, we find the nebulae scattered singly, in groups and occasionally in great clusters, but when very large volumes of space are considered, the tendency to cluster averages out and to the very limits of our telescopes the distribution is approximately uniform. If the observable region is divided into 100, 1000 or even 10,000 equal parts, the nebular contents of the various fractions are very closely similar. There is no evidence of a thinning out, no trace of a physical boundary. The realm of the nebulae, we must conclude, stretches on and on, far beyond the frontiers.

Observations give not the slightest hint of a super-system of nebulae. Hence, for purposes of speculation, we may invoke the principle of the uniformity of nature and suppose that any other equal portion of the universe, chosen at random, will exhibit much the same general characteristics as the region we can explore with our telescopes. As a working hypothesis, serviceable until it leads to contradictions, we may venture the assumption that the realm of the nebulae is the universe—that the Observable Region is a fair sample and that the nature of the universe may be inferred from the observed characteristics of the sample.

26.2.2   Characteristics of the Observable Region

The characteristics of the Observable Region as a whole forms the main subject of the present discussion, but a brief appendix will be added, indicating the kind of information concerning the universe we may hope to infer from the sample.

We are situated, by definition, at the Center of the Observable Region. Our immediate neighborhood—the system of the planets—we know rather intimately, but our knowledge fades rapidly with increasing distance. We know something about the stars, a little about the nearer nebulae, almost nothing about the more remote nebulae save their directions, their apparent luminosities and the nature of the light which they emit. Information concerning the Observable Region as a whole is thus restricted to the most general features only—the distribution of nebulae and the more conspicuous characteristics of their spectra. These data, together with the general laws of nature, which we assume to hold everywhere, are our present clues to the nature of the universe.

Let us start with distribution. The nebulae are beacons scattered through space. In order to determine their distribution it is necessary to know their intrinsic luminosity, *i.e.*, their candle power—both the average luminosity and the range. If some nebulae were intrinsically a million times brighter than others, as is the case with stars, the problem of the distribution would be extremely difficult, for apparent faintness would then be a very poor indication of distance. Fortunately, the nebulae are all of the same order of intrinsic luminosity. This information was derived as follows.

26.2.3   Distances of Nebulae

The nebulae are stellar systems, and some of them are so near that a few of the individual stars can be detected with the modern reflectors. Stars are the fundamental criteria of nebular distances. We know something about stars and wherever we find them we can generally recognize their types, estimate their candlepowers, and so derive their distance from their apparent faintness. In a dozen of the nearest nebulae, various types of stars are clearly recognized and distances are rather accurately determined.

For instance, some 40 cepheid variable stars are found in M31, the great spiral nebula in Andromeda. Such stars in our own system average about 3000 times as bright as the sun. In the spiral they appear about 150,000 times fainter than the faintest star that can be seen with the naked eye. A simple calculation indicates that the spiral must lie at a distance of nearly a million light years. Other types of stars which can be recognized in the nebula indicate the same order of distance, and hence we consider that the results are reliable. Such accuracy, however, is attained only for a very few of the nearest nebulae.

In several dozen other nebulae we can detect a few stars, but cannot recognize their types. Nevertheless, we have many reasons for supposing that there is an upper limit of stellar luminosity and that this limit, about 60,000 times as bright as the sun, is in general attained and seldom surpassed in all the great systems of stars. Hence we may assume that the brightest stars in all nebulae are 60,000 times as bright as the sun and estimate their distances from their apparent faintness. The results may not be accurate individually, but they are reliable for statistical purposes.

Finally, the great cluster in Virgo, a compact group of several hundred nebulae, is so near that a few stars can be seen in a few of its members. These stars indicate the distance of all the several hundred nebulae in the cluster. The 20 known clusters, moreover, are all similar groups, and their relative distances were already known. Hence the distance of the Virgo cluster indicates the distances of them all.

In this way it has been possible to assemble a sample collection of several thousand nebulae whose distances and hence whose real luminosities and dimensions are known. An analysis of the sample collection indicates at once that the nebulae are all of the same order of luminosity. They average about 80 million times as bright as the sun. The brightest are about ten times brighter than the average and the faintest are about ten times fainter, but the majority fall within the narrow limits from a half to twice the average of them all. The mean of any considerable number, say 100 nebulae, chosen at random will be closely similar to the general average. For statistical purposes, where large numbers are concerned, we may assume that the nebulae are equally luminous and that their apparent faintness indicates their distances.

26.2.4 Distribution of Nebulae

The distribution of nebulae can therefore be studied by counting the numbers of nebulae to successive limits of apparent faintness. The results represent numbers of nebulae in spheres of successively greater radii, and the differences give the numbers of nebulae in successive spherical shells. In this way the numbers of nebulae per unit volume, *i.e.*, the density distribution of nebulae, has been explored throughout the observable region. To a first approximation the density distribution is uniform. Each volume of space, represented by a sphere with a radius of 10 million light years, contains on the average about 2500 nebulae. Each nebulae is about 80 million times as bright as the sun and perhaps 800 million times as massive. On the average, the nebulae are about one and a half million light years apart (Fig. 26.3).

The uniform distribution of nebulae means that, on the grand scale, the density of matter in space is uniform and we can calculate the density. There is, on the average, 1 nebula, 800 million times as massive as the sun, for every billion billion cubic light

Fig. 26.3 Spiral nebula in Ursa Major (M. 101). The region of space which can be explored with existing telescopes is a vast sphere, 600 million light years in diameter, throughout which 100 million nebulae are scattered. (Taken at Mt. Wilson Observatory)

years or roughly one sun per billion or thousand million cubic light years. In ordinary units this is equivalent to 1 gm per 10^{30} cc, and may be visualized as corresponding to a grain of sand in each volume of space equal to the volume of the earth. The nebulae are scattered very thinly, and space is mostly empty (Fig. 26.4).

In this calculation we consider only the matter concentrated in nebulae. There is doubtless matter scattered between the nebulae which has been ignored. How much, we do not know; we can only say that there is not sufficient to be detected—not enough to appreciably dim the most distant nebulae that can be observed.

26.2.5   Velocity–Distance Relation

The second characteristic of the Observable Region, the velocity–distance relation, introduces the subject of spectrum analysis. When a light source is viewed through a glass prism, the various colors of which the light is composed are spread out into an ordered sequence, represented, for instance, in the rainbow. The sequence never varies, each color has its place. Different colors represent light of different wave-lengths. From the short waves of violet light, the waves lengthen steadily through the spectrum to the long waves of the red at the other end.

Three kinds of spectra are generally distinguished. An incandescent solid, *e.g.*, electric light filament, radiates all possible colors or wave-lengths; hence its spectrum is continuous from violet to red.

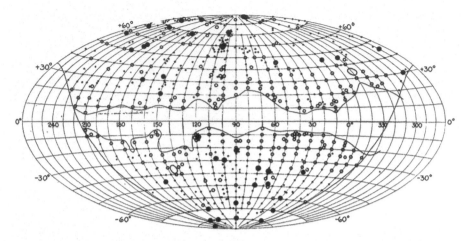

Fig. 26.4 Diagram showing the distribution of spiral nebulae. From counts made by Dr. Hubble on photographs distributed at intervals over the whole northern sky. The horizontal line is the plane of the galaxy, and the figures give galactic longitude. It should be noted that spirals avoid the region of the galaxy

An incandescent gas, *e.g.*, a neon tube, radiates only a few isolated colors, hence its spectrum consists of isolated bright spots or lines distributed in a certain pattern. This is called an emission spectrum. The pattern of bright lines is characteristic of the particular gas involved, and unknown gases are very readily identified from their spectra alone.

Finally, there are absorption spectra. When an incandescent solid, giving, of course, a continuous spectrum, is surrounded by a cooler gas, *e.g.*, a star surrounded by an atmosphere, the gas absorbs just those colors which it would radiate if itself incandescent. This absorption produces dark spaces or lines in the otherwise continuous spectrum of the background. The patterns of these dark lines identify the gases in the atmospheres of the stars.

The study of absorption spectra is the dominating feature of modern astronomy. They furnish an astonishing amount of information concerning the physical condition of stars and even of planets and of nebulae. Either directly or indirectly they indicate surface temperatures of stars, surface luminosities, total luminosities, distances, velocities in the line of sight (Fig. 26.5).

The significance of spectra may be indicated by a homely demonstration. From Mt. Wilson the lights of some sixty cities and towns are visible, spread over the valley below. A direct photograph with a camera shows swarms of lights similar to a field of stars, but tells nothing as to the nature of the lights. When a glass prism is placed in front of the camera lens, each light is spread out into a spectrum. Then the differences appear. The filament lamps show continuous spectra, arc lights show the emission spectra of carbon vapor, neon signs show two or three isolated colors. In the same way a direct photograph of the sky shows a field of stars. Except for their different luminosities the stars all appear alike. A photograph of the same field,

Fig. 26.5 One of the most beautiful of the spiral nebulae (M. 81 in Ursa Major). Its light takes 1,600,000 years to reach us. The central region is unresolved but in the outer portions swarms of stars are visible. These are similar to the very bright stars in our own galactic system. (Taken at Mt. Wilson Observatory)

with a prism in front of the lens, shows each star drawn out into its spectrum, and differences in the nature of the light are at once apparent.

Yellow stars like the sun show the absorption of hydrogen and of iron vapor in their atmospheres and, near the violet end, a pair of strong dark lines due to calcium absorption. These latter, the H and K lines of calcium, are the most conspicuous feature of the spectra and are unmistakable wherever they are found. On the same scale, the spectra of nebulae resemble those of yellow stars. The H and K lines are readily recognized, and certain hydrogen and iron lines as well.

The spectra of nebulae, however, exhibit a peculiar characteristic in that the details—the dark lines—are not in their usual positions. The lines are all displaced toward the red end of the spectrum and the displacements increase with the faintness of the nebulae observed. The observations are summed up in the statement—the fainter the nebula, the larger the red-shift (Fig. 26.6).

Now apparent faintness of nebulae is confidently interpreted in terms of distance; hence we can restate the observational results in the form—red-shifts increase with distance. Precise investigations indicate that the relation is linear—red-shifts are equal to distances times a certain constant (Fig. 26.7).

The relation was first established about five years ago among the brighter nearer nebulae for which Dr. Slipher, of the Lowell Observatory, had assembled his collection of spectra representing the pioneer work in the field. Since then the list of spectra has been more than trebled by Mr. Humason, using the large reflectors on Mt. Wilson. With the 150 red-shifts now available, the distance-relation has been

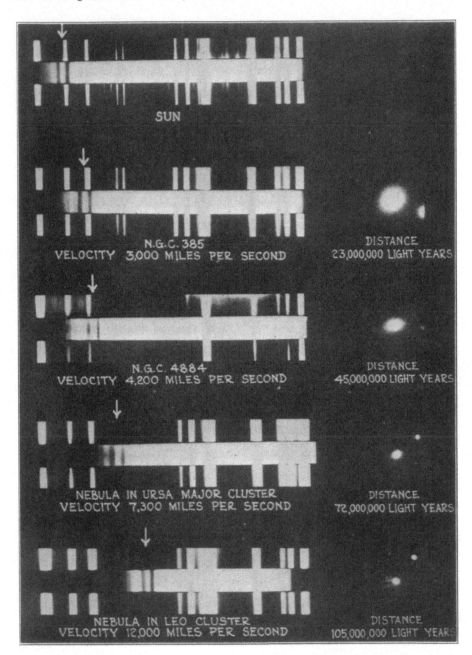

Fig. 26.6 Widened low-dispersion spectra. With direct photographs of distant extra-galacctic nebulae showing large red shift and giving estimated recession velocities. (Taken at Mt. Wilson Observatory)

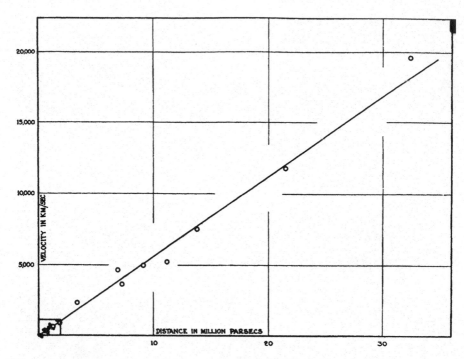

Fig. 26.7 Diagram showing relation between measured velocity and distance of spiral nebulae. The velocity is measured from spectrograms taken with the 100-inch reflector at Mt. Wilson Observatory and the distance is determined from the brightness of the nebulae as shown in direct photographs

confirmed and extended to the limit at which spectra can be recorded with existing instruments. Out to 150 million light years, the red-shifts increase at a uniform rate.

The significance of this strange characteristic of our sample of the universe depends upon the interpretation of red-shifts. The phenomena may be described in several equivalent ways—the light is redder, the light waves are longer, the vibrations are slower (the pitch is lower), the light quanta have lost energy.

Many ways of producing such effects are known, but of them all only one will produce large red-shifts without introducing other effects which should be conspicuous but actually are not found. This one known permissible explanation interprets red-shifts, as due to actual motion away from the observer. Rapid motion of recession drags out the light waves and, as it were, lowers the pitch. Red-shifts, we can say, are due either to actual motion or to some hitherto unrecognized principle of physics. Theoretical investigators almost universally accept the red-shifts as indicating motion recession of the nebulae, and they are fully justified in their position until evidence to the contrary is forthcoming.

On this interpretation the nebulae are rushing away from us, and the farther away they are, the faster they are traveling. The velocities increase by roughly 100 miles/s for each million light[2] years of distance. The present distribution of nebulae can be

---

[2] A repeat occurrence of the word "light" appearing in the original text has been omitted here.—[K.K.]

represented on the assumption that they were once jammed together in our particular region of space, and at a particular instant, about 2000 million years ago, they started rushing away in all directions at various velocities. The slower nebulae, on this assumption, are still in our neighborhood, but the faster nebulae are now far away. The faster they are traveling the farther they have gone. The time scale seems suspiciously short—a small fraction of the estimated age of some stars— and the apparent discrepancy suggests the advisability of further discussion of the interpretation of red-shifts as evidence of motion.

The largest red-shift actually recorded represents a velocity of about 15,000 miles/s at a distance of roughly 150 million light years. But nebulae can be photographed out to distances twice or thrice the distances to which their spectra can be recorded. Hence, if the observed relation holds to the very frontiers of the observable region, we should encounter red-shifts corresponding to velocities of 30,000 or 40,000 miles/s, say one fifth the velocity of light.

Such red-shifts are so enormous that we may expect appreciable indirect effects on colors and apparent luminosities. These effects are now under investigation. The field is new, but it offers very definite prospects not only of testing the form of the velocity–distance relation beyond the reach of the spectrograph, but even of critically testing the interpretation of red-shifts as actual motion. With this possibility in view, the cautious observer refrains from committing himself to the present interpretation and employs the colorless term "apparent velocity."

26.2.6   The Observable Region as a Sample of the Universe

Now, in conclusion, let us see what sort of information concerning the universe may be inferred from the observed characteristics of the sample. The sample is homogeneous and isotropic and the nebulae appear to be rushing away from our particular position. These meager data, together with the general laws of nature, are all we have to guide us.

Mathematics deals with possible worlds, *i.e.*, logically consistent systems. Science attempts to determine the actual world in which we live. So, in cosmology, mathematics presents us with an infinite array of possible universes. The explorations of science are eliminating type after type, class after class, and already the residue has dwindled to more or less comprehensible dimensions.

Considerations of the laws of nature led to the theory of relativity, developed by the genius of Einstein and now generally accepted. According to relativity, the large-scale geometry of space is determined by the contents of space. This dependence is expressed in Einstein's famous cosmological equation and modern cosmology is largely a series of attempts to solve the equation. It is a relation between symbols, and solutions are not possible until the symbols are interpreted. Observations as yet furnish only partial interpretations, and hence the various solutions already proposed incorporate a certain amount of guessing.

The most reasonable speculations run about as follows. Since the contents of space is distributed uniformly, the geometry of space must exhibit the same uniformity. Of all possible geometries, only three types fulfil this requirement completely. One is Euclidean geometry, so familiar to most of us that we ignore the very existence of others. Another is Riemannian geometry, and this it is which the consensus of opinion accepts as the most useful for describing the large scale features of the universe.

Riemannian space with constant positive curvature follows more or less directly from the observed characteristics of our sample. This space is usually described as the three-dimensioned analogue of the sphere. Just as the surface of a sphere with its uniform positive curvature has a finite area but no boundaries, so the three-dimensional analogue, with its uniform positive curvature, has a finite volume but no boundaries. In other words, we live in a finite universe. The volume is determined by the radius of curvature (again like the area of the surface of a sphere) and the radius of curvature is determined by the amount of matter, *i.e.*, the density of matter, in space. The density actually observed, 10^{-30} gm/cc, suggests a radius of about 3000 million light years and a volume of the order of two or three million times the volume of the Observable Region. Such a universe would contain about 500 million nebulae. This is an instantaneous picture, representing the situation for the past 200 or 300 million years.

The conception of a homogeneous universe with a definite volume and definite contents seems moderately comfortable until we remember the red-shifts. The conception is derived from relativity, and relativity assumes that the universe will appear much the same no matter where the observer happens to be situated. Since the nebulae appear to be rushing away from our particular position, they will appear to be rushing away from any other position in which an observer is located. This apparent anomaly is explained on the theory that the universe is expanding. The radius of curvature is a function of the time and is now increasing. The volume is increasing and the density is diminishing. The conventional analogy in two dimensions is again with the surface of a sphere, say a rubber balloon which is being inflated. From each point on the surface, all other points are retreating and, within certain limits, the farther away they are, the faster they recede.

The expanding universe, with its momentary dimensions as previously described, is the latest widely accepted development in cosmology. Various refinements as to the nature of the expansion have been discussed at length, but always with the aid of additional assumptions concerning the validity of which there is no consensus of opinion. Even the present position depends absolutely upon the interpretation of red-shifts as Doppler effects representing actual motions.

Further radical advances in cosmology will probably await the accumulation of more observational data—the elimination of more types of possible worlds. The data will come either from more detailed investigations of the present Observable Region or from a significant enlargement of the region itself.

The latter alternative will be achieved with the 200-inch reflector in course of construction for the California Institute of Technology, with the assistance and co-operation of the Carnegie Institution of Washington. This great telescope, in the hands of experienced research men in the two institutions, is expected to enlarge the available sample of the universe some ten times in a single step and will increase in a corresponding measure the chances that our sample is fair and significant.

I believe the 200-inch will definitely answer the question of the interpretation of red-shifts, whether or not they represent actual motions, and if they do represent motions—if the universe is expanding—the 200-inch may indicate the particular type of expansion.

This prospect is the climax of the story. Our present information concerning the universe is necessarily vague. It is new and raw and will mature only with time and continued study. The great significant feature is that the first steps have actually been achieved—that in our generation, for the first time, the structure of the universe is being investigated by direct observations.

26.3   Study Questions

QUES. 26.1 How bright are the nebulae? How is the brightness of a nebula related to its distance? What does the brightness of a nebula reveal about its size and composition?

QUES. 26.2 How are nebular forms classified? And how are their forms related to their internal motions?

QUES. 26.3 What is the distribution of nebulae throughout the universe?

a) How is the large-scale distribution of nebulae computed? And what renders such computations difficult? How is the depth distribution computed? To what conclusion is one led by such computations?
b) What is average separation between nebulae? What, then, is the approximate density of space, and how does this compare to the density of air? And how distant are the faintest nebulae?

QUES. 26.4 What is the relationship between nebular velocity and distance?

a) How does one measure the velocity of nebulae? How does this method differ from that of measuring their spatial distribution?
b) What spectral lines are of particular interest to astronomers? Why? And what do they reveal?
c) How did Hubble interpret the measured velocity–distance relation? How did Hubble generalize this relationship? Upon what fundamental assumption(s) does Hubble base his speculations?
d) Is the conclusion that space is expanding purely empirical, or does it rely upon a theoretical framework?

26.4   Exercises

Ex. 26.1 (HUBBLE'S LAW). Assuming that Hubble's law provides an accurate description of the velocity of all observable nebula, at what speed is M31 moving relative to Earth? At what speed are the most distant observable nebulae moving? Does this sound reasonable?

26.5   Vocabulary

1. Nebulae
2. Uniformity
3. Conspicuous
4. Spectrograph
5. Homogeneous
6. Isotropic
7. Spectrum
8. Incandescent
9. Red-shift
10. Empirical
11. Infer
12. Luminosity
13. Candlepower
14. Wavelength
15. Quanta
16. Recession
17. Discrepancy
18. Cosmology
19. Conception
20. Anomaly

Chapter 27
Measuring the Potentially Infinite

> *The world is not a dungeon, not even a nicely-decorated*
> *dungeon; it is a boundless perspective, marked out with bright*
> *guideposts.*
>
> —Georges Lemaître

27.1 Introduction

Georges Lemaître (1894–1966) was born in Charlerio, Belgium. He studied religion, humanities and classical languages at a Jesuit school before enrolling at the Catholic University of Louvain in 1911. His engineering studies were interrupted by the World War, in which he served as a soldier in the Belgian army. Returning to the university, he turned his attention to physics and mathematics, earning a doctor of sciences in mathematics in 1920. Lemaître was ordained into the priesthood in 1923, later serving as the president of the Pontifical Academy of Sciences from 1960 until his death. During the time between his ordination and his appointment as university professor in 1927, Lemaître travelled to Cambridge, where he studied under Arthur Eddington, and to the United States, where he received his Ph.D. under the guidance of Harlow Shapley. He was increasingly drawn to recent developments in cosmology and astrophysics, and in 1927 Lemaître published a paper in which he suggested that the universe originated from the radioactive disintegration of a single primeval atom and that its size was increasing according to what is now known as Hubble's law. The ideas presented in this paper were initially rejected by many prominent cosmologists since they suggested that the universe had a beginning—a notion which contradicted the prevailing static view of the universe. Nonetheless, Lemaître's ideas profoundly shaped modern Big Bang cosmology. The reading selections that follow were taken from the 1950 English translation by Betty H. Korff and Serge A. Korff of Lemaître's 1946 publication entitled *L'hypothèse de l'atome primitif*. Lemaître begins this book on *The Primeval Atom* by inquiring as to the actual size and shape of the universe taken as a whole. In attempting to answer this ancient question, considered over two millennia earlier by both Archimedes and Aristotle, Lemaître draws upon the recent work of Poincaré, Einstein, Hubble, De Sitter and Friedmann (Fig. 27.1).[1]

---

[1] Aristotle discussed the size and shape of the world (the universe) from a geocentric perspective in Book I, Chaps. 5–7, of *On the Heavens*. Archimedes attempted to calculate the number of sand

K. Kuehn, *A Student's Guide Through the Great Physics Texts*,
Undergraduate Lecture Notes in Physics, DOI 10.1007/978-1-4939-1360-2_27,
© Springer Science+Business Media, LLC 2015

Fig. 27.1 Thousands of galaxies scattered like grains of sand throughout a tiny region of space in the constellation Fornax. This ultra-deep field image was created with Hubble Space Telescope data from 2003 to 2004 (courtesy of NASA)

27.2   Reading: Lemaitre, *The Primeval Atom*

Lemaître, G., *The Primeval Atom: An Essay on Cosmogony*, D. Van Nostrand Company, New York, 1950. Chap. I: The Size of Space.[2]

27.2.1   Introduction

One of the oldest writings which have come down to us on the subject which I have the honor of discussing with you was left by the great geometrist of antiquity, Archimedes of Syracuse. His interesting work, *Arenarium* or *The Sand Reckoner*, begins with these words addressed to King Gelon who then shared power with his father, Hieron of Syracuse:

> There are some, King Gelon, who think that the number of the sand is infinite in multitude; and I mean by the sand not only that which exists about Syracuse and the rest of Sicily, but also that which is found in every region whether inhabited or uninhabited. Again, there are some who, without regarding it as infinite, yet think that no number has been named which is great enough to exceed its multitude. And it is clear that they who hold this view, if they imagined a mass made up of sand in other respects as large as the mass of the earth filled

---

grains required to fill the world based on Aristarchus' heliocentric model of the universe in his work entitled "The Sand Reckoner", included in Heath, T. (Ed.), *The Works of Archimedes*, Dover Publications, Mineola, NY, 2002, pp. 221–232.

[2] Conference held January 31, 1929, at "La Société Scientifique de Bruxelles" and first published in *La Revue des Questions Scientifiques*, March, 1929.

up to a height equal to that of the highest mountains, would be many times further still from recognizing that any number could be expressed which exceeded the multitude of the sand so taken. But I will try to show you by means of geometric proofs, which you will be able to follow, that, of the numbers named by me and given in the work which I sent to Zeuxippus, some exceed not only the number of the mass of sand equal in magnitude to the earth filled up in the way described, but also that of a mass equal in magnitude to the universe.

What is the volume of the world? How many grains of sand, how many cubic meters would have to pile up before all space would be filled, measured? Such is the question which I propose to examine with you and to which recent progress in astronomy and geometry enables us to give some answer.

The precise object which Archimedes had in view, in writing his *Arenarium*, no longer presents any difficulty, since the numbering system to which we are accustomed allows us to write at will any number, however large.

Nevertheless, the first problem which we must raise is the same one which Archimedes propounded at the beginning of his *Arenarium*: "Some think that the number of the sand is infinite in multitude." Let us alter the wording of the question a little and ask whether the number of stars is infinite.

27.2.2   Is the World Infinite?

What is an infinite multitude? It is clear that, however large a number may be I can conceive of another still larger. It is thus that the idea of infinity is introduced into mathematics. It is not less clear that any determined number cannot be the last one.

When we say that the regular sequence of numbers is infinite, we consider this sequence as continuing to be formed. As soon as we imagine the formation process of the numbers to have stopped, the sequence is ended. But our power to create numbers is not exhausted; the number which is actually finite is potentially infinite.

This concept of the potentially infinite is only a point of departure for pure mathematics. Pure mathematics deals with constructions of the mind which need not be set in actuality. Mathematics has framed transfinite numbers with regard to which the infinity of the series of numbers is almost nothing, and which evidently cannot help us to resolve our problem of the size of actual space.

The potentially infinite which we have just considered is obtained as an ordinal number; it simply means that the sequence of natural numbers is unlimited. On the other hand, one may wish to define the infinite number from the point of view of the cardinal number.

If one compares two collections of individuals, which are considered as equivalent entities, one will say that the two collections are equivalent if one can associate their respective members in such a way that each member of the first collection fits one, and only one, member of the second collection, and vice versa. Mathematicians say that the two equal collections have the same power, and preferentially use the last expression, in order to soften the paradoxical appearance of some of their results. We shall not follow them in this, since our aim is to select from their speculations whatever is directly available for studying the physical world and leave out those which would apply only to ideal collections.

There is a corresponding number to all the equivalent collections, a cardinal number, which is simply a typical collection, used to characterize the whole set of equivalent collections.

Is a collection necessarily finite? Nothing in the definition allows us to assert that. Let us see how one can discriminate between finite and infinite collections. A collection is said to be a part of another one when we get the first one by overlooking some member of the second. A collection is infinite when it is equal to one of its parts. Such is the definition of Weierstrasse, the only one, as far as I know, which has successfully withstood criticism.

It is easy to show that infinite collections are conceivable. If we consider the collection of all whole numbers, we get an infinite collection. The mathematicians deal only with the simplest one. As a part of this collection, we can consider the one which is obtained by omitting the odd numbers. The ensemble of even numbers is a part of the ensemble of whole numbers; and nevertheless, this part is equal to the whole, since a perfect correspondence can be established between the total of even numbers and the entire total of even and odd numbers. It is sufficient to assign the even number which is twice as great as a correspondent to each whole number. To each even number, there corresponds a single number, odd or even, its half, and vice versa. According to the definition of the equality of two collections, the total one and the partial one are equal; there are as many even numbers as there are whole numbers, whether even or odd.

These definitions allow one to answer the question which has already been raised: Is the number of stars finite?

From the point of view of potential infinity, we must say that the number of possible stars is infinite. The number of stars which could have come into being is infinite, but is the number of stars which really do exist now, or which will exist at some given time, finite?

From the point of view of the cardinal number, we can say that the question of whether the number of stars is finite or infinite reduces to knowing if the axiom that the whole is not equal to one of its parts applies to the assembly of stars, the word "equal" being used in its usual and obvious sense which we defined a moment ago. It seems, then, that unless one of our most immediately evident axioms no longer applies beyond Sirius or Aldebaran, we must conclude that the number of stars is finite.

Can we deduce from this that the volume occupied by the stars is finite? The conclusion is not quite straightforward, because if the stars are quite actual, the division of space which permits the evaluation of its volume is only potential.

Nevertheless, it is clear that, if the laws of geometry apply in the stellar world—at least, without too radical modifications—we may deduce, from the finite number of the stars, the finiteness of a convex polyhedron which encloses them. The essential hypothesis which allows such a deduction is that the principle of Archimedes applies to the distance between two stars, that is, if a finite length is carried, end to end, on a line joining these stars, the distance will be covered in a finite number of operations. If the stars are in relation to the distance, the volume which they occupy is finite.

What we have just said evidently applies to all the distinct entities of which the world is composed, Archimedes' grains of sand; the atoms or molecules, not numberless but very numerous, which wander from star to star; or even the grains of light which wander between them and which finally reach our telescopes. Everything in act is finite, potential being is the only legitimate field of application of the infinite or transcendent numbers of the mathematicians.

27.2.3   The Distance of the Stars

Now that we realize that reality is finite, we must try to determine how large it is. A few lines after the passage which I cited in the beginning, Archimedes continues: "But Aristarchus of Samos brought out a book consisting of certain hypotheses, in which the premises lead to the result that the universe is many times greater than that now so called."

It is through this passage that we know that the Greeks had already conceived of the hypothesis which must make us imply that the world is much greater than had formerly been believed.

Archimedes continues: "His hypotheses are that the fixed stars and the sun remain unmoved." Notice that he does not say that the sun remains unmoved, which would make no sense, but that the sun remains unmoved in relation to the stars; ". . . that the earth revolves about the sun in the circumference of a circle, the sun lying in the middle of the orbit, and that the sphere of the fixed stars, situated about the same center as the sun, is so great that the circle in which he supposes the earth to revolve bears such a proportion to the distances of the fixed stars as the center of the sphere to its surface."

What has the sphere of fixed stars got to do with this? If the earth rotates inside the sphere of fixed stars, it would result that the shape of the constellations will vary through an effect of perspective, each constellation appearing large as the earth approaches it more closely. Therefore, the hypothesis of Aristarchus implies that the stars are very far away, since this effect of perspective is not noticeable, or that the terrestrial orbit is like a point, in relationship to the sphere of fixed stars, with the result that the aspect of the sky remains practically the same as if it were effectively observed from the center.

Such is the obvious meaning, but the idea is rather inexactly expressed; the circle at the circumference of which the earth evolves has the same relationship to the distance of the fixed stars as the center of a sphere has to its surface. Therefore Archimedes refuses to understand and strongly resists such lack of precision.

"But it is evident," he says, "that this is impossible; for since the center of a sphere has no size, one cannot admit that it has any relationship with the surface of this sphere." That would involve admitting that the distance of the stars is infinite, since the relationship of the surface of the sphere to its center, that is, to zero, is clearly infinity. Archimedes cannot think of putting infinity into actuality.

Moreover, he adds that, according to the manner in which Aristarchus unfolds his hypothesis, he (Aristarchus) conceived that the relationship between the earth and the sphere on which the earth proceeds is the same as the relationship between that sphere and the sphere of the fixed stars.

That was not such a bad guess. The nearest stars are 12 times the distance fixed by Aristarchus. Is such a strain for the imagination quite necessary? The sun is 150 million km away, that is, 23,000 times the radius of the earth; we must admit that the stars are still 23,000 times further distant. That is provoking for the imagination. The imagination of astronomers has become accustomed to that, since those times. The unit of distance which is currently used to measure the distances of the stars, called a "parsec," is equal to 206,000 times the distance of the sun and we shall presently be talking of millions of parsecs.

But it can be understood that the astronomers awaited positive proofs before accepting the hypothesis of Copernicus, which was very tempting in some ways but which forced them to consider surprising distances.

The determination of star-distances by observing the effect of perspective due to the motion of the earth is called a "measurement of parallax." The measurements of Tycho Brahe allowed him to state that the stars were more than a million times farther away than the sun, 100 times farther than the furthest planet then known, Saturn. This statement induced him to reject the hypothesis of Copernicus.

The first efforts to measure parallaxes by means of telescopes were introduced as proofs of the motion of the earth. In 1674, the astronomer Hooke tells of his methods and his mishaps in a paper entitled: "An Attempt to Prove the Motion of the Earth." His telescope, pointed vertically, consisted of a lens fastened in an opening in the roof of the house, and a plumb line which hung down to the basement allowed the eyepiece to be correctly placed. During the fourth observation, the object-glass became loose and smashed.

When Römer, in 1704, thought he had found evidence of a difference of parallax between the stars Vega and Sirius, his really inexact observations were published by one of his co-workers under the significant title of "Copernicus Triumphans" ("The Triumph of Copernicus").

These researches culminated in a totally unexpected way with the discovery of aberration by Bradley in 1728. Aberration is an apparent displacement of stars resulting from the fact that light travels with a finite speed; the amplitude of the displacement is equal to the ratio between the speed of the earth and that of light. It amounts to one ten-thousandth of a radian or 21 arcsec. To reject the hypothesis of Copernicus, it would be necessary to imply that all stars simultaneously describe circular orbits in parallel planes, in one year. It is not possible to have a more direct proof of the system of Copernicus.

Measurements of parallaxes would have to await the powerful means of observation of today's astronomy, meridional telescopes, and, chiefly, photographic astronomy. No one yet doubted the system of Copernicus, but observation withheld another subject for astonishment.

There were plenty of stars at a dozen times the distance guessed by Aristarchus, a little farther than a parsec, about 206,000 times the radius of the terrestrial orbit; but there were very few, a 30th, at from 1 to 5 parsec; at the most 200 at less than 10 parsec; the others, the vast majority, did not show any perceptible parallax and were too far away to exhibit the effect of perspective due to the movement of the earth. The sphere of fixed stars spread out, far away, into space, while a few outposts swung with the motion of the earth. The others, the huge army, sparkled afar, as an apparent challenge to the patient pride of man.

How does the imagination of the poets compare with the reality of the heavens? The world is not a dungeon, not even a nicely decorated dungeon; it is a boundless perspective, marked out with bright guideposts which seem to have been placed at the farthest distance where they may still help us to answer the riddle, or, rather, to value and admire the work of beauty which has been prepared for us by the "God of the Armies."

What would we know of the sky if there were not a few stars of sensible parallax? If the world had been made on a scale only ten times greater, it would surely be beyond our grasp.

These neighboring stars provided us with a foothold in the stellar world; they supplied a firm basis for further findings. The movement, not of the earth around its sun, but of the whole solar system amidst the assembly of the stars, revealed the main features of the view of the heavens. The light of the stars, spread out by the prism, has shown us spectroscopic lines and described a chemical structure which is the same as that here below. Stellar chemistry has sometimes even outrun that of the laboratory, entering the name of the sun, "helium," in the list of elements. Stellar spectra have revealed the temperature of heavenly furnaces with the same reliability as has the engineer with a hearth. Finally, some special features of the spectra of the stars show us the strength of gravity at their surface and their absolute brightness. And while fine and ingenious interference devices made it possible to measure directly the diameter of these giant stars, Betelgeuse and Antares, which could hide the entire orbit of the earth inside their gaseous structure which is more rarefied than the vacuum of our Geissler tubes, astrophysicists did not need to modify the dimensions which had already been assigned by their customary estimates.

Pairs of stars, pulled one toward the other by gravity, were found to obey the same laws as our planets, and accordingly could even be weighed just as easily as the Sun, the Moon, or Jupiter. And these facts, added to all the others, have made it possible to account for the interior mechanism which provides stability to these globes of fire and, quoting Eddington, to understand what a simple thing a star is.

27.2.4   The Border of Naught

What is there beyond this army of stars which stretches out to a distance of several 1000 parsec, in the form of a flattened structure, ten times wider than it is thick?

We shall see, in a moment, that astronomy was again led to extend greatly the scale of cosmic formation, revealing, at prodigious distances, these enormous agglomerations of stars which are called the extragalactic nebulae. But further on, shall we reach, dare we hope to reach, a final nebula, a final star guarding the extreme limit of the world? To fix our ideas, let us talk of the ultimate stars, understanding by this word the last discrete entities at the extreme border of the universe. We can think of a convex polyhedron, with corners at the last stars, enclosing all the others. Provided that these last stars are in a relation of distance with the others, we are assured, as it has been explained before, that the volume of this polyhedron is finite.[3] This polyhedron encloses all the stars or all the particles of which matter is formed, whatever their nature may be. Outside this polyhedron, there is nothing. The universe would be a bubble of matter, dipped into a sea of nothingness.

What is space without matter? This question confronts us with one of the most difficult and most discussed problems: what is space? Without venturing into this dangerous territory, we can state that the principal philosophical systems, in spite of fundamental variance on so many points, agree in defining space in connection with matter. In our traditional philosophy, space is an abstraction of the extension of bodies, the latter is an "accident" of the corporeal substance; it is conceivable only where there is matter; without substance, the localizing "accident" becomes inconceivable. The system of Kant reaches the same conclusion in another way. Space is the form of phenomena, and one cannot conceive of it without phenomena. For one and the other of these two philosophies, one can truthfully say that space is in bodies; space which is absolutely empty could only be nothingness, and therefore does not exist.

The border of naught is a derisive wall, against which, it seems, the mind must stumble in its utmost endeavor to conquer the world. Are the stars nothing but a bright screen, hiding horrid naught? The thinking reed of Pascal prevails upon the rock by which it is bruised because it is aware of the rock; we rule the sky as we understand its harmony. Should we only be able to vanquish the universe, part by part, and should our mind be obliged to confess its impotency to face and understand the world as a whole?

It remains for me to tell you how one can avoid such a pessimistic conclusion and conceive of an intelligible form for the entirety of the world, and with what proofs, or, rather, with what beginnings of proof, with what hopes of proof, one can support this conception.

27.2.5   Space Which is Finite and Without Boundary

Real space, the universe, is finite; if not, the whole is equal to the part. Does a finite volume necessarily have a border? It is difficult for us to doubt it. But if we answer "Yes," is it our brain which impels us to this conclusion, or is it our imaginative, habitual concept of space? I should like to show you that it is our fancy, and to reject its testimony when it concerns an object so largely inadequate to the usual cases in which it can function and be informed.

Is it surprising that our imagination should be at fault when confronted with the whole of space? Should we be amazed? We are inside space, we lack distance from

---

[3] Always under the assumption of the hypothesis that the axiom, "The whole is unequal to a part," applies to the entirety of the universe.

which to contemplate it in its entirety. But we shall have prevailed over it if our mind reaches a correct valuation of our geometrical intuition and the field in which it can be legitimately applied, and if it can select the building stone with which, surpassing the imagination and silencing it, the mind can erect the edifice of the universe.

The space which is within reach of our direct experience—the Earth, the solar system—let us say, a parsec cubed, represents such small bricks with which to build the universe! Millions of these bricks will have to be piled up in order to fill up space.

Did you ever think of the strange shape which the palace of the League of Nations would have if its façade extended from Paris to Nankin? Such a pile of bricks would certainly be amazing!

Let us pile up bricks of space, let us say that each one is a cubic parsec; each brick has some volume and it has a boundary surface. The wall which these bricks build has a volume which is the sum of the composing volumes, because volumes are quantities which are essentially additive. But the boundary surfaces are not to be added to one another; the surface of the wall is not the sum of the surfaces of its composing bricks. The surfaces of the bricks which are in contact cancel one another; the outside surfaces alone make the wall. Let us go on piling up our bricks of space until all space is full; the surfaces which are in contact cancel one another. Will not the last brick cancel the last surface?

Your imagination rebels; remember that it has nothing to say, it has some value for the bricks, it cannot contemplate the edifice in its entirety. Syllogism only should be uttered. Is there any logical connection between these two concepts, a finite volume, and the occurrence of a boundary?

Some comparison may help us to understand. But every comparison is defective at some point, it is a game whose rules must be known and respected.

Let us compare space with the surface of the Earth. A volume is not a surface but it can be compared with it. A surface has a size, its area, analogous to the bulk of the volume; it has a border which divides the volume from exterior space. But the comparison is only valid if we do not go outside the surface which we compare to space, because although we can go outside a surface, we cannot leave space through some fourth dimension in order to look at it from the outside. Such are the rules of the game.

Every country has a border, just as each brick of space has a boundary surface. Let us juxtapose countries, let us federate Europe as a republic; borders cancel out but the total area is maintained. Let us annex the Atlantic Ocean and the Americas, Africa and Asia, the Indian, Arctic and Pacific Oceans, *etc.* When we place the last brick—Australia, let us say—the last border will vanish. The surface of the Earth as a whole has no boundary; why should it not be the same for the whole of space?

Let us not forget the rule of the game, let us not look at the question from the standpoint of the Moon.

27.2.6 Properties of Finite, Homogeneous Space, Without Boundary

I hope that I have grasped the possibility of space having a finite volume without having a boundary, and that you understand the importance of such a conception for comprehending the universe as a whole. We must now investigate some of the properties of such a space and the opportunity which may arise for an experimental check of whether our space does possess them.

We have curbed our imagination and we will not relax the reins. We shall let it run into each brick of space, but reason alone shall be allowed to go further.

Thus, we can well conceive what a straight line is. A brick of space is enough to define a segment of it and we are free to use the customary processes of alignment, to check by inversion, *etc.* Nothing prevents the production of a straight line from one brick to another, and we know that such a process cannot end because we shall nowhere reach a boundary of space. Now to the reasoning, to tell us what will come out of this process, the elements of which have been supplied by geometric intuition but which that intuition is unable to grasp in its entirety.

Here is what will happen. Every line—therefore a straight line, as well—when it is indefinitely prolonged, will finally come back, as closely as may be desired, to any point from which it has started.

For instance, let us show that the line will come again within less than a meter of one of its points. Imagine a sphere, one meter in diameter, the center of which runs along the line. This sphere traces a cylinder in space, a cylinder whose volume is equal to the section of the sphere multiplied by the length passed through. The volume of the cylinder thus swept out will finally become larger than any given quantity, hence larger than the finite volume of space. That is not possible unless some part of the cylinder is repeated, one portion of space being swept twice. Therefore, the line, even if it is straight, must pass again at less than a meter from a former point.

We can go further if we admit that space is homogeneous, that is, that all its points and all its straight lines have the same properties. Then it is clear that if the straight line passes again within less than a meter of its points, a small change in its initial direction will make it repass exactly through one of its points and therefore through all of them. If one straight line is closed, homogeneity makes them all closed. The straight line is thus a closed line; departing from one point, it returns to that point after having described a certain length, the tour of the universe which, because of homogeneity, must be the same, no matter what the starting point is.

27.2.7 An Euclidean Map of Elliptical Space

I suppose that your imagination is tugging at its bridle, so I am going to endeavor, not to satisfy it (which is impossible) but to have it appeased.

For a long time, mathematicians have considered the straight line as a closed line, but, in doing so, they believed that they were introducing a mathematical fiction which was apt in explaining their theorems but which did not express a reality of the physical world. Let us consider, with them, an Euclidian plane, that is, let us loose our imagination beyond the admitted limits.

In this plane, let us consider a straight line, naturally infinite since it is Euclidian, and a point outside this line. Then, pivoting around the point, a second straight line intersecting the first one. When the mobile straight line turns, the point of intersection runs over, let us say, on the right side; at the moment when the two straight lines are parallel, it disappears, the lines no longer intersect one another and then the point reappears at the left-hand side to return to its starting point.

Mathematicians say that when the straight lines were parallel, they should intersect, but at a fictitious point which they called the point at infinity of the straight line. Essentially, they insist that there is only one point at infinity on a straight line and not, as might perhaps seem natural, a point at infinity on the right-hand side and a point at infinity on the left-hand side. There is but one point at infinity and this point, which is neither on the right nor on the left, closes up the straight line. A point running over to the right reaches the point at infinity which is also on the left, and goes on through the left until it returns to its starting point.

If we wish to apply these concepts to our homogeneous space, finite and without boundary, we must forget that the point at infinity is a fictitious point and consider it as any other point; moreover, we must admit that it is not at infinity and that the length of the tour of the straight line has a finite size.

There is an important branch of geometry, projective geometry, which just sets aside that which annoys us. In fact, projective geometry ignores all material relations and concerns itself only with those properties which do not involve any measurement, for instance, the fact that points are on the same straight line or that several straight lines pass through the same point and, further on, it agrees to make no distinction between the point at infinity and the ordinary points.

Therefore, we can make use of our habits of thinking relative to projective space to give some idea of space which is homogeneous and finite but without boundary. This is perhaps of such a nature as to soothe the imagination, at least that of the mathematicians.

We can go further. Since our imagination is Euclidian, let us conceive of a projective space with, of course, its points at infinity which close the space, but also with its Euclidian metric. Just as the geographer represents the spheroid Earth on a plane map, using scales which are defined in each point and in each direction, so can we introduce, into our projective, Euclidian space a new metric which will be that of real space, which will be defined by a mapping in Euclidian space.

Let us take some definite point in the center of this map and let us propose to graduate a straight line passing through this point. At the center, let us raise a perpendicular on a straight line of unit length, and let us draw straight lines from the extremity of this perpendicular at equal angles. Let us agree that the segments intercepted by the sides of the equal angles on the line to be graduated represent equal lengths of actual space. The whole tour of the straight line will correspond to two right angles, the length represented by any segment will be to the tour of the straight line as the angle subtended by the segment is to two right angles.

This definition can be extended without difficulty to the case of a segment away from the center of the map. This segment defines, with the center, a plane surface whereby we draw out from the center a perpendicular of unit length; the angle subtended by the segment, with corner at the extremity of the perpendicular, will again be the measure of the length represented by the segment.

It is possible to show that space thus represented is homogeneous. The proof is immediate for a plane surface passing through the center of the map. From the point of projection used for its graduation, let us construct a sphere of unit radius. The straight lines projecting from the center of this sphere of segments representing equal lengths form equal angles and therefore cut the sphere at equidistant points. The metric properties of the plane surface which is represented are therefore the same as the metric properties of the sphere. Both surfaces are therefore homogeneous.

Homogeneous space, finite and without boundary, for which we have defined an Euclidian map, is called elliptical space, introduced by Cayley–Klein. An elliptical plane surface is wholly represented, without distortion, on an Euclidian half-sphere; a half-sphere and not an entire sphere, because two points on the sphere correspond to a unique point on the plane surface, the two extremities of a diameter. The elliptical plane is represented twice on the sphere, the two antipodal points representing only one point in real space.

The map of elliptical space which we have just described has the advantage of exhibiting the closed character of the straight line. Furthermore, it preserves straight lines and plane surfaces; it has the inconvenience of representing the finite volume of space in the totality of infinite, Euclidian space. One could easily alter the map in such a way that the volumes would be represented in it without alteration. All real space will then be represented by an Euclidian sphere of the same volume. There will be a border of the map; but the points represented on this border are not, in reality, distinguished in any way from the other points. They are a peculiarity of the map, not the space.

When we follow the round-the-world trip of Magellan on a planisphere, we go along, ever further west, until we reach the left edge of the map, yet we know that we have not reached the end of the Earth, but only a point which is represented twice at two opposite borders of the map.

In the same way, while we follow the tour of the universe along a straight line of elliptical space on our Euclidian map, we are not at the border of naught when we reach the extreme edge of the map, but at a point which is represented twice at two antipodal points of the sphere inside which we have represented finite space without boundary.

We cannot visualize the whole of space, but we have been able to make a map of it in the space of our imagination. The map has a border which, for a moment, we mistook for the border of nothingness. No doubt, it does not now surprise us any more than the frame of a Mercator planisphere. We realize that it is possible to read out all the properties of elliptical space in all detail on the map. Are we, perhaps, ready to wonder if the space in which we live may not be elliptical?

27.2.8   The Possibility of an Experimental Check on the Ellipticity of Real Space

If space is elliptical shall we ever know it?

Here we must meet an objection raised by Poincaré which has been widely spread through his books. A demonstration could be attempted, in trying to prove that the sum of the angles of some triangle of astronomic dimensions is very little larger than the two right angles. Such a demonstration would be hopeless, because we have no means of ascertaining that the sides of this triangle are really straight lines. This oddly restricts the chances of proof. We must rely only on geometric properties which last even when the figures are stretched, keeping only qualitative relation. Such properties comprise what is called the "analysis situs."

Are there properties of "analysis situs" which allow discrimination between Euclidian space and elliptical space? If so, we have some hope that the question might be settled by experiment.

In Euclidian space, every closed line splits the area of the plane into two positions, internal and external, so that it is not possible to proceed continuously from one to the other without crossing the dividing line. This is a characteristic of "analysis situs" which is not right in elliptical space. Actually, in elliptical space, the straight line is a closed line and two straight lines meet in a single point; after a complete tour along one of these lines, we shall come back from the opposite side before meeting the other line. The straight line is therefore a closed line which does not make a cut in the plane. Likewise, planes are closed surfaces which do not divide space into separate regions.

An experimental proof is therefore possible. Let us describe some fictitious experiment which may help to visualize how real proof might possibly occur.

Let us imagine a world in which, when Sirius is rising in the East, a similar Sirius is setting in the West; where each constellation has a duplicate, located at the opposite point of the celestial sphere, with the same binary stars revolving with the same period, with like color, *etc.*, and situated at distances whose sum is the same for all. Should we not agree that it is the same Sirius that we see in the East and in the West? We should be compelled to acknowledge it and to realize that the two rays of light which it sent us in opposite directions are two parts of one and the same straight line, closed fast at the two opposite sides of the star.

This is but a fiction, but in making clear that a world could be described where the ellipticity of space would be plain, it emphasizes the possibility of a proof. Mankind was sure of the roundness of the Earth before it could be circumnavigated; will our sight ever make the complete tour of space? It is not very probable. Nevertheless, we may dare to hope that less direct signs, we cannot guess which, will some day establish our convictions.

27.2.9   Space and Matter: Einstein's Relation

According to the theory of relativity, one of the effects of gravity is to alter the properties of space. Near a mass like the Sun, or a star, geometry is no longer Euclidian. Space is said to have some curvature, which is a quantitative estimate of the variance of the geometrical properties of this space from those of Euclidian space.

The equations which govern this dependence between geometric and gravitational properties contain a parameter, which remains insignificant on a small scale, such as for the motion of the planets, but which becomes very effective when one studies the universe as a whole. Accordingly, it has been called the "cosmological constant."

Ignoring local geometric perturbations due to aggregation of matter into stars, one can study the properties of the universe as a whole under the proposition that space is widely homogeneous. This hypothesis requires that matter will also be distributed in a widely homogeneous way.

If equal volumes are considered, volumes which are small with reference to the total volume of space but which contain, nevertheless, a very large number of local condensations, the mass contained in each of these volumes is about the same. From the cosmic point of view, one can therefore consider that matter has a constant density.

Then there exists a relationship between the cosmological constant and the density of matter. Their amounts are proportional. As long as there is matter, the cosmological constant cannot vanish and it is necessarily positive. The value of the cosmological constant determines the geometric properties of space; a positive value of the cosmological constant demands, as a consequence, that space be of an elliptical type, and it determines, in addition, the length of the tour of the universe. This tour is thus inversely proportional to the square root of the density of matter. This remarkable relation, discovered by Einstein, provides us with some estimate of the size of space.

Indeed, the density of matter is a quantity which we know from experiments made in our neighborhood. We mean by that word some piece of space, now large indeed, as large as our research power may allow, and nevertheless a piece of space which may still be very small in view of the whole of space.

Here are the facts we know about the density of matter in our neighborhood.

27.2.10   The World of Nebulae

Recently we described the local system of stars which surrounds us as a disc-shaped structure some 1000 parsec in diameter. The symmetry plane of this disc is inscribed in the sky as the Milky Way whose structure and distance are not yet too well known. Beyond this disc, called the local system and in continuity with it, lie assemblies of stars, the so-called globular clusters, which are distributed in a flattened system some 6000 parsec in diameter, our local system being near the edge. This assembly of celestial bodies is called the galaxy. Outside the assemblies of stars and of nebulae of an amount not much smaller than the galaxy are the extra-galactic nebulae. The nearest representative members of these large cosmic assemblies are the Andromeda nebula and the Magellanic clouds. The existence in these nebulae of stars which seem to show the same characteristics as those of the local system, mainly of novae and chiefly of some variable stars called cepheid variables, provides an estimate of their distance. The Magellanic clouds are about 3000 parsec away, the Andromeda nebula is ten times farther. Beyond, small nebulae can be seen which are like a miniature of the neighboring ones which have been studied in detail. They look like these same nebulae observed from a greater distance. A statistical study of the nebulae hitherto photographed enabled Hubble, if not to prove, at least to point out a number of convergent indications showing that nebulae are less unlike in size and brightness than stars are, and that their spatial distribution is roughly uniform. Nevertheless, some signs of clustering organization have been detected. This will undoubtedly provide new means of investigations without altering the main results already obtained.

From Hubble's study, it results that the faintest nebulae on Mount Wilson photographs are at some 44 million parsec distant and that their total number within the range of our telescopes is about 2 million. It is thus possible to get some estimate of the mean distance of the nebulae. We may say that a cube of 500,000 parsec in size contains, on the average, one nebula.

Furthermore, nebulae have intrinsic brightness and, therefore, masses which are much alike. Our galaxy may be considered as a typical nebula, somewhat bigger than the average. Our knowledge of the stars in the local system has been brought far enough to provide a fair estimate of the total mass of our galaxy. Thus the available facts enable us to get an idea of the density of the universe in our neighborhood, which is within reach of our observations and hence to infer, from Einstein's equation, the size of space.

According to Einstein's equation, the half-tour of the universe, that is, the greatest distance that may exist in space, is inversely proportional to the square root of density. For a density of a sun per cubic parsec, the half-tour of the universe would be 2 million parsec. There is, on the average, one nebula in a cube of 500,000 parsec, and each nebula has a mass as great as that of 250 million suns. The density of matter is, therefore: $500^3/0.25 = 500$ million times less than one sun per one cubic parsec. Taking the square root, the half-tour of space is therefore 22,000 times 2 million, *i.e.*, 40 billion parsec or 1000 times the present range of our telescopes.

The uncertainties of these estimates must not be minimized. Nevertheless, they must not be exaggerated, either.

An error of 10 or 20 % in the mean distance of nebulae is not improbable, but an error of 100 or 200 % is scarcely conceivable; we should need an error of not less than 250 % to reduce the half-tour of the universe from 40 to 10 billion parsec.

The estimate of the mass of the nebulae is perhaps elusive, since masses of low luminosity or even some which are totally obscure would escape detection. But in order to reduce the tour of the universe to one-third, the mass would have to be ten times greater.

27.2.11   Does the Size of Space Vary?

A rather serious objection could be made to the theory expounded above, an objection which questions the very foundation of the theory. As we shall see, it could require a sensible correction and, besides, it would pave the way to trustworthy experimental verification.

The cosmic theory of Einstein, in addition to the hypothesis of the homogeneity of space, implied an hypothesis which may seem so natural that we must be forgiven for not having mentioned it at the outset, namely, the hypothesis that the tour of the universe does not vary with time, or, in other words, that the universe is static.

The great Dutch astronomer, De Sitter, has indeed unfolded a cosmic theory founded on the theory of relativity which has the inconvenience of suggesting that the universe contained no matter but which, on the other hand, accounted for an extremely interesting phenomenon.

The spectrum of the extra-galactic nebulae is like that of the spectrum of stars of an average type, such as that of the Sun; but when it is compared with the solar spectrum, one notices that all the lines are strongly shifted toward the red. This is true for almost all of the nebulae whose spectra are available (about 40-odd). Two or three of the nearest nebulae are the only exceptions. Such a displacement of the entire spectrum found a natural interpretation as an indication that the nebulae are receding from us at an enormous speed, 600 km/s as a mean. An experimental error is scarcely possible for such large speeds and the few exceptions make it even less probable.

On the other hand, if the nebulae recede from our galaxy, that seems to indicate that our galaxy is a central point in the universe, endowed with special properties. We are very reluctant to accept such a conclusion. It would seem very strange that the region of intelligence should be thus distinguished by material properties. We know that it is neither the center of the local system nor of the galaxy. It would seem astonishing that it should be the center of the system of the nebulae. We should like to assume that if our observation post had been located in some of the far-away nebulae, the appearances would not be essentially different from those which we observe from our galaxy; in particular, the spectra of the nebulae around us would present the same systematic displacement toward the red. Moreover, such an hypothesis is a simple extension of the hypothesis of the homogeneity of space.

Now, this is perfectly conceivable. We need only assume that nebulae remain lightly distributed in space, but that the very properties of this space do vary with time, the tour of the universe, in fact, increasing with time. Then the distance of two nebulae would remain the same fraction of the tour of the universe and would therefore increase with it. Any two nebulae should recede, one from the other. Everything happens as it might appear to two microbes located on the outside of a soap-bubble. While the bubble expands, each microbe could state that his neighbors are receding from him; he would be deceived into thinking that he is at the center, but that would be an illusion.

It is in this sense that De Sitter's universe must be interpreted. Weyl has shown that it can be described as an Euclidian space where the figure composed of material points should inflate while keeping its same proportions. Lanczos has given an analogous interpretation for a closed, expanding space.

The fact that several different interpretations are possible for the same universe comes from the fact that it is empty of matter. The presence of matter induces a natural partition of the universe into time and space; the partition first adopted by De Sitter did not acknowledge the condition of homogeneity in space, in which he introduced a nonhomogeneous field of gravity. The paradoxical results which he obtained are due to this oversight.

The success of De Sitter's universe arises solely from the fact that he considers the tour of the universe as variable. Friedmann has shown later how such a universe can be modified in order to take account of the presence in the universe of a certain amount of matter.

Friedmann's universe involves yet another unknown parameter; it can be determined if one makes the natural hypothesis that the observed phenomena are not essentially altered during a relatively short interval, if we mean short as regards what we know of the deviation of evolution of the stars.

One finds, then, that the tour of the universe must be reduced to one-fifth of the value found from Einstein's formula. The remotest objects would then be at a distance from us of about 200 times the present range of our largest telescopes.

We therefore may dare to hope that the study of extra-galactic nebulae, a study whose first results date from less than 10 years, will enable us to uncover positive proofs of the closed character of space and to verify, no doubt with some correction, the value which we can already assign to the size of space.

We cannot end this rapid review which we have made together of the most magnificent subject that the human mind may be tempted to explore without being proud of these splendid endeavors of Science in the conquest of the Earth, and also without expressing our gratitude to One Who has said: "I am the Truth," One Who gave us the mind to understand Him and to recognize a glimpse of His glory in our universe which He has so wonderfully adjusted to the mental power with which He has endowed us.

27.3   Study Questions

QUES. 27.1 Is the number of stars infinite?

a) In what units did Archimedes measure the volume of the world? Did he find it to be infinite, or finite, in size? What mathematical technique has greatly facilitated the calculation of very large numbers or magnitudes?

b) What quality renders two collections of individuals *equivalent*? How did Weierstrasse use the concept of equivalence to define an infinite collection? Is an infinite collections of individuals even conceivable? If so, how?

c) What is the difference between the *potential* and the *actual*? Is the number of stars infinite, in either sense? Also, is the volume occupied by the stars infinite? How does Lemaître approach this problem?

QUES. 27.2 What is the distance to the stars?

a) What was the opinion of Aristarchus regarding the motion of the sun and the earth? On what grounds did Archimedes criticize this opinion? In particular, what does Aristarchus' opinion imply regarding the distance to the nearest stars?
b) By what reasoning did Tycho Brahe reject the Copernican hypothesis? Was he justified?
c) Why did Bradley's observation of stellar aberration support the Copernican hypotheses whereas early observations of stellar parallax initially failed?
d) How far, then, are the nearest stars? And what other significant information about the stars may be gleaned from their emitted light?

QUES. 27.3 Can space exist apart from matter?

a) In traditional philosophy, what is the difference between a corporeal "substance" and its "accidents"? In what sense, is extension—the space occupied by a body—an accident? Does this conception of space agree with that of Kant?
b) Does Lemaître agree with Aristotle's dictum that "everything in act is finite"? How does he deploy this dictum?
c) Must a finite volume have a border? More specifically: is there a logical connection between a finite volume and the occurrence of a boundary? Or is this merely a common-sense (but unnecessary) connection?

QUES. 27.4 What is the shape of space?

a) What is meant by the term elliptical space? Specifically, in what sense is elliptical space like the surface of the earth?
b) Can one experimentally check the ellipticity of real space by measuring its geometrical properties? What practical difficulty did Poincaré point out which would render such checks impossible?
c) How, then, might such a check be made using *analysis situs*—that is, the topological properties of space? Are such experiments practical?

QUES. 27.5 How does the existence of matter itself affect the size and shape of the universe?

a) What is the cosmological constant? And how does its value determine the shape of space?
b) What did Hubble's observations reveal about the range and the number of nebulae? What does this imply about the density of matter in the universe?
c) How, in turn, can the density of matter be used to determine the size of the universe? Are such calculations reliable? Upon what data and assumptions do such calculations depend?
d) Does the size of space vary with time? How do you know?
e) What does the observed nebular recession seem to imply? What is objectionable about this? And how can this be reconciled with the assumption of the homogeneity of space?

27.4 Exercises

Ex. 27.1 (SAND RECKONING). The approximate diameter of a sand grain is half a millimeter. About how many sand grains would fit inside the earth, whose diameter is approximately 13,000 km? Using Lemaître's value for the size and shape of the universe, how many sand grains would fit inside the universe? Finally, are these finite numbers (according Weierstrasse's conception of infinity)?

Ex. 27.2 (SPHERICAL TRIGONOMETRY). Is the sum of the interior angles of a triangle drawn on the surface of a sphere equal to π? If not, then upon what does it depend, and what are the maximum and minimum values which it can take? Finally, for a sphere with a diameter of 1 km, what is the size of an equilateral triangle having a *spherical excess* of 1°?

27.5 Vocabulary

1. Antiquity
2. Infinite
3. Transfinite
4. Ordinal
5. Cardinal
6. Paradox
7. Speculation
8. Discriminate
9. Radical
10. Deduce
11. Convex
12. Polyhedron
13. Legitimate
14. Transcendent
15. Premise
16. Parallax
17. Aberration
18. Prodigious
19. Agglomeration
20. Vanquish
21. Impotence
22. Intelligible
23. Edifice
24. Testimony
25. Syllogism
26. Parsec

Chapter 28
The Birth of the Big Bang

The evolution of the world does not depend solely upon the law of universal attraction.

—Georges Lemaître

28.1   Introduction

In Chap. I of his book, *The Primeval Atom*, Lemaître suggested that our universe is not a flat, Euclidean space (as one might presume), but rather an elliptical, non-Euclidean space. Indeed, the nebulae which populate our universe bend space in their vicinity, according to Einstein's theory of general relativity, allowing its edges to meet up—just like the opposite edges of a mercator projection map meet up to form a single meridian. In this sense, both the surface of the earth and the space of the universe are *finite* yet *unbounded*. Now, in Chap. IV Lemaître turns to the question of cosmogony: how did the world begin? Before providing his own hypothesis, Lemaître reviews the cosmogonic theories of Buffon, Kant and Laplace. Which (if any) of these theories of cosmogony do you find most compelling? Why?

28.2   Reading: Lemaitre, *The Primeval Atom*

Lemaître, G., *The Primeval Atom: An Essay on Cosmogony*, D. Van Nostrand Company, New York, 1950. Chap. IV: Cosmogonic Hypothesis.[1]

28.2.1   Introduction

In a celebrated work entitled *La Science et l'Hypothese*, Henri Poincaré has emphasized the role which hypothesis plays in the development of science. Science progresses through making use of both observation and hypothesis, by confronting, and often, contrasting, facts with ideas, practice with theory, of minutely-analyzed detail or imaginative organization of a series of details, and of guessing its sequel.

---

[1] Lecture given at La Société Royale Belge des Ingénieurs et Industriaux, at Brussels, January 10, 1945, and published in *Ciel et Terre*, March–April, 1945.

K. Kuehn, *A Student's Guide Through the Great Physics Texts*,
Undergraduate Lecture Notes in Physics, DOI 10.1007/978-1-4939-1360-2_28,
© Springer Science+Business Media, LLC 2015

Confronted by the object of study constituted by the universe, it has been useful to the development of science for impatient and audacious minds to conceive and suggest cosmogonic hypotheses which seek to reconstitute the evolution which matter has undergone in order to form, ultimately, our solar system with its planets, or even the system of all the suns, called the galaxy, and the entirety of all the galaxies which make up the universe.

Nevertheless, it seems that the principal cosmogonic hypotheses have not been formulated for a utilitarian end. When one reads Laplace, Kant, or Buffon, one notices that these authors have experienced a particular pleasure in developing their systems, a sort of exaltation related to the enthusiasm of the poets; the pleasure of discovering an enigma, of perceiving a simplicity hidden under the apparent complexity of the world, also, without doubt, an aesthetic pleasure before grandiose beauty, perhaps also the pleasure of risk which their enterprise brings, since the progress of positive knowledge must ultimately control their intuitions by confirming them, unless it annuls them or even makes them seem almost ridiculous, after a while.

The cosmogonic problem was not posed in a very precise manner until Newton discovered the law of universal gravitation. The law of attraction being known, mechanics permits the state of a system to be calculated at any moment whatever, provided one knows the initial conditions, that is to say, the positions and velocities of all the points of the system at an instant which is considered to be initial. A cosmogonic hypothesis seeks to find initial conditions presenting some characteristic of simplicity and from which the present state of the world could have resulted through the action of the laws of mechanics.

Nevertheless, the evolution of the world does not depend solely upon the law of universal attraction. It necessarily brings into play, in a more or less essential manner, the other physical properties of matter. It is only quite recently that our physical knowledge has attained the extent of our astronomical knowledge. Until now, astronomy had always been far in advance of the other sciences. Celestial mechanics was founded when Lavoisier had not yet made his experiments on the oxidation of mercury which established the composition of air and removed from fire, one of the four elements of the ancients, its appearance of substance. We have difficulty in imagining the concepts of matter which were then current. Buffon wrote: "Upon, and near the surface of the Earth, there are substances which are 14,000–15,000 times more dense than others; the densities of gold and air are nearly in this proportion. But the interior parts of the Earth and the planets are more uniform." Kant will say: "Variety in kinds of elements is essential to the organization of chaos. . . This variety is undoubtedly infinite, since nature is shown to be limitless everywhere. . . The elements of a specific weight which is a thousand times greater are a thousand times, and perhaps millions of times, more disseminated than those which are a thousand times lighter. And since this difference of densities has no limits. . . ," etc.

It is with such a concept of matter that Kant dares to say: "Give me matter and I will construct a world out of it," that is to say, I shall show you how a world can come from it.

The discovery of universal attraction permitted Newton to determine the masses of the planets which possess known satellites, that is, the Earth, Jupiter, and Saturn. He deduced the relative densities of these planets and determined that the density of the planets was increasingly lower, the further away the planets were from the Sun and thus received less of its heat. This observation of Newton played a dominant role in the development of cosmogonic hypotheses. As long as it was believed that the elements which make up matter were extremely varied, it was necessary to imagine initial conditions which permitted the understanding that the lightest elements formed the outer planets, while the heaviest formed the inner planets. It was in this manner that the problem was posed for Buffon or Kant.

The oldest cosmogonic theory about which I wish to speak to you is that of Buffon. George-Louis Le Clerc, Count de Buffon, published the first of fifteen volumes of his *Natural History* in 1749, a volume dealing with "The Theory of the Earth." This date and this title should hold our attention. In fact, it was 6 years later, in 1755, that Emmanuel Kant, who had just been named "privat-docent" at the University of Königsberg at the age of 31, to teach philosophy, physics and mathematics, published his cosmogonic theory with the title: *General Natural History and Theory of the Heavens*. The allusion to Buffon's work, *Natural History*, "Theory of the Earth," is clear; besides, Kant's work was manifestly conceived as a reaction to that of Buffon.

28.2.2   Buffon's Cosmogony

For Buffon, the Earth and the other planets had been formed as the result of a glancing collision between a comet and the Sun. The astronomers of this period were very much impressed by comets, which they took to be stars whose mass was comparable to that of the planets. Kant agrees with Buffon on this point. He writes, speaking of the mass of comets, that one must believe that certain comets are heavier than Saturn and Jupiter. It is difficult to understand the reason why comets were taken so seriously. Perhaps it is due to the outstanding role which certain of them, Halley's comet, in particular, had played in the discovery of universal attraction.

Buffon recognizes that a powerful comet was necessary to his theory and, here, allow me a rather long quotation which illustrates very well the ideas of the period: "But if we consider the prodigious rapidity of the comets in their perihelion; the near approach they make to the Sun; density and strong cohesion of parts necessary to sustain, without destruction, the inconceivable heat they undergo; and the solid and brilliant nucleus which shines through their dark atmospheres; it cannot be doubted that comets are composed of matters extremely dense and solid; that they contain, in small limits, a great quantity of matter; and, consequently, that a comet of no enormous size may remove the Sun from his place and give a projectile motion to a mass of matter equal to the 650th part of his body. This remark corresponds with what we know concerning the respective densities of the planets, which always decrease in proportion to their distance from the Sun, having less force of heat to resist. Accordingly, Saturn is less dense than Jupiter and Jupiter much less than the Earth. Thus, if the density of the planets, as Newton alleges, were in proportion to the quantity of heat they can support, Mercury would be 7 times denser than the Earth and 28 times less dense than the Sun, and the comet of 1680, 28,000 times more dense than the Earth or 112,000 times denser than the Sun. Now, supposing the quantity of matter in this comet to be equal to the ninth part of the Sun, or, allowing it to be only the one-hundredth part of the bulk of the Earth, its quantity of matter would still be equal to a 900th part of the Sun: Hence a body of this kind, which would be but a small comet, might push off from the Sun a 900th or a 650th part, especially when the amazing rapidity of comets, in their perihelion, is taken into account."[2]

Buffon takes thoroughly into account the difficulties which his theory presents. How could matter, driven by the comet from the rim of the Sun, acquire a circular motion? He answers as best he can, without being really convincing.

He explains that the outer planets, formed from matter which was ejected with the greatest force, must be lighter because, he says: "the projectile force being proportional to the surface to which it is applied, the same stroke would make the larger and lighter parts of the solar matter move with more rapidity than the smaller and heavier."

Buffon's hypothesis has been revived during recent years. The comet is replaced by a star, no less hypothetical, of course. Instead of the glancing collision, one considers the tidal forces which the two stars, the Sun and the star which would pass near it, exercise upon one another; the ejection of a filament of matter could result from this, a filament which would then splinter, giving birth to the planets.

28.2.3   Kant's Cosmogony

Properly speaking, Buffon's hypothesis is not a cosmogonic hypothesis, since the Sun and the comet are taken for granted. It is only a theory of the formation of the planets. Kant's hypothesis is another matter entirely. Kant imagines primeval matter as being formed of particles of very different densities and distributed uniformly in space. This state of nature seems to him to be "the simplest which could have followed nothingness." Next, he imagined that there is a point where attraction acts more energetically than anywhere else, that is, that the homogeneity of the distribution is not perfect. Then it is toward this point that all the elementary particles which are distributed in this space are assembled. "The first effect of this general fall is the formation of a body at this center of attraction, which, so to speak, grows from an infinitely small nucleus by rapid strides; and in the proportion in which this mass increases, it also draws with greater force the surrounding particles to unite with it."

Such is the explanation of the formation of the Sun, and it would be all there is to say, without doubt, if there were not other forces in play besides that of gravitation. But nature holds in reserve other forces which are exercised, in particular, when matter is decomposed into very small parts. These are those forces of repulsion which are manifested in the elasticity of vapors, the diffusion of odors and the expansion of

---

[2] We quote according to: *The History and Theory of the Earth*, translated by William Smellie, member of the Antiquarian & Royal Societies of Edinburgh, published by W. Clowes, London.

gaseous matter. This force is designated by the untranslatable verb "zurückstossen." It means a reverse shock such as is exercised upon one another by two marbles which collide. The effect of these collisions is to produce a state of minimum interaction, one in which the particles no longer collide and thus describe circular vortices. The particles modify their fall toward the Sun, until the particles no longer criss-cross one another and continue their free circular motion.

All this is correct if it concerns inelastic collisions and not elastic collisions between the particles. But that is nowhere clearly stated.

The particles are assumed to be of very different densities, as we already know. The densest particles will experience the resistance of the others less strongly; therefore, they will be able to approach the sun more easily, while the light particles will have the circular motion imposed upon them at a greater distance. The planets formed very far from the Sun will thus be of lesser density than those which are formed near the Sun. In this manner, the law of densities formulated by Newton is found again.

Kant draws the most extraordinary consequences from this error, thus showing that when powerful intellects make mistakes, they do not go halfway.

Taking up again the idea of the *plurality of worlds* put forth by Fontanelle in 1686, Kant remarks that the inhabitants of other planets must be made of matter which is denser or lighter than terrestrial matter. He formulates the following general law: "The matter of which the inhabitants of diverse planets are formed, animals as well as plants, must, above all, be of a nature which is lighter and more subtle, the elasticity of the fibres, and, at the same time, the conformation of their bodies, must be more and more perfect as the stars become more distant from the Sun."

Since the faculties of thinking creatures are necessarily dependent on the matter which forms the machine inhabited by their minds, he is led to conclude "that it is more than likely that the excellence of intelligent creatures, the readiness of their thought, the vivacity of the notions which they receive of exterior impressions, as well as their faculty of associating them, finally, also, the nimbleness in the exercise of their activity, in a word, the entirety of their moral being must be subjected to a stated law, which is more and more perfect and excellent as their habitation becomes further from the Sun. This law being thus established with a degree of reasonableness which scarcely differs from a demonstrated truth, the imagination can have free play in the comparison of the qualities of these diverse inhabitants," and, in fact, Kant gives free play to his imagination and manifestly regrets that he was not lucky enough to have been born on Jupiter or Saturn.

Kant's theory is not only a theory of the formation of the solar system; it is a cosmogonic theory in the complete sense of the word, a theory of the universe. Kant notes that there is in the universe that which he calls a "systematic arrangement," in German, "systematische Verfassung." He understands by that a certain number of stars arranged around a common center about which they move without departing by very much from a plane passing through this center. Jupiter and his satellites form a systematic arrangement; it is the same with the Sun and the planets. Why should not the stars form a systematic arrangement on a grand scale? Here Kant anticipates later discoveries and divines the phenomenon which is known by the name of the rotation of the galaxy. He guesses that the Milky Way which makes the circle of the sky is the plane of a vast systematic arrangement when seen in projection.

Kant goes further. Why should there not be systematic arrangements of galaxies similar to our own and in which there would be certain of those nebulae described by Maupertuis in "Le Discours sur la Figure des Astres," which appeared in Paris in 1742? There would be more arrangements of these arrangements and so on, to infinity.

Nevertheless, Kant seems to have forgotten here an essential characteristic of his theory of a systematic arrangement which we set forth, a moment ago. Besides the Newtonian attraction, it is necessary to have some repulsive force analogous to the effects of the inelastic collisions which are exerted upon the particles of primeval matter. It is hard to see how something analogous to these collisions can be exerted upon the stars in order that they may form a systematic arrangement on a grand scale, as the primeval particles form the arrangement of the solar system on a small scale. Above all, one does not see how it would be possible to find an infinity of devices for contriving repulsive forces which would make possible the infinite succession of subordinate systematic arrangements. This reservation having been made, let us recognize that the idea does not lack scope. It lends itself to fine developments, of which here are several samples: "The creation, or better still, the fashioning of matter must have first started at a central point and then have been extended to all distances to fill infinite space, in the succession of eternity, of worlds and of systems of worlds. . . The indefinite, exterior region will still be the seat of disorder and chaos. . . While, on the one side, nature is aging around the center, on the other, it is always young and prolific in new creations. The world already formed is found to be limited, on the one hand, by the ruins of the world which has been destroyed and, on the other, by the chaos of unformed nature. . . "

28.2.4 Laplace's Cosmogony

The third hypothesis about which I wish to speak is that of Laplace. It is also the best known, therefore I shall be able to be more brief. It was outlined in the first edition of *L'Exposition du Système du Monde*, which dates from 1794, then was completed in the third edition of 1808 to form, in the following editions, "Note VII" which ends the work.

Thus we are 40 or 50 years later than Kant. Nevertheless, there is no doubt that Laplace was not acquainted with Kant's cosmogony. This fact arises, in large part, from the misadventure which befell Kant's work. The publisher went bankrupt, just at the moment when he was finishing the printing, the stock was seized, and the work was not put on sale. Kant published it, in part, in 1763, under a rather extraordinary title, which scarcely attracted the attention of astronomers: *The Only Possible Proof of the Existence of God*. In 1791, an edition of the first half of the work appeared, in the appendix to a work of Herschel, and it was not until 1797 that the work was published *in extenso*.

   Furthermore, there is hardly any relationship between the hypothesis of Laplace and that of Kant, and it is rather a mistake to speak of the Laplace–Kant hypothesis, as is often done. In Laplace's view, the initial nebula is a gas, whereas, for Kant, it is a swarm of particles. The nebula of Laplace resembles a giant star, such as Antares or Betelgeuse, the atmosphere of which extends over spaces comparable to that which the solar system occupies. Such a nebula, supposedly animated by an initial rotation, is condensed by cooling while giving off rings which will be able to be broken up, thus forming planets. The work of Laplace is handled with a sobriety and a precision which one expects from a mathematician of the very first rank; nothing is found here which resembles the philosophical fancy in the work of Kant. It was the origin of important mathematical works at the hands of numerous scholars, among whom it will suffice to mention Roche and, more recently, Poincaré and Jeans. The hypothesis itself has been amended and modified without any definite conclusion emerging from these researches. Besides, these theories do not provide any indication whatever of the origin of the nebular star, which is held to be primeval.

   This exposition, no doubt too summary, of cosmogonic hypotheses, which have been suggested since the discovery of universal attraction, leads us, with Poincaré and Jeans, to our own epoch. Now we must ask ourselves how the cosmogonic problem is posed at the present time.

28.2.5   The Primeval Atom

First of all, our knowledge of physical matter has been greatly extended, in recent times. The infinite diversity of the elements which Kant spoke of is reduced to several hundred simple bodies (counting the isotopes). These simple bodies have among them some links of relationship. Certain ones among them are radioactive, that is, they split while emitting rays which are sometimes, themselves, simple bodies, such as nuclei of helium. Uranium, for example, evolves by a cascade of transformations, involving radium and finishing up as lead, not without having ejected, in the course of these transformations, rays of various kinds. More recently, we have learned that radioactivity is a general property of matter; in addition to spontaneous radioactivity, there exists, in fact, artificial radioactivity which can be provoked by appropriate atomic bombardments. One can say that every chemical element is radioactive, as is uranium, or is the product of a radioactive transformation, as is lead.

   It is transformations of this nature that explain the heat of the Sun or of the stars. At the temperature of some 20 million degrees which obtains in the interior of the Sun, hydrogen can be introduced into the nuclei of lightweight atoms. These atoms will then disintegrate, but, at the end, we find that helium, not hydrogen, remains. Finally, hydrogen is transformed into helium, thanks to other atoms which have played the role of catalyst. Four atoms of hydrogen weigh a little more than one atom of helium. The difference is found in the form of thermal energy. We have here a magnificent example of the equivalence of mass and energy, propounded long ago by the theory of relativity; the disintegration of a gram of matter frees c^2 ergs or 9×10^{20} ergs. And that is why the Sun shines.

These transformations by which hydrogen is incorporated in lightweight atoms to make them into heavier atoms cannot go much further back than carbon. It may be shown that it is altogether impossible for the heaviest atoms to be formed in this manner under conditions of temperature and pressure analogous to those which obtain in the stars. On the contrary, it is perfectly possible that the matter which currently exists may be the residue of the radioactive disintegration of atoms which have disappeared. If radium had been conserved for thousands of years, it would have largely disappeared, leaving behind it lead and helium. The average life of radium is, in fact, about 1000 years. The radium which we find today still exists only because it is continually being formed as an intermediate product of the disintegration of uranium, and the average life of uranium is 4 billion years. If one had tried to conserve radium for many billions of years, it would have disappeared and would have given place to lead and to helium. Besides, it is through the lead content of uranium ore that one can obtain the most precise measurement of the period at which these ores were formed and thus determine, within about 2 billion years, the age of the terrestrial crust. Therefore, it is very likely that uranium itself was produced by the disintegration of an atom which is unknown because it has entirely disintegrated. Moreover, what we say about uranium can be said about all the elements, since all the elements are radioactive, or can be the products of radioactive transformations. Thus one sees that, in the present state of physics, the most simple point of departure for a cosmogonic theory is no longer a more or less uniform nebula but a single atom, the radioactive disintegration of which would have created the less stable atoms which exist today, through a series of successive splittings.

This hypothesis of the primeval atom permits an immediate explanation of certain facts whose appearance is rather surprising. In the first place, disintegration products of an atom are present in quite definite proportions. The composition of matter must, therefore, be quantitatively identical everywhere, and that is what is stated for terrestrial matter, the matter of meteorites, of the Sun and of the stars. The exceptions, of which the principal one is the hydrogen content, are explained by special circumstances and are therefore among those which confirm the rule.

In the second place, if the matter resulting from the disintegration of the primeval atom had been able to accumulate in order to form stars arranged in galaxies by a process which the theory will have to explain, it is clear that the radiation which was ejected at the time of successive transformations must have escaped, at least in part, from this process of agglomeration, to circulate freely in interstellar space. It must be possible to recognize this radiation, as much by its quality as by its total quantity. This radiation constitutes cosmic rays. In fact, these rays have an individual intensity which is at least a thousand times greater than that of radium rays, and, moreover, their total intensity is comparable to the totality of existing energy. This last point demands some explanation. The density of matter contained in the nebulae is estimated by Hubble at 10^{-30} gm/cc. That represents about an atom per cubic meter, and it is what one would obtain if the matter contained in all the stars of all the nebulae were distributed in space. Furthermore, the intensity of cosmic rays is measured in erg/cm^2. This measurement is transformed into erg/cc by dividing by the velocity $c = 3 \times 10^{10}$, then into gm/cc by dividing again by c^2. In this manner, one obtains 10^{-34} gm/cc, which is about one ten-thousandth the density of the matter of the stars. It is what would be expected if cosmic rays, together with the matter of the stars, were produced by radioactive transformations. In ordinary radioactive transformations, the energy of the rays is, in fact, of the order of percents of the energy remaining in the mass of the atom.

One sees how our physical knowledge has profoundly modified the given facts of the cosmogonic problem. But there is an even more profound modification which is a consequence of the very law of universal attraction, a modification which has been introduced by the theory of relativity or Einsteinian theory of gravitation. According to this theory, and in those conditions where the law of universal attraction is applicable, that is, for velocities which are small compared to the velocity of light, the bodies act upon one another not only in a manner inversely as the square of the distance but also proportionally to their distance. Thus there are two forces, instead of the single Newtonian attraction. The theory of relativity teaches us that the force which varies in inverse ratio to the square of the distance is an attraction, and furnishes us the value of this attraction, that is, the value of the constant of gravitation. The other force, which is proportional to the distance, can be an attraction or a repulsion, and it depends on a constant called the cosmological constant. It has been agreed to give this constant a positive value in the case of repulsion.

If the cosmological constant is positive, that is, if a force of repulsion exists, besides the attraction, it is possible that these two forces which thwart one another succeed in exactly neutralizing one another; in this way, matter will remain in equilibrium between these two forces. Then one obtains that which has been called a "universe of Einstein." The density by which matter must be distributed in order that there may thus be equilibrium is linked to the value of the cosmological constant by a simple relation.

In this manner, we are led to ask ourselves if there does not exist, in the universe, a realization of this equilibrium, and if astronomical observations would not permit us to recognize these realizations.

With regard to the hypothesis of Kant, we have seen that our galaxy forms a "systematic arrangement" where matter revolves about a center and remains densely concentrated in the neighborhood of a plane passing through this center. On the other hand, as Kant had already guessed, following Fontanelle, some nebulae which are called extra-galactic nebulae or galaxies and any divided into spiral or elliptical nebulae also form a systematic arrangement, in the Kantian sense, and it has been shown that several of them are effectively animated by a motion of rotation, the period of which has been deduced by spectrographic measurements. In fact, it is these observations which provide the estimate of the mass of the nebulae and Hubble's figure of 10^{-30} for the density of matter.

As Kant also guessed, systems of nebulae likewise exist which have been called groups or clusters of nebulae, but, contrary to what Kant could imagine, these groups or clusters of nebulae give no indication of a systematic arrangement. In certain of these clusters, the velocities of a great number of their nebulae have been measured, and these velocities appear to be distributed at random. They do not indicate, in any way, a rotation of the entire cluster. In these clusters, there is no definite concentration in the central region. Their form is capricious and it admits of extensions without any apparent order.

Such an aspect suggests the absence of a dominant force, rather like the actual clouds which exist in our atmosphere. This invites us to identify the clusters of nebulae with the regions of equilibrium between Newtonian attraction and cosmic repulsion, in the manner of Einstein's universe.

In order that this may be possible, it is necessary that the clusters have a density linked to the cosmological constant by the relation to which we referred, a moment ago. Therefore it is necessary that the clusters all have the same density in the nebulae. This density can be estimated by determining the mean distance of the nebulae in the cluster, that is, the diameter of the cluster divided by the cube root of the number of nebulae which it contains. In this way, numbers are found for the various clusters, averaging 200,000 light years.

28.2.6   The Three Periods of Expansion

The two ideas which have been formulated, thus far, are: first, the origin of matter as a single atom which disintegrated; second, the equilibrium of clusters of nebulae. These ideas can easily be combined into a single theory.

At the time of the atom's disintegration, matter was strongly condensed. The fragments of the atom were separated from one another with great speeds, speeds which were progressively slowed down by the force of gravitation which, for these large densities, amply overcame the cosmic repulsion. This period of rapid expansion was followed by a period of slowing-down in which the density attained the value of equilibrium. Then, on the average, repulsion overcame attraction and the expansion started again at an accelerated rate. Only regions where velocities and densities differed from the average were delayed in the state of equilibrium and finally constituted the clusters of nebulae. In comparing the mutual distances of isolated nebulae to those of nebulae which are located in the clusters, one can estimate that the universe has been dilated about ten times since its passage into equilibrium.

28.2.7   Experimental Verifications

This figure makes possible a quantitative verification of the theory. In fact, at the present time, cosmic repulsion greatly exceeds attraction, and we can therefore apply the formula established for the universe of De Sitter, that is, an empty universe. We can thus deduce, from the speed of expansion, the value of the cosmological constant. We therefore have two determinations of the cosmological constant which are entirely independent of one another; the first is that which results from the equilibrium of the clusters of nebulae and the rotational value of the masses of the nebulae; the second is the one which we have just mentioned. The agreement between these two determinations is excellent.

Thus, it may well be that we have found initial conditions which are ideally simple and sufficiently precise for the entire evolution of the world to result from them in the necessary manner. It should remain to deduce, from these initial conditions, detailed consequences which should result from them and to verify whether these deductions are in agreement with astronomical observations.

There is scarcely any doubt that such a deduction may be possible. But it is not an easy task, and it must be thoroughly recognized that this task has not yet been accomplished.

Without going into detail on the technical difficulties which are presented, and in recognizing the provisory character of what can be said from now on, I should like to point out to you the consequences of the theory which it seems already possible to perceive.

In the first place, what might well be the state of matter during the first period of rapid expansion? A sort of gas must have resulted from the splitting of the primeval atom which was broken into smaller and smaller pieces, a gas which was, of course, not much like present-day gases which are in statistical equilibrium and perfectly homogeneous. This must have been a kind of gas which was very much lacking in homogeneity. It could rather be described as a mixture of more homogeneous, gaseous clouds with random velocities, from which the primeval, radioactive rays became more and more disengaged and finally emerged as cosmic rays.

When these gaseous clouds approached the slowing down period of world expansion, during which repulsion and attraction balanced one another, circumstances became somewhat analogous to those which Kant considered. According to chance and fluctuations, condensations could be formed and could become amplified. The gaseous clouds, rushing toward some occasional center, would encounter one another. The energy of the impact would be transformed into light or radiant heat, and it will surely be a case of inelastic collisions. If the total moment of momentum was small, an elliptical nebula will result from the collision. If, on the contrary, the total moment of momentum was large, a true "systematic arrangement," a spiral nebula will result.

What will happen to the gaseous clouds which have met in this manner? We have supposed that these clouds were of such density that the cosmic repulsion and gravitation created a balance at this point, that is, at a density of about 10^{-27} gm/cc. As a result of the encounter of two clouds, the density will be at least doubled, thus gravitation will clearly overcome repulsion and we shall obtain a gaseous nebula which will be condensed toward its center, greatly resembling the nebula of Laplace. A star will evolve from it and, possibly, planets.

At the time when expansion resumed its speed, that is, up to the present time, we should thus have two condensed regions which were separating progressively from one another and which are themselves made up of stars mixed with the remains of the gaseous clouds. Accidentally, several hundred nebulae have remained in states of slowed-down equilibrium and they form a cluster of nebulae. The system of nebulae outside the clusters is distributed in a strongly homogeneous manner, but there must be great fluctuations in the distribution, representing all the degrees between the mean distribution and the properly-termed clusters. This corresponds very well with that which is revealed to us by the observations of Hubble and, especially, of Shapley.

It is clear that if this theory can be developed in a detailed manner, it will come into contact, at many points, with the facts as observed. This deduction can be difficult, but it does not leave room for any arbitrariness or for any adjustment which would permit the hypothesis to be saved if it entered into conflict with the facts. In particular, it must be possible to calculate a priori the law of the distribution of matter in the elliptical nebulae, a law which is known through observation. If these calculations, which have not yet been made, lead to a result which is in agreement with the observed facts, the hypothesis deserves, without doubt, to be considered as established. If, on the contrary, there is final and manifest disagreement, it will be necessary to seek something else.

There are also other consequences of the theory about which it is interesting to speak, although they do not lead to confirmations of this theory. On the contrary, the theory being granted, they furnish us with information which experiment alone cannot attain.

28.2.8   Geometry

Astronomers are capable of recognizing, on their photographic plates, nebulae whose distance is half a billion light-years. What is there beyond? To try to answer this question, within the framework of the theory, I am obliged to call upon the most difficult aspects of the theory of relativity. No doubt, I shall be able to do no more than touch upon the subject at this time, but I would leave you with an idea of the primeval atom which is too incomplete if I failed to discuss the geometric consequences which it implies. The theory of relativity shows that the presence of matter is linked to geometric properties which are described by the name of the curvature of space. A space of positive, uniform curvature is called a Riemannian space, or, again, an elliptical space. If, in such a space, one considers concentric spheres of increasing radii, one ends with a sphere of maximum radius which fills all space. Each point of this sphere encounters the point diametrically opposite. Expressed differently, on the periphery of the sphere, the antipodal points which we consider as separate points are, in reality, identical. These are, if you like, two representations which we make for ourselves of the same, single point in elliptical space. Two radii, diametrically opposed, are thus soldered to one another by the single point represented by their two extremities. Together, they form a straight line which is a closed line, the length of which is called the tour of space. One third of the tour of space is called the radius of space. The theory of relativity permits the radius of space to be calculated if the density and velocities of matter are known. One finds, thus, that, at the moment of the slowing-down of expansion and of the passage through equilibrium, the radius of space was equal to a billion light-years. At the present time, it is about ten times greater. The half-billion directly accessible to astronomical observation therefore represents a portion which is not trifling.

All this supposes that there is matter beyond the observed region, so that the curvature of space which exists in this region continues beyond, in such a manner as to close space. This can be inferred from the presence of cosmic rays. In fact, these rays have run along through space since the formation of the stars, and calculation shows that they have thus traveled over the tour of space several times. In this manner, they extend the given facts of observation beyond visual observations and guarantee to us that elliptical space is full of matter which, if it has not exactly the same density, has at least a density of the same order of magnitude as that which is observed in our own neighborhood.

Finally, if we go backward in thought along the course of time and seek to imagine for ourselves the geometric conditions which prevailed in the first period of hurried expansion when the fragments of the primeval atom were being repeatedly broken off, we find, for the radius of space, values which are smaller and smaller, finally reaching an initial limiting value of zero.

In this manner, we see that at each stage in its evolution, the primeval atom—or the products of its disintegration—uniformly filled all space, with an ever-increasing radius from its zero origin.

We can compare space–time to an open, conic cup. One progresses from the past to the future up to the generating lines of the cone, one runs along the tour of space when circulating along the parallel, horizontal circles. The bottom of the cup is the origin of atomic disintegration; it is the first instant at the bottom of space–time, the now which has no yesterday because, yesterday, there was no space.

We must conclude: I can think of nothing better to repeat than, in a somewhat altered form, the words of Kant: "Give me an atom and I will construct a universe out of it."

28.3 Study Questions

QUES. 28.1 Can Newton's theory of universal gravitation alone account for the structure, formation and evolution of the universe?

a) What is Newton's universal theory of gravitation? How did Newton's theory precipitate speculations regarding the origin and evolution of the planets, the galaxy and the universe?

b) What assumption regarding the nature of matter is common to the cosmogonic theories of Buffon and Kant?

c) Is it fruitful to speculate about he origin and evolution of the planets, the galaxy and the universe itself? If so, upon what should such speculation be based? And what is the end, or goal, of such speculations?

QUES.28.2 How do the cosmogonic theories of Buffon, Kant and Laplace differ?

a) What is Buffon's theory of cosmogony? Upon what assumption was it based and what difficulties did it encounter? Strictly speaking, is his really a "cosmogonic theory"?
b) What is Kant's theory of cosmogony? How does it account for the formation of the sun and the distribution of the planets? And what conclusion does he then draw regarding the inhabitants of said planets? Does Lemaître take this view seriously?
c) What does Kant infer from the systematic arrangement of the solar system? Is this a reasonable inference? Is it correct? What reservation does Lemaître express regarding Kant's cosmogony?
d) What is Laplace's theory of cosmogony? How is it similar to, and different from, that of Kant? On what grounds does Lemaître criticize this theory?

QUES. 28.3 Is Lemaître's theory of cosmogony superior to those of Buffon, Kant and Laplace?

a) In what way does Lemaître's understanding of the nature of matter—and radioactivity in particular—influence his cosmogony?
b) How does Lemaître's theory of cosmogony account for the formation of the elements? Are there any natural limits to the formation of heavy elements by means of nuclear transformations?
c) How does Lemaître's theory account for the presence, and even the density, of interstellar cosmic radiation?
d) What, according to Lemaître, are the three periods of expansion of the universe?
e) Given the existence of universal gravitation, why might the universe (or certain regions of the universe) be found in a state of equilibrium? Are there any regions which are actually found in such a state?
f) How might Lemaître's theory explain the systematic arrangement of spiral nebulae, the formation of elliptic nebulae, and the formation of stars and planets? And how might the truth (or falsehood) of Lemaître's theory of cosmogony be experimentally tested?

QUES. 28.4 What are the geometric consequences which follow from the theory of relativity?

a) What is meant by an elliptical, or Riemannian, space? What causes it to curve, and how is its radius defined?
b) How can the radius of an elliptical space be calculated? Using this method, what is the radius of our universe at the moment of equilibrium?
c) What lies beyond the observed region of our universe? Is our universe closed? How might its curvature be inferred?
d) What might one conclude by extrapolating backwards from the presently observed expansion? Is it reasonable to do so?

28.4   Exercises

Ex. 28.1 (MASS–ENERGY EQUIVALENCE). According to the theory of relativity, mass, m, and energy, E, are equivalent: they are related by Einstein's famous formula

$$E = mc^2, \tag{28.1}$$

where c is the speed of light in a vacuum. Using Eq. (28.1), what is the energy equivalent of a 1 kg lump of mud? Compare this value with the energy released when 1 kg of an explosive, such as gunpowder, burns.[3]

Ex. 28.2 (RADIOACTIVE DECAY). What are the decay products, and the half-life, of the most common radium isotope? What fraction of such a sample of radium will remain after ten half-lives have expired? (ANSWER: 0.098 %)

Ex. 28.3 (THE SIZE OF THE UNIVERSE AND BIG BANG COSMOLOGY). Look up a current estimate of the size and density of the universe. Are these in agreement with those of Lemaître? Is this surprising? More generally, how does modern Big Bang cosmology differ from that proposed in *The Primeval Atom*?

28.5   Vocabulary

1. Cosmogonic
2. Audacious
3. Utilitarian
4. Enigma
5. Aesthetic
6. Grandiose
7. Perihelion
8. Primeval
9. Terrestrial
10. Vivacity
11. Subordinate
12. Prolific
13. Sobriety
14. Radiation
15. Agglomeration
16. Thwart
17. Extra-galactic
18. Capricious
19. Equilibrium
20. Provisory
21. Manifest

---

[3] For a more detailed discussion of mass-energy equivalence see Chap. 32 of volume II, and especially Ex. 32.1.

Chapter 29
The Primeval Atom

The purpose of any cosmogonic theory is to seek out ideally simple conditions which could have initiated the world and from which, by the play of recognized physical forces, that world, in all its complexity, may have resulted.

—Georges Lemaître

29.1 Introduction

In Chap. IV of *The Primeval Atom*, Georges Lemaître outlined his own theory of cosmogony, contrasting it with those previously conceived by thinkers such as Buffon, Kant and Laplace. Now, in Chap. V, he presents his theory "in deductive form", explaining in detail its points of contact with modern cosmological observations. As you study this text, consider the question: is Lemaître's theory plausible?

29.2 Reading: Lemaitre, *The Primeval Atom*

Lemaître, G., *The Primeval Atom: An Essay on Cosmogony*, D. Van Nostrand Company, New York, 1950. Chap. V: The Primeval Atom.[1]

29.2.1 Introduction

The primeval atom hypothesis is a cosmogonic hypothesis which pictures the present universe as the result of the radioactive disintegration of an atom.

I was led to formulate this hypothesis, some 15 years ago, from thermodynamic considerations while trying to interpret the law of degradation in the frame of quantum theory. Since then, the discovery of the universality of radioactivity shown by artificially provoked disintegrations, as well as the establishment of the corpuscular

---

[1] Lecture given at the annual session of the "Société Helvétoque des Sciences Naturelles" at Freibourg, in September 1945, and published in the "Proceedings" of this Society.

K. Kuehn, *A Student's Guide Through the Great Physics Texts*,
Undergraduate Lecture Notes in Physics, DOI 10.1007/978-1-4939-1360-2_29,
© Springer Science+Business Media, LLC 2015

nature of cosmic rays, manifested by the force which the Earth's magnetic field exercises on these rays, made more plausible an hypothesis which assigned a radioactive origin to these rays, as well as to all existing matter.

Therefore, I think that the moment has come to present the theory in deductive form. I shall first show how easily it avoids several major objections which would tend to disqualify it from the start. Then I shall strive to deduce its results far enough to account, not only for cosmic rays, but also for the present structure of the universe, formed of stars and gaseous clouds, organized into spiral or elliptical nebulae, sometimes grouped in large clusters of several thousand nebulae which, more often, are composed of isolated nebulae, receding from one another according to the mechanism known by the name of the expanding universe.

For the exposition of my subject, it is indispensable that I recall several elementary geometric conceptions, such as that of the closed space of Riemann, which led to that of space with a variable radius, as well as certain aspects of the theory of relativity, particularly the introduction of the cosmological constant and of the cosmic repulsion which is the result of it.

29.2.2   Closed Space

All partial space is open space. It is comprised in the interior of a surface, its boundary, beyond which there is an exterior region. Our habit of thought about such open regions impels us to think that this is necessarily so, however large the regions being considered may be. It is to Riemann that we are indebted for having demonstrated that total space can be closed. To explain this concept of closed space, the most simple method is to make a small-scale model of it in an open space. Let us imagine, in such a space, a sphere in the interior of which we are going to represent the whole of closed space. On the rim surface of the sphere, each point of closed space will be supposed to be represented twice, by two points, A and A', which, for example, will be two antipodal points, that is, two extremities of the same diameter. If we join these two points A and A' by a line located in the interior of the sphere, this line must be considered as a closed line, since the two extremities A and A' are two distinct representations of the same, single point. The situation is altogether analogous to that which occurs with the Mercator projection, where the points on the 180th meridian are represented twice, at the eastern and western edges of the map. One can thus circulate indefinitely in this space without ever having to leave it.

It is important to notice that the points represented by the outer surface of the sphere, in the interior of which we have represented all space, are not distinguished by any properties of the other points of space any more than is the 180th meridian for the geographic map. In order to account for that, let us imagine that we displaced the sphere in such a manner that point A is superposed on B, and the antipodal point A' on B'. We shall then suppose that the entire segment AB and the entire segment $A'B'$ are two representations of a similar segment in closed space. Thus we shall have a portion of space which has already been represented in the interior of the initial sphere which is now represented a second time at the exterior of this sphere. Let us disregard the interior sphere as useless; a complete representation of the space in the interior of the new sphere will remain. In this representation, the closed contours will be soldered into a point which is twice represented, namely, by the points B and B', mentioned above, instead of being welded, as they were formerly, to point A and A'. Therefore, these latter are not distinguished by an essential property.

Let us notice that when we modify the exterior sphere, it can happen that a closed contour which intersects the first sphere no longer intersects the second, or, more generally, that a contour no longer intersects the finite sphere at the same number of points. Nevertheless, it is evident that the number of points of intersection can only vary by an even number. Therefore, there are two kinds of closed contours which cannot be continually distorted within one another. Those of the first kind can be reduced to a point. They do not intersect the outer sphere or they intersect it at an even number of points. The others cannot be reduced to one point, we call them *odd contours* since they intersect the sphere at an odd number of points.

If, in a closed space, we leave a surface which we can supposed to be horizontal, in going toward the top we can, by going along an odd contour, return to our point of departure from the opposite direction without having deviated to the right or left, backward or forward, without having traversed the horizontal plane passing through the point of departure.

29.2.3   Elliptical Space

That is the essential of the topology of closed space. It is possible to complete these topological ideas by introducing, as is done in a geographical map, scales which vary from one point to another and from one direction to another. That can be done in such a manner that all the points of space and all the directions in it may be perfectly equivalent. Thus, Riemann's homogeneous space, or elliptical space, is obtained. The straight line is an odd contour of minimum length. Any two points divide it into two segments, the sum of which has a length which is the same for all straight lines and which is called the tour of space.

All elliptical spaces are similar to one another. They can be described by comparison with one among them. The one in which the tour of the straight line is equal to $\pi = 3.1416$ is chosen as the standard elliptical space. In every elliptical space, the distanced between two points are equal to the corresponding distances in standard space, multiplied by the number R which is called the radius of elliptical space under consideration. The distances in standard space, called space of unit radius, are termed angular distances. Therefore, the true distances, or linear distances, are the product of the radius of space times the the the angular distances.

29.2.4   Space of Variable Radius

When the radius of space varies with time, space of variable radius is obtained. One can imagine that material points are distributed evenly in it, and that spatio–temporal observations are made on these points. The angular distance of the various observers remains invariant, therefore the linear distances vary proportionally to the radius of space. All the points in space are perfectly equivalent. A displacement can bring any point into the center of the representation. The measurements made by the observers are thus also equivalent, each one of them makes the same map of the universe.

If the radius increases with time, each observer sees all points which surround him receding from him, and that occurs at velocities which become greater as they recede further. It is this which has been observed for the extra-galactic nebulae that surround us. The constant ratio between distance and velocity has been determined by Hubble and Humason. It is equal to $T_H = 2 \times 10^9$ years.

If one makes a graph, plotting as abscissa the values of time and as ordinate the value of radius, one obtains a curve, the sub-tangent of which at the point representing the present instant is precisely equal to T_H.

29.2.5   The Primeval Atom

These are the geometric concepts that are indispensable to us. We are now going to imagine that the entire universe existed in the form of an atomic nucleus which filled elliptical space of convenient radius in a uniform manner.

Anticipating that which is to follow, we shall admit that, when the universe had a density of 10^{-27} gm/cc, the radius of space was about a billion light-years, that is, 10^{27} cm. Thus the mass of the universe is 10^{54} gm. If the universe formerly had a density equal to that of water, its radius was then reduced to 10^{18}, say, one light-year. In it, each proton occupied a sphere of 1 Å, say 10^{-8} cm. In an atomic nucleus, the protons are contiguous and their radius is 10^{-13}, thus about 100,000 times smaller. Therefore, the radius of the corresponding universe is 10^{13} cm, that is to say, an astronomical unit.

Naturally, too much importance must not be attached to this description of the primeval atom, a description which will have to be modified, perhaps, when our knowledge of atomic nuclei is more perfect.

Cosmogonic theories propose to seek out initial conditions which are ideally simple, from which the present world, in all its complexity, might have resulted, through the natural interplay of known forces. It seems difficult to conceive of conditions which are simpler than those which obtained when all matter was unified in an atomic nucleus. The future of atomic theories will perhaps tell us, some day, how far the atomic nucleus must be considered as a system in which associated particles still retain some individuality of their own. The fact that particles can issue from a nucleus, during radioactive transformations, certainly does not prove that these particles pre-existed as such. Photons issue from an atom of which they were not constituent parts, electrons appear there, where they were not previously, and the theoreticians deny them an individual existence in the nucleus. Still more protons or alpha particles exist there, without doubt. When they issue forth, their existence becomes more independent, nevertheless, and their degrees of freedom more numerous. Also, their existence, in the course of radioactive transformations, is a typical example of the degradation of energy, with an increase in the number of independent quanta or increase in entropy.

That entropy increases with the number of quanta is evident in the case of electromagnetic radiation in thermodynamic equilibrium. In fact, in black body radiation, the entropy and the total number of photons are both proportional to the third power of the temperature. Therefore, when one mixes radiations of different temperatures and one allows a new statistical equilibrium to be established, the total number of photons has increased. The degradation of energy is manifested as a pulverization of energy. The total quantity of energy is maintained, but it is distributed in an ever larger number of quanta, it becomes broken into fragments which are ever more numerous.

If, therefore, by means of thought, one wishes to attempt to retrace the course of time, one must search in the past for energy concentrated in a lesser number of quanta. The initial condition must be a state of maximum concentration. It was in trying to formulate this condition that the idea of the primeval atom was germinated. Who knows if the evolution of theories of the nucleus will not, some day, permit the consideration of the primeval atom in a single quantum?

29.2.6   The Formation of Clouds

We picture the primeval atom as filling space which has a very small radius (astronomically speaking). Therefore, there is no place for superficial electrons, the primeval atom being nearly an *isotope of a neutron*. This atom is conceived as having existed for an instant only, in fact, it was unstable and, as soon as it came into being, it was broken into pieces which were again broken, in their turn; among these pieces electrons, protons, alpha particles, *etc.*, rushed out. An increase in volume resulted, the disintegration of the atom was thus accompanied by a rapid increase in the radius of space which the fragments of the primeval atom filled, always uniformly. When these pieces became too small, they ceased to break up; certain ones, like uranium, are slowly disintegrating now, with an average life of four billion years, leaving us a meager sample of the universal disintegration of the past.

In this first phase of expansion of space, starting asymptotically with a radius practically zero, we have particles of enormous velocities (as a result of recoil at the time of the emission of rays) which are immersed in radiation, the total energy of which is, without doubt, a notable fraction of the mass energy of the atoms.

The effect of the rapid expansion of space is the attenuation of this radiation and also the diminution of the relative velocities of the atoms. This latter point requires some explanation. Let us imagine that an atom has, along the radius of the sphere in which we are representing closed space, a radial velocity which is greater than the velocity normal to the region in which it is found. Then this atom will depart faster from the center than the ideal material particle which has normal velocity. Thus the atom will reach, progressively, regions where its velocity is less abnormal, and its proper velocity, that is, its excess over normal velocity, will diminish. Calculation shows that proper velocity varies in this way in inverse ratio to the radius of space. We must therefore look for a notable attenuation of the relative velocities of atoms in the first period of expansion. From time to time, at least, it will happen that, as a result of favorable chances, the collisions between atoms will become sufficiently moderate so as not to give rise to atomic transformations or emissions of radiation, but that these collisions will be elastic collisions, controlled by superficial electrons, so considered in the theory of gases. Thus we shall obtain, at least locally, a beginning of statistical equilibrium, that is, the formation of gaseous clouds. These gaseous clouds will still have considerable velocities, in relation to one another, and they will be mixed with radiations that are themselves attenuated by expansion.

It is these radiations which will endure until our time in the form of cosmic rays, while the gaseous clouds will have given place to stars and to nebulae by a process which remains to be explained.

29.2.7   Cosmic Repulsion

For that explanation, we must say a few words about the theory of relativity. When Einstein established his theory of gravitation, or generalized theory, he admitted, under the name of the principle of equivalence, that the ideas of special relativity were approximately valid in a sufficiently small domain. In the special theory, the differential element of space–time measurements had for its square a quadratic form with four coordinates, the coefficients of which had special constant values. In the generalization, this element will still be the square root of a quadratic form, but the coefficients, designated collectively by the name of *metric tensors*, will vary from place to place. The geometry of space–time is then the general geometry of Riemann at three plus one dimensions. The spaces with variable radii are a particular case in this general geometry, since the theory of spatial homogeneity or of the equivalence of observers is introduced here.

It can be that this geometry differs only apparently from that of special relativity. This is what happens when the quadratic form can be transformed, by a simple change in coordinates, into a form having constant coefficients. Then one says with Riemann that the corresponding variety (that is, space–time) is flat or Euclidian. For that, it is necessary that certain expressions, expressed by components of a tensor with four indices called Riemann's tensor, vanish completely at all points. When it is not so, the tensor of Riemann expresses the departure from flatness. Riemann's tensor is calculated by the average of second derivatives of the metric tensor. Starting with Riemann's tensor with four indices, it is easy to obtain a tensor which has only two indices like the metric tensor; it is called the contracted Riemann tensor. One can also obtain a scalar, the totally contracted Riemann tensor.

In special relativity, a free point describes a straight line with uniform motion, that is the principle of inertia. One can also say that, in an equivalent manner, it describes a geodesic of space–time. In the generalization, it is again presumed that a free point describes a geodesic. These geodesics are no longer representable by a uniform, rectilinear motion, they now represent a motion of a point under the action of the forces of gravitation. Since the field of gravitation is caused by the presence of matter, it is necessary that there by a relation between the density of the distribution of matter and Riemann's tensor which expresses the departure from flatness. The density is, in itself, considered as the principal component of a tensor with two indices called the *material tensor*; thus one obtains as a possible expression of the material tensor $T_{\mu\nu}$ as a function of the metric tensor $g_{\mu\nu}$ and of the two tensors of Riemann, contracted to $R_{\mu\nu}$ and totally contracted to R,

$$T_{\mu\nu} = aR_{\mu\nu} + bRg_{\mu\nu} + cg_{\mu\nu}, \tag{29.1}$$

where a, b, and c are three constants.

But this is not all; certain identities must exist between the components of the material tensor and its derivatives. These identities can be interpreted, for a convenient choice of coordinates, a choice which corresponds, moreover, to the practical conditions of observations, as expressing the principles of conservation, that of energy and that of momentum. In order that such identities may be satisfied, it is no longer possible to choose arbitrarily the values of the three constants. b must be taken as equal to $-a/2$. Theory cannot predict either their magnitude or their sign. It is only observation which can determine them.

The constant a is linked to the constant of gravitation. In fact, when theory is applied to conditions which are met in the applications (in particular, the fact that astronomical velocities are small in comparison to the speed of light) and when one profits from these conditions by introducing coordinates which facilitate comparison with experiment, one finds that the geodesics differ from rectilinear motion by an acceleration which can be interpreted as an attraction in inverse ratio to the square of the distances, and which is exercised by the masses represented by the material tensor. This is simply the principal effect foreseen by the theory; this theory predicts small departures which, in favorable cases, have been confirmed by observation.

A good agreement with planetary observations is obtained by leaving out the term in c. That does not prove that this term may not have experimental consequence. In fact, in the conditions which were employed to obtain Newton's law as an approximation of the theory, the term in c would furnish a force varying, not in the inverse square ratio of the distance, but proportionally to this distance. This force therefore could have a marked action at very great distances although, for the distances of the planets, its action would be negligible. Also, the relation c/a, designated customarily by the letter *lambda*, is called the cosmological constant. When Λ is positive, the additional force proportional to the distance is called *cosmic repulsion*.

The theory of relativity has thus unified the theory of Newton. In Newton's theory, there were two principles posed independently of one another: universal attraction and the conservation of mass. In the theory of relativity, these principles take a slightly modified form, while being practically identical to those of Newton in the case where these have been confronted with the facts. But universal attraction is now a result of the conservation of mass. The size of the force, the constant of gravitation, is determined experimentally.

The theory again indicates that the constancy of mass has, as a result, besides the Newtonian force of gravitation, a repulsion proportional to the distance of which the size and even the sign can only be determined by observation and by observation requiring great distances.

Cosmic repulsion is not a special hypothesis, introduced to avoid the difficulties which are presented in the study of the universe. If Einstein has re-introduced it in his work on cosmology, it is because he remembered having arbitrarily dropped it when he had established the equations of gravitation. To suppress it amounts to determining it arbitrarily by giving it a particular value: zero.

29.2.8   The Universe of Friedmann

The theory of relativity allows us to complete our description of space with a variable radius by introducing here some dynamic considerations. As before, we shall represent it as being in the interior of a sphere, the center of which is a point which we can choose arbitrarily. This sphere is not the boundary of the system, it is the edge of the map or of the diagram which we have made of it. It is the place at which the two opposite, half-straight lines are soldered into a closed straight line. Cosmic repulsion is manifested as a force proportional to the distance to the center of the diagram. As for the gravitational attraction, it is known that, in the case of distribution involving spherical symmetry around a point, and that is certainly the case here, the regions farther away from the center than the point being considered have no influence upon its motion; as for the interior points, they act as though they were concentrated at the center. By virtue of the homogeneity of the distribution of matter, the density is constant, the force of attraction which results is thus proportional to the distance, just as is cosmic repulsion.

Therefore, a certain density exists, which we shall call the density of equilibrium or the *cosmic density*, for which the two forces will be in equilibrium.

These elementary considerations permit recognition, in a certain measure, of the result which the calculation gives and which is contained in Friedmann's equation:

$$\left(\frac{dR}{dt}\right)^2 = -1 + \frac{2M}{R} + \frac{R^2}{T^2}. \tag{29.2}$$

The last term represents cosmic repulsion (it is double the function of the forces of this repulsion). T is a constant depending on the value of the cosmological constant and being able to replace this. The next-to-last term is double the potential of attraction due to the interior mass. The radius of space R is the distance from the origin of a point of angular distance $\sigma = 1$. If one multiplied the equation by σ^2, one would have the corresponding equation for a point at any distance.

That which is remarkable in Friedmann's equation is the first term -1. The elementary considerations which we have just advanced would allow us to assign it a value which is more or less constant; it is the constant of energy in the motion which takes place under the action of two forces. The complete theory determines this constant and thus links the geometric properties to the dynamic properties.

29.2.9   Einstein's Equilibrium

Since, by virtue of equations, the radius R remains constant, the state of the universe in equilibrium, or Einstein's universe, is reached. The conditions of the universe in equilibrium are easily deduced from Friedmann's equation:

$$R_E = \frac{T}{\sqrt{3}},$$

$$\rho_E = \frac{3}{4\pi}\frac{1}{T^2}, \tag{29.3}$$

$$M = \frac{T}{\sqrt{3}}.$$

In these formulas, the distances are calculated in light-time, which amounts to taking the velocity of light c as equal to unity, but, in addition, the unit of mass is chosen in such a way that the constant of gravitation may also be equal to unity. It is easy to pass on to the numerical values in C.G.S. by re-establishing in the formulas the constants c and G in such a manner as to satisfy the equations of dimension. In particular, if one takes T as being equal to 2×10^9 years, as we shall suppose in a moment, one finds that the density ρ_E is equal to 10^{-27} gm/cc.

These considerations can be extended to a region in which distribution is no longer homogeneous and where even the spherical symmetry is no longer verified, provided that the region under consideration be of small dimension. In fact, it is known that, in a small region, Newtonian mechanics is always a good approximation. Naturally, it is necessary, in applying Newtonian mechanics, to take account of cosmic repulsion but, aside from this easy modification, it is perfectly legitimate to utilize the intuition acquired by the practice of classic mechanics and its application to systems which are more or less complicated. Among other things, it can be noted that the equilibrium of which we have just spoken is unstable and that the equilibrium can even be disturbed in one sense, in one place, and in the opposite sense in another region.

Perhaps it is necessary to mention here that Friedmann's equation is only rigorously exact if the mass M remains constant. While one takes account of the radiation which circulates in space and also of the characteristic velocities of the particles which cross one another in the manner of molecules in a gas and, as in a gas, give rise to pressure, it is necessary to consider the work of this pressure during the expansion of space, in the evaluation of the mass or the energy. But it is apparent that such an effect is generally negligible, as detailed researches elsewhere have shown.

29.2.10   The Significance of Clusters of Nebulae

We are now in a position to take up again the description which we had begun of the expansion of space, following the disintegration of the primeval atom. We had shown how, in a first period of rapid expansion, gaseous clouds must have been formed, animated by great, characteristic velocities. We are now going to suppose that the mass M is slightly larger than $T/\sqrt{3}$.

The second member of Friedmann's equation will thus be able to become smaller, but it will not be able to vanish. Thus, we may distinguish three phases in the expansion of space. The first rapid expansion will be followed by a period of deceleration, during the course of which attraction and repulsion will virtually bring themselves into equilibrium. Finally, repulsion will definitely prevail over attraction, and the universe will enter into the third phase, that of the resumption of expansion under the dominant action of cosmic repulsion.

Let us consider the phase of slow expansion in more detail. The gaseous clouds are undoubtedly not distributed in a perfectly uniform manner. Let us consider a region sufficiently small, and that only from the point of view of classic mechanics, the conflict between the forces of repulsion and attraction which almost produces equilibrium. We easily see that as a result of local fluctuations of density there will be regions where attraction will finally prevail over repulsion, in spite of the fact that we have supposed that, for the universe in its entirety, it is the contrary which takes place. These regions in which attraction has prevailed will thus fall back upon themselves, while the universe will be entering upon a period of renewed expansion. We shall obtain a universe formed of regions of condensations which are separated from on another. Will not these regions of condensations be elliptical or spiral nebulae? We shall come back to this question in a moment.

Let us note that, although it is of rare occurrence, it will be possible for large regions where the density or the speed of expansion differ slightly from the average to hesitate between expansion and contraction, and remain in equilibrium, while the universe has resumed expansion. Could these regions not be identified with the clusters of nebulae, which are made up of several hundred nebulae located at relative distances from one another, which are a dozen times smaller than those of isolated nebulae? According to this interpretation, these clusters are made up of nebulae which are retarded in the phase of equilibrium; they represent a sample of the distribution of matter, as it existed everywhere, when the radius of space was a dozen times smaller than it is at present, when the universe was passing through equilibrium.

29.2.11   The Findings of De Sitter

This interpretation gives the explanation for a remarkable coincidence upon which De Sitter insisted strongly, in the past. Calculating the radius of the universe in the hypothesis which bears his name, that is, ignoring the presence of matter and introducing into the formulas the value T_H given by the observation of the expansion, he obtained a result which scarcely differs from that which is obtained, in Einstein's totally different hypothesis of the universe, by introducing into the formula the observed value of the density of matter. The explanation of this coincidence is, according to our interpretation of the clusters of nebulae, that, for a value of the radius which is a dozen times the radius of equilibrium, the last term in Friedmann's formula greatly prevails over the others. The constant T which figures in it is therefore practically equal to the observed value T_H: but since, in addition, the clusters are a fragment of Einstein's universe, it is legitimate to use the relationship existing between the density and the constant T for them. For $T = T_H$ one finds, as we have seen, that the density in the clusters must be 10^{-27} gm/cc, which is the value given by observation. This observation is based on counts of nebulae and on the estimate of their mass indicated by their spectroscopic velocity of rotation.

In addition to this argument of a quantitative variety, the proposed interpretation also takes account of important facts of a qualitative order. It explains why the clusters do not show any marked central condensations and have vague forms, with irregular extensions, all things which it would be difficult to explain if they formed dynamic structures controlled by dominant forces, as is manifestly the case for the star clusters or the elliptical and spiral nebulae. It also takes into account a manifest fact which is the existence of large fluctuations of density in the distribution of the nebulae, even outside the clusters. This must be so, in fact, if the universe has just passed through a state of unstable equilibrium, a whole gamut of transition between the properly termed clusters which are still in equilibrium, while passing through regions where the expansion, without being arrested, has nevertheless been retarded, in such a manner that these regions have a density which is greater than the average.

This interpretation permits the value of the radius at the moment of equilibrium to be determined at a billion light-years, and thus 10^{10} light-years for the present value of the radius. Since American telescopes prospect the universe as far as half a billion light-years, one sees that this observed region already constitutes a sample of a size which is not at all negligible compared to entire space; hence, it is legitimate to hope that the values of the coefficient of expansion T_H and of the density, obtained for this restricted domain, are representative of the whole.

The only indeterminate which exists is that which is relative to the degree of approximation with which the situation of equilibrium has been approached. It is on this value which the estimate of the duration of expansion depends. Perhaps it will be possible to estimate this value by means of statistical considerations regarding the relative frequency of the clusters, compared to the isolated nebulae.

29.2.12   The Proper Motion of Nebulae

Now we must come back to the question of the formation of nebulae from the regions of condensation. We have seen that the characteristic velocities, or the relative velocities of gaseous clouds, which cross one another in the same place, must have been very large. Since certain of them, because of a density which is a little too large, form a nucleus of condensation, they will be able to retain the clouds which have about the same velocity as this nucleus. The proper velocity of the cloud so formed will hence be determined by the velocity of the nucleus of condensation. The nebulae formed by such a mechanism must have large relative velocities. In fact, that is what is observed in the clusters of nebulae. In the one which has been best studies, that of *Virgo*, the dispersion of the velocities about the mean velocity is 650 km/s. The proper velocity must have been the proper velocity of all the nebulae at the moment of passage through equilibrium. For isolated nebulae, this velocity has been reduced to about one twelfth, as a result of expansion, by the same mechanism which we have explained with reference to the formation of gaseous clouds.

29.2.13   The Formation of Stars

The density of the clouds is, on the average, the density of equilibrium 10^{-27}. For this density of distribution, a mass such as the Sun would occupy a sphere of 100 light-years in radius. These clouds have no tendency to contract. In order that a contraction due to gravitation can be initiated, their density must be notably increased. This is what can occur if two clouds happen to collide with great velocities. Then the collision will be an inelastic collision, giving rise to ionization and emission of radiation. The two clouds will flatten one another out, while remaining in contact, the density will be easily doubled and condensation will be definitely initiated. It is clear that a solar system or a simple multiple star may arise from such a condensation, through known mechanisms. That which characterizes the mechanism to which we are led is the greatness of the dimensions of the gaseous clouds, the condensation of which will form a star. This circumstance takes account of the magnitude of the angular momentum, which is conserved during the condensation and whose value could only be nil or negligible if the initial circumstances were adjusted in a wholly improbably manner. The least initial rotation must give rise to an energetic rotation in a concentrated system, a rotation incompatible with the presence of a single body but assuming either multiple stars turning around one another or, simply, one star with one or several large planets turning in the same direction.

29.2.14   The Distribution of Densities in Nebulae

Here is the manner in which we can picture for ourselves the evolution of the regions of condensation. The clouds begin by falling toward the center, and by describing a motion of oscillation following a diameter from one part and another of the center. In the course of these oscillations, they will encounter one another with velocities of several hundreds of km/s and will give rise to stars. At the same time, the loss of energy due to these inelastic collisions will modify the distribution of the clouds and stars already formed in such a manner that the system will be further condensed.

It seems likely that this phenomenon could be submitted to mathematical analysis. Certain hypotheses will naturally have to be introduced, in such a way as to simplify the model, so as to render the calculation possible and also so as artificially to eliminate secondary phenomena. There is scarcely any doubt that there is a way of thus obtaining the law of final distribution of the stars formed by the mechanism described above. Since the distribution of brilliance is known for the elliptical nebulae and from that one can deduce the densities in these nebulae, one sees that such a calculation is susceptible of leading to a decisive verification of the theory.

One of the complications to which I alluded, a moment ago, is the eventual presence of a considerable angular momentum. In excluding it, we have restricted the theory to condensations respecting spherical symmetry, that is, nebulae which are spherical or slightly elliptical. It is easy to see what modification will bring about the presence of considerable angular momentum. It is evident that one will obtain, in addition to a central region analogous to the elliptical nebulae, a flat system analogous to the ring of Saturn or the planetary systems, in other words, something resembling the spiral nebulae. In this theory, the spiral or elliptical character of the nebula is a matter of chance; it depends on the fortuitous value of the angular momentum in the region of condensation. It can no longer be a question of the evolution of one type into another. Moreover, the same thing obtains for stars where the type of the star is determined by the accidental value of its mass, that is, of the sum of the masses of the clouds whose encounter produced the star.

29.2.15 Distribution of Supergiant Stars

If the spirals have this origin, it must follow that the stars are formed by an encounter of clouds in two very distinct processes. In the first place, and especially in the central region, the clouds encounter one another in their radial movement, and this is the phenomenon which we have invoked for the elliptical nebulae. Kapteyn's preferential motion may be an indication of it. But besides this relatively rapid process, there must be a slower process of star formation, beginning with the clouds which escaped from the central region as a result of their angular momentum. These will encounter one another in a to-and-fro motion, from one side to the another of the plane of the spiral. The existence of these two processes, with different ages, is perhaps the explanation of the fact that supergiant stars are not found in the elliptical nebulae or in the nucleus of spirals, but that one observes them only in the exterior region of the spirals. In fact, it is known that the stars radiate energy which comes from the transformation of their hydrogen into helium. The supergiant stars radiate so much energy that they could only maintain this output during a hundred million years. It should be understood, thus, that, for the oldest stars, the supergiants may be extinct for lack of fuel, whereas they still shine where they have been recently formed.

29.2.16   The Uniform Abundance of the Elements

But it is doubtless not worthwhile to allow ourselves to be prematurely led to the attempted pursuit of the theory in such detail, but rather to restrict ourselves, for the moment, to the more general consequences of the hypothesis of the primeval atom. We have seen that the theory takes account of the formations of stars in the nebulae. It also explains a very remarkable circumstance which could be demonstrated by the analysis of stellar spectra. It concerns the quantitative composition of matter, or the relative abundance of the various chemical elements, which is the same in the Sun, in the stars, on the Earth and in the meteorites. This fact is a necessary consequence of the hypothesis of the primeval atom. Products of the disintegration of an atom are naturally found in very definite proportions, determined by the laws of radioactive transformations.

29.2.17   Cosmic Rays

Finally, we said in the beginning that the radiations produced during the disintegrations, during the first period of expansion could explain cosmic rays. These rays are endowed with an energy of several billion electron-volts. We know no other phenomenon currently taking place which may be capable of such effects. That which these rays resemble most is the radiation produced during present radioactive disintegrations, but the individual energies brought into play are enormously greater. All that agrees with rays of superradioactive origin. But it is not only by their quality that these rays are remarkable, it is also by their total quantity. In fact, it is easy, form their observed density which is given in erg/cm, to deduce their density of energy by dividing by c, then their density in gm/cc by dividing by c^2. Thus one finds 10^{-34} gm/cc, about one ten-thousandth the present density of the matter existing in the form of stars. It seems impossible to explain such an energy which represents one part in 10,000 of all existing energy, if these rays had not been produced by a process which brought into play all existing matter. In fact, this energy, at the moment of its formation, must have been at least ten times greater, since a part of it was able to be absorbed and the remainder has been reduced as a result of the expansion of space. The total intensity observed for cosmic rays is therefore just about that which might be expected.

29.2.18   Conclusion

The purpose of any cosmogonic theory is to seek out ideally simple conditions which could have initiated the world and from which, by the play of recognized physical forces, that world, in all its complexity, may have resulted.

I believe that I have shown that the hypothesis of the primeval atom satisfies the rules of the game. It does not appeal to any force which is not already known. It accounts for the actual world in all its complexity. By a single hypothesis it explains stars arranged in galaxies within an expanding universe as well as those local exceptions, the clusters of nebulae. Finally, it accounts for that mighty phenomenon, the ultrapenetrating rays. They are truly cosmic, they testify to the primeval activity of the cosmos. In their course through wonderfully empty space, during billions of years, they have brought us evidence of the superradioactive age, indeed they are a sort of fossil rays which tell us what happened when the stars first appeared.

I shall certainly not pretend that this hypothesis of the primeval atom is yet proved, and I would be very happy if it has not appeared to you to be either absurd or unlikely. When the consequences which result from it, especially that which concerns the law of the distribution of densities in the nebulae, are available in sufficient detail, it will doubtless be possible to declare oneself definitely for or against.

29.3 Study Questions

QUES. 29.1 In what sense is a Riemannian space similar to the surface of a globe?

a) What distinguishes total space from all partial spaces? Why does Lemaître claim that all partial spaces are open spaces?

b) What is a Mercator projection map? Are any points missing on a Mercator map? Are any points duplicates?

c) What is a closed space, as opposed to an open space? How can one construct a closed three-dimensional space? And how is this like a Mercator map?

d) What additional condition renders a closed space an elliptical closed space? Specifically, how are all elliptical spaces alike?

e) In an elliptical space, how many points must be placed on a straight line in order to divide it into two segments? What is notable about the sum of the length of these two segments?

f) What is the radius of an elliptical space? And why does Lemaître refer to a distance in elliptical space as an angular distance?

g) In what sense are all points on an elliptical space equivalent? Is there any evidence that our universe is, in fact, an elliptical space? If so, is its radius constant or is it changing?

QUES. 29.2 Was the initial state of the universe simpler or more complex than it is today?

a) What does Lemaître assume regarding the initial state of the universe? What would be the radius of such a universe, and how does he arrive at this value?

b) As radioactive decay proceeds, what happens to the number of quanta, or particles in the universe?

c) Is energy conserved during radioactive decay? What, then, does it mean to say that energy is "degraded"? In particular, does the entropy of the universe increase or decrease as a result of radioactive decay?

QUES. 29.3 How did gaseous nebulae arise?

a) To what does Lemaître compare the primeval atom? Was this primeval atom
   stable? And how did the radioactive decay proceed thereafter?
b) How does the speed of the decay products change as the radius of space expands?
   And by what mechanism do they eventually achieve equilibrium?
c) Are the gaseous clouds composed of matter or cosmic rays? What eventually
   arises from the gaseous clouds?

QUES. 29.4 How does Einstein's theory of gravity differ from Newton's?

a) What is the nature of gravity according to Newton? And what is the trajectory of
   a free particle in the absence of gravity?
b) According to Einstein's theory, what is the relationship between matter, space and
   gravity? In particular, how does matter affect space? And how does the curvature
   of space, in turn, affect the trajectory of a free particle?
c) How many distinct principles did Newton's theory of gravity tacitly assume? In
   what sense is Einstein's theory simpler than Newton's?
d) What additional component of force is implied by Einstein's theory of gravity?
   What is the nature of this force—is it attractive or repulsive? What is the term
   used for this force?

QUES. 29.5 Is our universe in equilibrium?

a) How is Einstein's equilibrium universe represented mathematically? What do
   each of the symbols in these equations signify?
b) Is an equilibrium universe stable or unstable with respect to small perturbations?
   Is it possible for some regions to be expanding while others are contracting? What
   remarkable coincidence(s) might this explain?

QUES. 29.6 What is the origin of stars, and of the elements?

a) How can stars form within gaseous clouds? How might one test this hypothesis?
   How does Lemaître then explain the formation of spiral nebulae? Of supergiant
   stars within these nebulae?
b) What is the relative abundance of chemical elements in the Sun, the stars and the
   earth? How is this determined? And can Lemaître's theory account for this fact?

QUES. 29.7 What is the origin of the cosmic rays? What are they? Where are they
found? What is remarkable about these rays? In particular, how does their density
compare to that of matter? And what does this suggest?

QUES. 29.8 Is Lemaître's cosmogonic theory true?

a) What is the role of cosmogonic theories? Does Lemaître's theory satisfy the "rules
   of the game"? Do any other theories?
b) Upon how many assumptions does Lemaître's rely? And what specific facts does
   it purport to explain?
c) Is Lemaître's theory plausible? What observations might render it more so?
d) What role does chance play in Lemaître's theory? Is chance a valid mode of
   explanation?

29.4 Exercises

Ex. 29.1 (MERCATOR, RIEMANN AND HUBBLE). Mercator's projection map of the surface of Earth was constructed so as to preserve the angle of intersection between any two lines drawn on the surface of Earth. For example, latitude and longitude lines always meet at right angles on Earth so they also meet at right angles on Mercator's map. This so-called *conformality* also ensures that in the immediate neighborhood of any particular point the scale (*e.g.* 1 inch = 100 miles) is the same in all directions. But conformality is a strictly local property; the scale at the equator and near the poles is vastly different. Mercator's map is particularly suitable for navigation, since lines of constant bearing appear as straight lines.

a) Are there any points on Mercator's map which are depicted twice? Are there any which are not depicted at all? Why is this?
b) Into how many line segments is the equator divided by two distinct points placed on the equator? Does the sum of these line segments depend upon the specific location of these two points?
c) What is the relationship between the length of a tour of the earth around the equator and the radius of the earth itself? Does this mean that a traveler can infer the radius of the earth without ever leaving its surface?
d) Does the radius of the earth provide a natural length scale? If so, how might you define a new unit of length so that a tour around the equator is just Π in these natural units?
e) Suppose that you were one of eight people standing, equidistant from each another, along the earth's equator. Also, suppose that the earth's radius were increasing at a constant rate of one radius per hour. Are the seven other people moving away from you at the same speed? If not how do their speeds depend on their distances from you? Would the other seven agree on this ratio or law?

29.5 Vocabulary

1. Cosmogonic
2. Disintegration
3. Degradation
4. Universality
5. Corpuscular
6. Plausible
7. Deductive
8. Exposition
9. Cosmic
10. Indispensable
11. Impel
12. Antipodal
13. Meridian
14. Solder
15. Topology
16. Invariant
17. Abscissa
18. Ordinate
19. Angstrom
20. Contiguous
21. Photon
22. Proton
23. Alpha particle
24. Radioactive

25. Quanta
26. Entropy
27. Electromagnetic
28. Thermodynamic
29. Germinate
30. Superficial
31. Isotope
32. Neutron
33. Asymptotic
34. Attenuation
35. Diminution
36. Equilibrium
37. Nebulae
38. Space–time
39. Coefficient
40. Tensor
41. Homogeneity
42. Euclidean

43. Riemannian
44. Geodesic
45. Rectilinear
46. Conservation
47. Facilitate
48. Astronomical
49. Negligible
50. Manifest
51. Unity
52. Gamut
53. Condensation
54. Nucleus
55. Dispersion
56. Ionization
57. Fortuitous
58. Preferential
59. Supergiant
60. Electron-volt

Bibliography

Aristotle, On the Heavens, in *Aristotle: I, Great Books of the Western World*, vol. 8, edited by Robert Maynard Hutchins, Encyclopedia Britannica, 1952.

Chapman, D. (Ed.), *Observer's Handbook*, Royal Astronomical Society of Canada, 2013.

Copernicus, N., *On the Revolutions of the Heavenly Spheres*, Great Minds, Prometheus Books, Amherst, NY, 1995.

Densmore, D. (Ed.), *Euclid's Elements*, second ed., Green Lion Press, Santa Fe, NM, 2003.

Dolling, L. M., Gianelli, A. F., and Statile, G. N. (Eds.), *The Tests of Time: Readings in the Development of Physical Theory*, Princeton University Press, Princeton, NJ, 2003.

Donahue, W. H. (Ed.), *Selections from Kepler's Astronomia Nova*, Green Lion Press, Santa Fe, NM, 2004.

Einstein, A., *Relativity*, Great Minds, Prometheus Books, Amherst, NY, 1995.

Einstein, A., *The Einstein Reader*, Citadel Press, 2006.

Finocchiaro, M. A. (Ed.), *The Essential Galileo*, Hackett Publishing Company, Indianapolis, Indiana, 2008.

Gilbert, W., *De Magnete*, Dover Publications, New York, 1958.

Gingerich, O., *The Book Nobody Read: Chasing the Revolutions of Nicholaus Copernicus*, Penguin Books, 2004.

Heath, T. (Ed.), *The Works of Archimedes*, Dover Publications, Mineola, NY, 2002.

Heilbron, J. L., *The Sun in the Church*, Harvard University Press, Cambridge, Massachusetts, 1999.

Herschel, J. F. W., *Outlines of Astronomy*, Longmans, Green, and Co., 1893.

Hubble, E., The realm of the nebulae, *The Scientific Monthly*, 39(3), 193–202, 1934.

Kepler, J., *Epitome of Copernican Astronomy & Harmonies of the World*, Great Minds, Prometheus Books, Amherst, NY, 1995.

Leavitt, H., Periods of 25 variable stars in the small magellanic cloud, *Harvard College Observatory Circular*, 173, 1912.

Lemaître, G., *The Primeval Atom: An Essay on Cosmogony*, D. Van Nostrand Company, New York, 1950.

Lester, T., *The fourth part of the world: the race to the ends of the earth, and the epic story of the map that gave America its name*, Free Press, New York, NY, 2009.

Mueller, I. (Ed.), *Simplicius On Aristotle's "On the Heavens 2.10-14"*, Cornell University Press, Ithaca, NY, 2005.

Munitz, M. K. (Ed.), *Theories of the Universe*, The Free Press of Glencoe, 1957.

Neugebauer, O., *A History of Ancient Mathematical Astronomy*, Springer-Verlag, New York, Heidelberg, Berlin, 1975.

Pedersen, O., *A Survey of the Almagest*, revised ed., Springer, New York, 2010.

Ptolemy, C., *Claudii Ptolemaei Opera Quae Exstant Omnia*, vol. I, Lipsiae in aedibus B. G. Teubneri, 1898.

Rome, A., L'Astrolabe et le Météoroscope d'aprés le commentaire de Pappus sur le 5e livre de l'Almageste, *Annales de la Société Scientifique de Bruxelles*, *47*(2), 77–102, 1927.

Rosenkranz, Z., *The Einstein Scrapbook*, The Johns Hopkins University Press, 2002.

Sinnott, R. W., *Sky and Telescope's Pocket Sky Atlas*, New Track Media, LLC, Cambridge, Massachusetts, 2006.

Slipher, V., Nebulae, *Proceedings of the American Philosophical Society*, *56*(5), 403–409, 1917.

Swerdlow, N. M., A Star Catalogue Used by Johannes Bayer, *Journal for the History of Astronomy*, *17*, 189, 1986.

Toomer, G. J. (Ed.), *Ptolemy's Almagest*, Princeton University Press, Princeton, NJ, 1998.

Voelkel, J. R., *Johannes Kepler and the New Astronomy*, Oxford Portraits in Science, Oxford University Press, New York, Oxford, 1999.

Waldseemüller, *Introduction to Cosmography*, 31–81 pp., Ann Arbor: University Microfilms, Princeton, NJ, 1966.

Wallace, W. A., The Problem of Causality in Galileo's Science, *The Review of Metaphysics*, *36*(3), 607–632, 1983.

Wallis, F. (Ed.), *Bede: The Reckoning of Time*, Liverpool University Press, 1999.

Index

K. Kuehn, *A Student's Guide Through the Great Physics Texts,*
Undergraduate Lecture Notes in Physics, DOI 10.1007/978-1-4939-1360-2,
© Springer Science+Business Media, LLC 2015

CPSIA information can be obtained at www.ICGtesting.com
Printed in the USA
LVOW01*1802161015

458592LV00001B/27/P